Electric Drives and Electromechanical Systems

Electric Drives and Electromechanical Systems

Richard Crowder

Amsterdam • Boston • Heidelberg • London • New York • Oxford
Paris • San Diego • San Francisco • Singapore • Sydney • Tokyo

Butterworth-Heinemann is an imprint of Elsevier

Butterworth-Heinemann is an imprint of Elsevier
Linacre House, Jordan Hill, Oxford OX2 8DP, UK
The Boulevard, Langford Lane, Kidlington, Oxford OX5 1GB, UK
84 Theobald's Road, London WC1X 8RR, UK
Radarweg 29, PO Box 211, 1000 AE Amsterdam, The Netherlands
30 Corporate Drive, Suite 400, Burlington, MA 01803, USA
525 B Street, Suite 1900, San Diego, CA 92101-4495, USA

First edition 2006

British Library Cataloguing in Publication Data
A catalogue record for this book is available from the British Library

Library of Congress Cataloging-in-Publication Data
A catalog record for this book is availabe from the Library of Congress

ISBN–13: 978-0-7506-6740-1
ISBN–10: 0-7506-6740-0

For information on all Butterworth-Heinemann publications
visit our web site at books.elsevier.com

Printed and bound in Great Britain

06 07 08 09 10 10 9 8 7 6 5 4 3 2 1

Contents

Preface

Electrical drives play an important role as electromechanical energy converters a wide range of applications, for example machine tools in manufacturing industries, photocopies, CD player, electric windows in the car, prosthetic hands and other medical devices; some are obvious other not so, until the they fail. It is critically important that the correct drive is matched to the application with due regard to its requirements. With the recent developments in power semiconductors and microprocessors with signal processing capabilities, the technology of the modern drive system has changed dramatically in recent years. However, the selection of a drive system relies on a systems approach – without which, it is highly probable that either the mechanical, electrical or electronic elements will not be fully considered.

A complete drive system consists of many different components, hence this book has been structured to present a logical discussion, on a wide range of topics relating to selection *of the complete* motor-drive system. It does not, however, extend to a detailed consideration of control and electromagnetic theory; if the reader wishes to pursue this path many excellent books are available, some of which are highlighted in the bibliography.

The structure of the book is as follows. Chapter 1 gives a brief overview of the problems that need to be solved, with particular emphasis on a wide range of electromechanical applications, including machine tools, robotics and related high performance applications. Chapters 2 and 3 concentrate on the problem of motor-drive selection, and give an insight into the decisions required during this procedure. It is hoped that this will lift the veil on what is thought by many to be a black art, or on what more commonly falls into the gap between the responsibilities of electrical, electronic, and mechanical engineers. Chapter 3 concludes with suitable algorithms to size a wide range of applications. Chapter 4 considers the types, selection and installation of velocity and position transducers, the correct selection of which will have a significant impact on the overall performance of the system. In order to illustrate the various points in the chapters, use has been made of a range of numerical examples, and hopefully these will show how the theory can be applied.

While the initial chapters concentrate on the mechanical aspects of a drive application, the second part of the book concentrates on the main classes of drives, which are available, and are used, to drive the applications discussed in Chapter 1.

The technologies considered include: the brushed motor (Chapter 5), brushless motors (Chapter 6), vector controlled induction motors (Chapter 7), and the stepper motor (Chapter 8). In addition a of motor-drives fall outside this rather arbitrary classification system, and these are considered in Chapter 9. It should be recognised that some of the larger drive systems have been omitted, due to the application domain being restricted to small or medium sized applications. Within each of theses chapters there is a review of the relevant theory, and an examination of the specific drive and control requirements.

Finally the book concludes with Chapter 10, which briefly reviews the theory and architecture of current controllers, including the programmable logic controller (PLC). Due to the increasing reliance on decentralised control within many application domains, a review of network technologies and their current standards are presented.

The production of a book such as this is not a solitary affair although it tends to be at times. I must acknowledge the help and assistance given by my colleagues in industry and academia; finally I would particularly like to acknowledge my wife Lucy and my daughter Emma for their continued support throughout the writing period.

Southampton
May 2005

List of principal symbols

The following are the principal symbols used in this book. In general they are also defined within the text. Lower-case symbols normally refer to instantaneous values, and upper-case symbols to constants. In practice, a symbol may refer to more than one quantity; however, within a particular context the meaning should be clear.

B	Damping constant
B_g	Air-gap flux
d	Duty cycle
e_p	Per-phase r.m.s. e.m.f.
e_{ss}	Steady-state error
E_a, E_b, E_c	Direct-current (d.c.) brushless motor back e.m.f.
E_g	Tachogenerator voltage
E_k	Kinetic energy
E_m	d.c. brushed-motor back e.m.f.
E_r	Regenerative energy
f_e	Supply frequency
f_p	Sample frequency
f_s	PWM switching frequency
F	Force
F_f	Frictional force
F_L	External linear force
I	Moment of inertia, Current
I_{eff}	Inertia referred to the input of a gearbox
I_L	Load inertia
I_m	Motor inertia
I_{tot}	Total inertia of a system
I_a	Armature current
I_f	Field current
I_m	Induction-motor magnetisation current
I_r	Induction-motor rotor current referred to the stator
I'_r	Induction-motor rotor current
I_R	Regenerative current

I_S	Induction-motor stator current
K	Stiffness
K_e	Voltage constant for a d.c. brushed motor
K_g	Tachogenerator voltage constant
K_p	Proportional gain
K_s	Effective stiffness of a joint
K_t	d.c. brushed-motor torque constant
K_T	Induction-motor torque constant
K'_T	d.c. brushless-motor torque constant
L_a	Armature inductance
L_s	Induction motor stator inductance
n	Gear ratio
n^*	Optimum gear ratio
N	Rotary speed, teeth on a gear
N_L	Lead-screw input speed
N_p	Turns per pole
N_T	Number of teeth on belt-drive pulley
p	Number of pole pairs
P	Power
P_i	Input power
P_o	Output power
R_a	Armature resistance
R_L	Load resistance
R'_r	Referred motor resistance
R_s	Stator resistance
R_T	Number of teeth on a stepper-motor stack
s	Slip
t_d	Dwell time
t_m	Time to complete a move
t_z	Time to reach standstill under regeneration
t_o	Time to reach zero terminal voltage under regeneration
T	Torque
T_{cm}	Continuous torque requirement
T_e	Electrical torque output
T_f	Friction torque
T_i	Input torque
T_L	Load torque
T_M	Motor torque
T_o	Output torque
T_{peak}	Peak torque
T_s	Torque stiffness
T_w	Windage torque
T_0	Stall torque
v_a^s, v_b^s, v_c^s	Voltages in a stationary three-axis frame

v_d, v_q	Voltages in a rotating two-axis frame
v_d^s, v_q^s	Voltages in a stationary two-axis frame
V_a, V_b, V_c	Brushless-motor terminal voltages
V_c	PWM control voltage, Cutting speed
V_{cpk}	PWM peak control voltage
V_L	Linear speed
V_m	d.c. motor terminal voltage
V_r	Induction-motor induced rotor voltage
V_s	Supply voltage
α	Acceleration
β	Torque angle
Δ	Peak current deviation in a pulse width modulated drive
ϵ	Efficiency
ζ	Damping
θ	Rotary position
θ_e	Static-position error
μ	Coefficient of friction
ρ	Load factor of a PWM amplifier
ϕ	Flux linkage
ω	Rotational speed
ω_e	Synchronous speed
ω_i	Input speed
ω_{int}	Initial speed prior to application of regeneration
ω_m	Mechanical speed
ω_n	Natural frequency
ω_o	Output speed
ω_r	Rotor electrical speed
ω_s	Slip frequency
ω_0	No-load speed

Chapter 1

Electromechanical systems

In the design of any complex system, all the relevant design details must be considered to ensure the development of a successful product. In the development of motion systems, problems in the design process are most likely to occur in the actuator or motor-drive system. When designing any actuation system, mechanical designers work with electrical and electronic systems engineers, and if care is not taken, confusion will result. The objective of this book is to discuss some of the electric motor-drive systems in common use, and to identify the issues that arise in the selection of the correct components and systems for specific applications.

A key step in the selection of any element of a drive system is a clear understanding of the process being undertaken. Section 1.1 provides an overview to the principles of industrial automation, and sections 1.2 and 1.3 consider machine tools and industrial robotics, respectively. Section 1.4 considers a number of other applications domains.

1.1 Principles of automation

Within manufacturing, automation is defined as the technology which is concerned with the application of mechanical, electrical, and computer systems in the operation and control of manufacturing processes. In general, an automated production process can be classified into one of three groups: fixed, programmable, or flexible.

- *Fixed automation* is typically employed for products with a very high production rate; the high initial cost of fixed-automation plant can therefore be spread over a very large number of units. Fixed-automation systems are used to manufacture products as diverse as cigarettes and steel nails. The significant feature of fixed automation is that the sequence of the manufacturing operations is fixed by the design of the production machinery, and therefore the sequence cannot easily be modified at a later stage of a product's life cycle.

- *Programmable automation* can be considered to exist where the production equipment is designed to allow a range of similar products to be produced. The production sequence is controlled by a stored program, but to achieve a product change-over, considerable reprogramming and tooling changes will be required. In any case, the process machine is a stand-alone item, operating independently of any other machine in the factory; this principle of automation can be found in most manufacturing processes and it leads to the concept of islands of automation. The concept of programmable automation has its roots in the Jacquard looms of the nineteenth century, where weaving patterns were stored on a punched-card system.

- *Flexible automation* can be considered to be an enhancement of programmable automation in which a computer-based manufacturing system has the capability to change the manufacturing program and the physical configuration of the machine tool or cell with a minimal loss in production time. In many systems the machining programs are prepared at a location remote from the machine, and they are then transmitted as required over a computer-based local-area communication network.

The basic design of machine tools and other systems used in manufacturing processes changed little from the eighteenth century to the late 1940s. There was a gradual improvement during this period as the metal cutting changed from an art to a science; in particular, there was an increased understanding of the materials used in cutting tools. However, the most significant change to machine-tool technology was the introduction of *numerical-control* (NC) and *computer-numerical-control* (CNC) systems.

To an operator, the differences between these two technologies are small: both operate from a stored program, which was originally on punched tape, but more recently computer media such as magnetic tapes and discs are used. The stored program in a NC machine is directly read and used to control the machine; the logic within the controller is dedicated to that particular task. A CNC machine tool incorporates a dedicated computer to execute the program. The use of the computer gives a considerable number of other features, including data collection and communication with other machine tools or computers over a computer network. In addition to the possibility of changing the operating program of a CNC system, the executive software of the computer can be changed, which allows the performance of the system to be modified at minimum cost. The application of NC and CNC technology permitted a complete revolution of the machine tool industry and the manufacturing industries it supported. The introduction of electronic systems into conventional machine tools was initially undertaken in the late 1940s by the United States Air Force to increase the quality and productivity of machined aircraft parts. The rapid advances of electronics and computing systems during the 1960s and 1970s permitted the complete automation of machine tools and the parallel development of industrial robots. This was followed during the 1980s by the integration

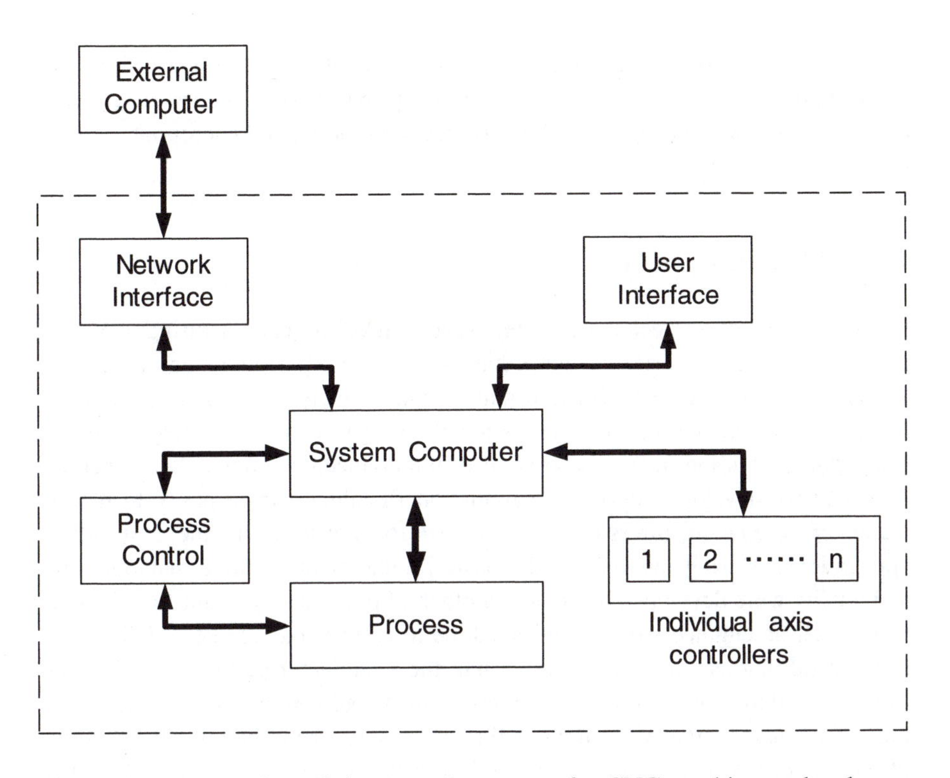

Figure 1.1. The outline of the control structure for CNC machine tool, robot or similar multi-axis system. The number of individual motion axes, and the interface to the process are determined by the system's functionality.

of robots, machine tools, and material handling systems into computer-controlled factory environments. The logical conclusion of this trend is that individual product quality is no longer controlled by direct intervention of an operator. Since the machining parameters are stored either within the machine or at a remote location for direct downloading via a network (see Section 10.4) a capability exists for the complete repeatability of a product, both by mass production and in limited batches (which can be as small as single components). This flexibility has permitted the introduction of management techniques, such as just-in-time production, which would not have been possible otherwise.

A typical CNC machine tool, robot or multi-axis system, whatever its function, consists of a number of common elements (see Figure 1.1). The axis position, or the speed controllers, and the machining-process controller are configured to form a hierarchical control structure centred on the main system computer. The overall control of the system is vested in the system computer, which, apart from sequencing the operation of the overall system, handles the communication between the operator and the factory's local-area network. It should be noted that industrial robots, which are considered to be an important element of an automated factory, can be considered to be just another form of machine tool. In a machine tool or industrial robot or related manufacturing systems, controlled motion (position and speed) of the axes is necessary; this requires the provision of actuators, either linear or rotary, associated power controllers to produce motion, and appropriate sensors to measure the variables.

1.2 Machine tools

Despite advances in technology, the basic stages in manufacturing have not changed over the centuries: material has to be moved, machined, and processed. When considering current advanced manufacturing facilities it should be remembered that they are but the latest step in a continuing process that started during the Industrial Revolution in the second half of the eighteenth century. The machine-tool industry developed during the Industrial Revolution in response to the demands of the manufacturers of steam engines for industrial, marine, and railway applications. During this period, the basic principles of accurate manufacturing and quality were developed by, amongst others, James Nasmyth and Joseph Whitworth. These engineers developed machine tools to make good the deficiencies of the rural workers and others drawn into the manufacturing towns of Victorian England, and to solve production problems which could not be solved by the existing techniques. Increased accuracy led to advantages from the interchangeability of parts in complex assemblies. This led, in turn, to mass production, which was first realised in North America with products (such as sewing machines and typewriters) whose commercial viability could not be realised except by high-volume manufacturing (Rolt, 1986). The demands of the market place for cost reductions and the requirement for increased product quality has led to dramatic changes in all aspects of manufacturing industry, on an international scale, since 1970. These changes, together with the introduction of new management techniques in manufacturing, have necessitated a considerable improvement in performance and costs at all stages of the manufacturing process. The response has been a considerable investment in automated systems by manufacturing and process industries.

Machining is the manufacturing process in which the geometry of a component is modified by the removal of material. Machining is considered to be the most versatile of production processes since it can produce a wide variety of shapes and surface finishes. To fully understand the requirements in controlling a machine tool, the machining process must be considered in some detail. Machining can be classified as either *conventional machining*, where material is removed by direct physical contact between the tool and the workpiece, or *non-conventional machining*, where there is no physical contact between the tool and the workpiece.

1.2.1 Conventional machining processes

In a conventional machining operation, material is removed by the relative motion between the tool and the workpiece in one of five basic processes: turning, milling,

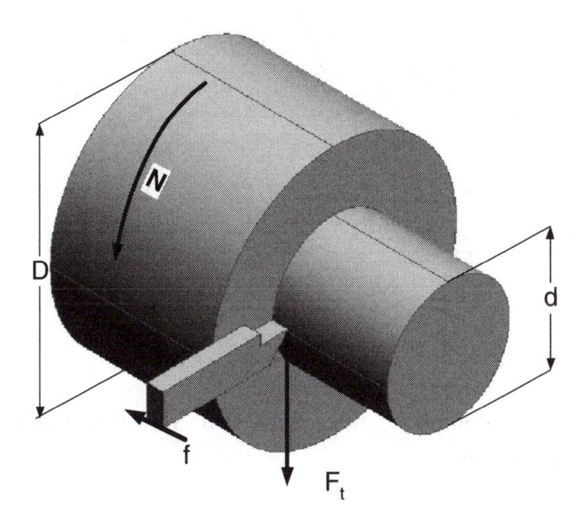

Figure 1.2. The turning process, where a workpiece of initial diameter D is being reduced to d; F_t is the tangential cutting force, N is the spindle speed, and f the feed rate. In the diagram the depth of the cut is exaggerated.

drilling, shaping, or grinding. In all machining operations, a number of process parameters must be controlled, particularly those determining the rate of material removal; and the more accurately these parameters are controlled the higher is the quality of the finished product (Waters, 1996). In sizing the drives of the axes in any machine tool, the torques and speed drives that are required in the machining process must be considered in detail. Figure 1.2 illustrates a turning operation where the tool is moved relative to the workplace. The power required by the turning operation is of most concern during the roughing cut (that is, when the cutting depth is at its maximum), when it is essential to ensure that the drive system will produce sufficient power for the operation. The main parameters are the tangential cutting force, F_t, and the cutting speed, V_c. The cutting speed is defined as the relative velocity between the tool and the surface of the workpiece (m min^{-1}). The allowable range depends on the material being cut and the tool: typical values are given in Table 1.1. In a turning operation, the cutting speed is directly related to the spindle speed, N (rev min^{-1}), by

$$V_c = D\pi N \tag{1.1}$$

The tangential force experienced by the cutter can be determined from knowledge of the process. The specific cutting force, K, is determined by the manufacturer of the cutting tool, and is a function of the materials involved, and of a number of other parameters, for example, the cutting angles. The tangential cutting force is given by

Table 1.1. Machining data

Material	Cutting Speed, V_c	Specific Cutting Force, K	Material Removal Rate, R_p
Low carbon steel	90-150	2200	25
Cast iron	60-90	1300	35
Aluminium	230-730	900	80

$$F_t = \frac{Kf(D-d)}{2} \tag{1.2}$$

Knowledge of the tangential forces allows the power requirement of the spindle drive to be estimated as

$$Power = \frac{V_c F_t}{60} \tag{1.3}$$

In modern CNC lathes, the feed rate and the depth of the cut will be individually controlled using separate motion-control systems. While the forces will be considerably smaller than those experienced by the spindle, they still have to be quantified during any design process. The locations of the radial and axial the forces are also shown in Figure 1.2; their magnitudes are, in practice, a function of the approach and cutting angles of the tool. Their determination of these magnitudes is outside the scope of this book, but it can be found in texts or manufacturers' data sheets relating to machining processes.

In a face-milling operation, the workpiece is moved relative to the cutting tool, as shown in Figure 1.3. The power required by the cutter, for a cut of depth, W_c, can be estimated to be

$$Power = \frac{dfW_c}{R_p} \tag{1.4}$$

where R_p is the quantity of material removed in m^3 min^{-1} kW^{-1} and the other variables are defined in Figure 1.3. A number of typical values for R_p are given in Table 1.1. The determination of the cutting forces is outside the scope of this book, because the resolution of the forces along the primary axes is a function of the angle of entry and of the path of the cutter relative to the material being milled. A value for the sum of all the tangential forces can, however, be estimated from the cutting power; if V_c is the cutting speed, as determined by equation (1.1), then

$$\sum F_t = \frac{60000 \times Power}{V_c} \tag{1.5}$$

The forces and powers required in the drilling, planing, and grinding processes can be determined in a similar manner. The sizes of the drives for the controlled axes in all types of conventional machine tools must be carefully determined to ensure that the required accuracy is maintained under all load conditions. In addition,

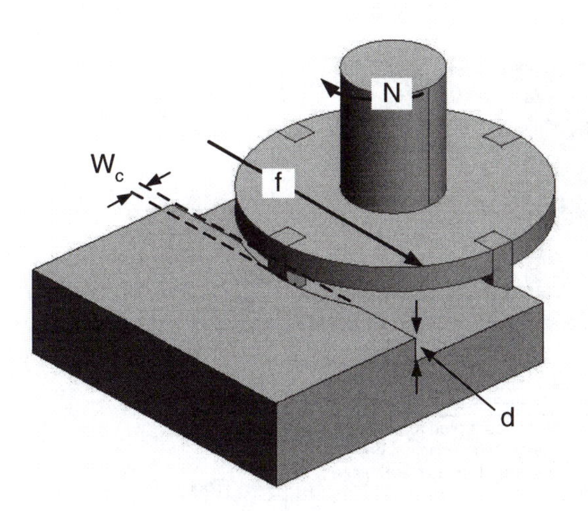

Figure 1.3. The face-milling process where the workpiece is being reduced by d: f is the feed rate of the cutter across the workpiece, W_c is the depth of the cut and N is the rotary speed of the cutting head.

a lack of spindle or axis drive power will cause a reduction in the surface quality, or, in extreme cases, damage to the machine tool or to the workpiece.

1.2.2 Non-conventional machining

Non-conventional processes are widely used to produce products whose materials cannot be machined by conventional processes, for example, because of the workpiece's extreme hardness or the required operation cannot be achieved by normal machine processes (for example, if there are exceptionally small holes or complex profiles). A range of non-conventional processes are now available, including

- laser cutting and electron beam machining,
- electrochemical machining (ECM),
- electrodischarge machining (EDM),
- water jet machining.

In laser cutting (see Figure 1.4(a)), a focused high-energy laser beam is moved over the material to be cut. With suitable optical and laser systems, a spot size with a diameter of 250 μm and a power level of 10^7 W mm^{-2} can be achieved. As in conventional machining the feed speed has to be accurately controlled to achieve

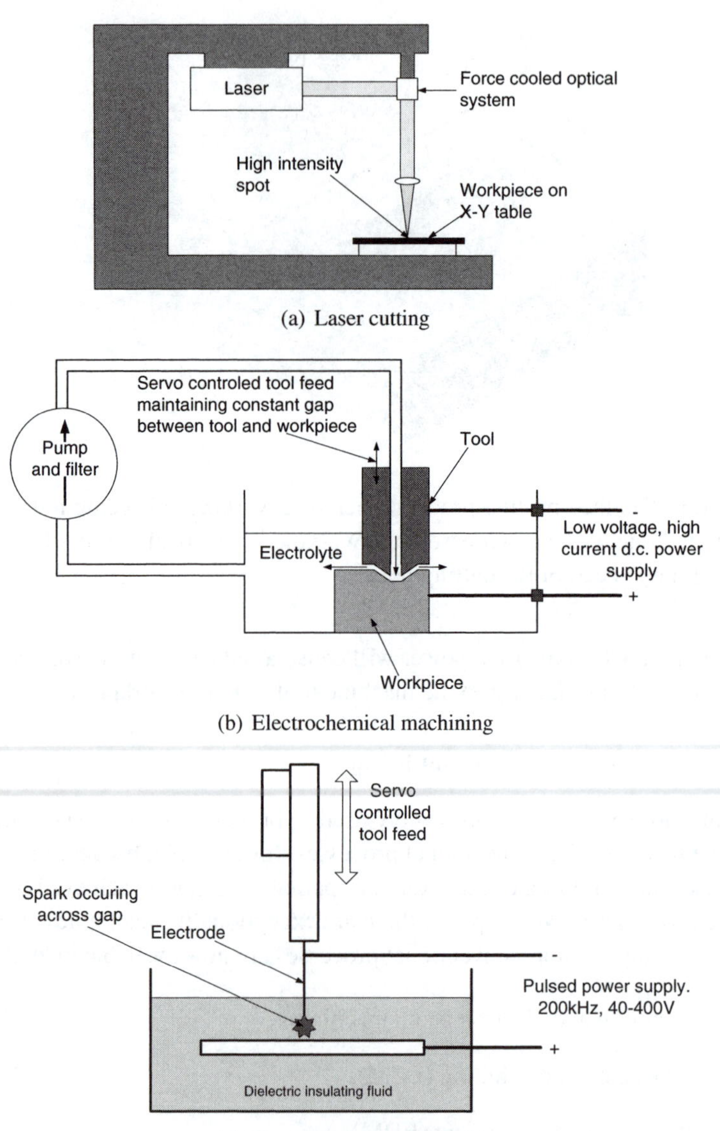

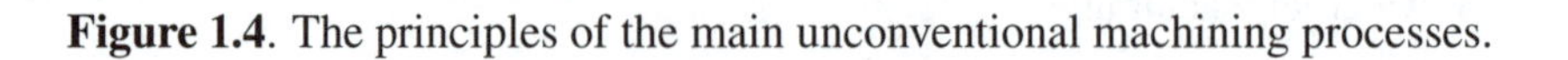

Figure 1.4. The principles of the main unconventional machining processes.

the required quality of finish: the laser will not penetrate the material if the feed is too fast, or it will remove too much material if it is too slow. Laser cutting has a low efficiency, but it has a wide range of applications, from the production of cooling holes in aerospace components to the cutting of cloth in garment manufacture. It is normal practice, because of the size and delicate nature of laser optics, for the laser to be fixed and for the workpiece to be manoeuvred using a multiaxis table. The rigidity of the structure is critical to the quality of the spot, since any vibration will cause the spot to change to an ellipse, with an increase in the cutting time and a reduction in the accuracy. It is common practice to build small-hole laser drills on artificial granite bed-plates since the high density of the structure damps vibration.

In electron beam machining, a focused beam of electrons is used in a similar fashion to a laser, however the beam is generated and accelerated by a cathode–anode arrangement. As the beam consists of electrons it can be steered by the application of a magnetic field. The beam beam can be focused to 10 to 200 μm and a density of 6500 GW mm^{-2} At this power a 125 μm diameter hole in a steel sheet 1.25 mm thick can be cut almost instantly. As in the case of a laser, the bean source is stationary and the workpiece is moved on an X–Y table. The process is complicated by the fact that it is undertaken in a vacuum due to the nature of the electron beam. This requires the use of drives and tables that can operate in a vacuum, and do not contaminate the environment.

Electrochemical machining can be considered to be the reverse of electroplating. Metal is removed from the workpiece, which takes up the exact shape of the tool. This technique has the advantage of producing very accurate copies of the tool, with no tool wear, and it is widely used in the manufacture of moulds for the plastics industry and aerospace components. The principal features of the process are shown in Figure 1.4(b). A voltage is applied between the tool and the workpiece, and material is removed from the workpiece in the presence of an electrolyte. With a high level of electrolyte flow, which is normally supplied via small holes in the tooling, the waste product is flushed from the gap and held in solution prior to being filtered out in the electrolyte-supply plant. While the voltage between the tool and the workpiece is in the range 8–20 V, the currents will be high. A metal removal rate of 1600 mm^{3}min^{-1} per 1000 A is a typical value in industry. In order to achieve satisfactory machining, the gap between the tool and the workpiece has to be kept in the range 0.1–0.2 mm. While no direct machining force is required, the feed drive has to overcome the forces due to the high electrolyte pressure in the gap. Due to the high currents involved, considerable damage would occur if the feed-rate was higher than the required value, and the die and the blank tool collided. To ensure this does not occur, the voltage across the gap is closely monitored, and is used to modify the predefined feed rates, and, in the event of a collision, to remove the machining power.

In electrodischarge machining (see Figure 1.4(c)), a controlled spark is generated using a special-purpose power supply between the workpiece and the electrode. As a result of the high temperature (10 000 °C) small pieces of the workpiece and the tool are vaporised; the blast caused by the spark removes the waste so that it can be flushed away by the electrolyte. The choice of the electrode (for example, copper, carbon) and the dielectric (for example, mineral oil, paraffin, or deionised water) is determined by the material being machined and the quality of the finish required. As material from the workpiece is removed, the electrode is advanced to achieve a constant discharge voltage.

Due to the nature of the process, the electrode position tends to oscillate at the pulse frequency, and this requires a drive with a high dynamic response; a hydraulic drive is normally used, even if the rest of the machine tool has electric drives. A number of different configuration can be used, including wire machining, small-hole drilling, and die sinking. In electrodischarge wire machining, the electrode is a moving wire, which can be moved relative to the workpiece in up to five axes; this allows the production of complex shapes that could not be easily produced by any other means.

Water jet machining involves the use of a very-high pressure of water directed at the material being cut. The water is typically pressurised between 1300–4000 bar. This is forced through a hole, typically 0.18–0.4 mm diameter, giving a nozzle velocity of over 800 m s^{-1}. With a suitable feed rate, the water will cleanly cut through a wide range of materials, including paper, wood and fibreglass. If an abrasive powder, such as silicon carbide, is added to the water a substantial increase in performance is possible though at a cost of increased nozzle ware. With the addition of an abrasive powder, steel plate over 50 mm thick can easily be cut. The key advantages of this process include very low side forces, which allows the user to machine a part with walls as thin as 0.5 mm without damage, allowing for close nesting of parts and maximum material usage. In addition the process does not generate heat hence it is possible to machine without hardening the material, generating poisonous fumes, recasting, or distortion. With the addition of a suitable motion platform, three dimensional machining is possible, similar to electrodischarge wire cutting.

1.2.3 Machining centres

The introduction of CNC systems has had a significant effect on the design of machine tools. The increased cost of machine tools requires higher utilisation; for example, instead of a manual machine running for a single shift, a CNC machine may be required to run continually for an extended period of time. The penalty for this is that the machine's own components must be designed to withstand the extra wear and tear. It is possible for CNC machines to reduce the non-productive time in the operating cycle by the application of automation, such as the loading and unloading of parts and tool changing. Under automatic tool changing a number of tools are stored in a magazine; the tools are selected, when they are required, by a program and they are loaded into the machining head, and as this occurs the system will be updated with changes in the cutting parameters and tool offsets. Inspection probes can also be stored, allowing in-machine inspection. In a machining centre fitted with automatic part changing, parts can be presented to the machine on pallets, allowing for work to be removed from an area of the machine without stopping the machining cycle. This will give a far better usage of the machine, including unmanned operation. It has been estimated that seventy per cent of all manufacturing is carried out in batches of fifty or less. With manual operation (or even with programmable automation) batches of these sizes were uneconomical; however, with the recent introduction of advanced machining centres, the economic-batch size is equal to one.

1.3 Robots

The development of robots can be traced to the work in the United States at the Oak Ridge and Algonne National Laboratories of mechanical teleoperated manipulators for handling nuclear material. It was realised that, by the addition of powered actuators and a stored program system, a manipulator could perform the autonomous and repetitive tasks. Even with the considerable advances in sensing systems, control strategies, and artificial intelligence, the standard industrial robots are not significantly different from the initial concept. Industrial robots can be considered to be general-purpose reprogrammable machine tools moving an end effector, which holds either components or a tool. The functions of a robot are best summarised by considering the following definition of an industrial robot as used by the Robotic Industries Association (Shell and Hall, 2000, p499):

> *An industrial robot is a reprogrammable device designed to both manipulate and transport parts, tools, or specialised manufacturing implements through programmed motions for the performance of specific manufacturing tasks.*

While acceptable, this definition does exclude mobile robots and non-industrial applications. Arkin (1998) on the other hand proposes a far more general definition, namely:

> *An intelligent robot is a machine able to extract information from its environment and use knowledge about its world to move safely in a meaningful and purposive manner.*

1.3.1 Industrial robots

Depending on the type of robot and the application, the mechanical structure of a conventional robot can be divided into two parts, the main manipulator and a wrist assembly. The manipulator will position the end effector while the wrist will control its orientation. The structure of the robot consists of a number of links and joints; a joint allows relative motion between two links. Two types of joints are used: a revolute joint to produce rotation, and a linear or prismatic joint. A minimum of six joints are required to achieve complete control of the end effector's position and orientation. Even though a large number of robot configurations are possible, only five configurations are commonly used in industrial robotics:

- *Polar* (Figure 1.5(a)). This configuration has a linear extending arm (Joint 3) which is capable of being rotated around the horizontal (Joint 2) and vertical axes (Joint 3). This configuration is widely used in the automotive industry due to its good reach capability.

- *Cylindrical* (Figure 1.5(b)). This comprises a linear extending arm (Joint 3) which can be moved vertically up and down (Joint 2) around a rotating column (Joint 1). This is a simple configuration to control, but it has limited reach and obstacle-avoidance capabilities.

- *Cartesian and gantry* (Figure 1.5(c)). This robot comprises three orthogonal linear joints (Joints 1–3). Gantry robots are far more rigid than the basic Cartesian configuration; they have considerable reach capabilities, and they require a minimum floor area for the robot itself.

- *Jointed arm* (Figure 1.5(d)). These robots consists of three joints (Joints 1–3) arranged in an anthropomorphic configuration. This is the most widely used configuration in general manufacturing applications

- *Selective-compliance-assembly robotic arm (SCARA)* (Figure 1.5(e)). A SCARA robot consists of two rotary axes (Joints 1–2) and a linear joint (Joint 3). The arm is very rigid in the vertical direction, but is compliant in the horizontal direction. These attributes make it suitable for assembly tasks.

A conventional robotic arm has three joints; this allows the tool at the end of the arm to be positioned anywhere in the robot's working envelope. To orientate the tools, three additional joints are required; these are normally mounted at the end of the arm in a *wrist* assembly (Figure 1.5(f)). The arm and the wrist give the robot the required six degrees of freedom which permit the tool to be positioned and orientated as required by the task.

The selection of a robot is a significant problem for a design engineer, and the choice depends on the task to be performed. One of the earliest applications of robotics was within a foundry; such environments were considered to be hazardous to human operators because of the noise, heat, and fumes from the process. This is a classic application of a robot being used to replace workers because of environmental hazards. Other tasks which suggest the use of robots include repetitive work cycles, the moving of difficult or hazardous materials, and requirements for multishift operation. Robots that have been installed in manufacturing industry are normally employed in one of four application groups: materials handling, process operations, assembly, or inspection. The control of a robot in the performance of a task necessitates that all the joints can be accurately controlled. A basic robot controller is configured as a hierarchial structure, similar to that of a CNC machine tool; each joint actuator has a local motion controller, with a main supervisory

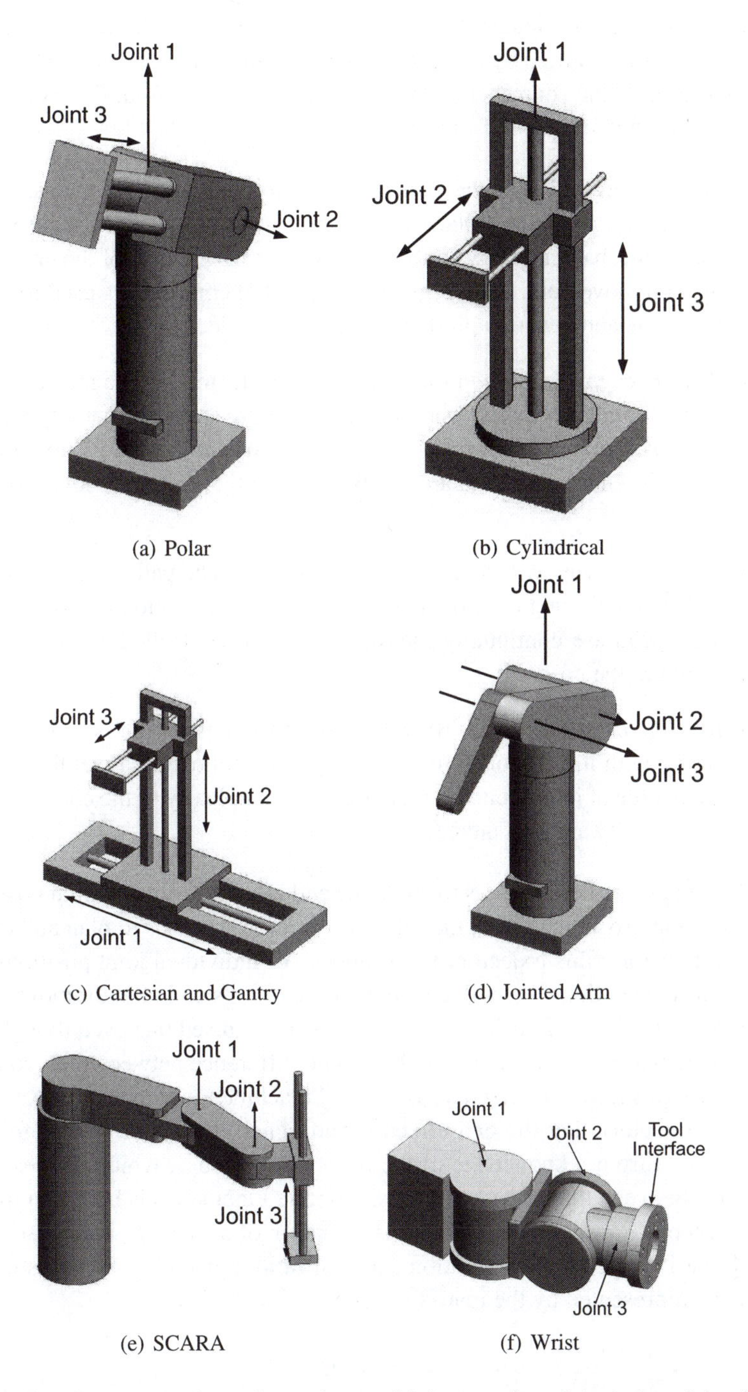

(a) Polar

(b) Cylindrical

(c) Cartesian and Gantry

(d) Jointed Arm

(e) SCARA

(f) Wrist

Figure 1.5. The standard configurations of joints as found in industrial robots, together with the wrist.

controller which coordinates the motion of each joint to achieve the end effector trajectory that is required by the task. As robot control theory has developed so the sophistication of the controller and its algorithms has increased. Controllers can be broadly classified into one of four groups:

- *Limited sequence control.* This is used on low-cost robots which are typically designed for pick-and-place operation. Control is usually achieved by the use of mechanical stops on the robot's joint which control the end positions of each movement. A step-by-step sequential controller is used to sequence the joints and hence to produce the correct cycle.
- *Stored program with point-to-point control.* Instead of the mechanical stops of the limited-sequence robot, the locations are stored in memory and played back as required. However, the end effector's trajectory is not controlled; only the joint end points are verified before the program moves to the next step.
- *Stored program with continuous-path control.* The path control is similar to a CNC contouring controller. During the robot's motion the joint's position and speed are continually measured and are controlled against the values stored in the program.
- *Intelligent-robot control.* By the use of sensors, the robot is capable of interacting with its environment for example, by following a welding seam. As the degree of intelligence is increased the complexity of the control hardware and its software also increase.

The function of the robot is to move the end effector from an initial position to a final position. To achieve this, the robot's control system has to plan and execute a motion trajectory; this trajectory is a sequence of individual joint positions, velocities, and accelerations that will ensure that the robot's end effector moves through the correct sequence of positions. It should be recognised that even though robotic manipulators are being considered, there is no difference between their control and the control of the positioning axes of a CNC machine tool.

The trajectory that the end effector, and hence each joint, has to follow can be generated from a knowledge of the robot's kinematics, which defines the relationships between the individual joints. Robotic kinematics is based on the use of homogeneous transformations. A transformation of a space H is represented by a 4×4 matrix which defines rotation and translation; given a point u, its transform V can be represented by the matrix product

$$V = Hu \tag{1.6}$$

Following an identical argument, the end of a robot arm can be directly related to another point on the robot or anywhere else in space. Since a robot consists of a number of links and joints, it is convenient to use the homogeneous matrix, ${}^0T_i$

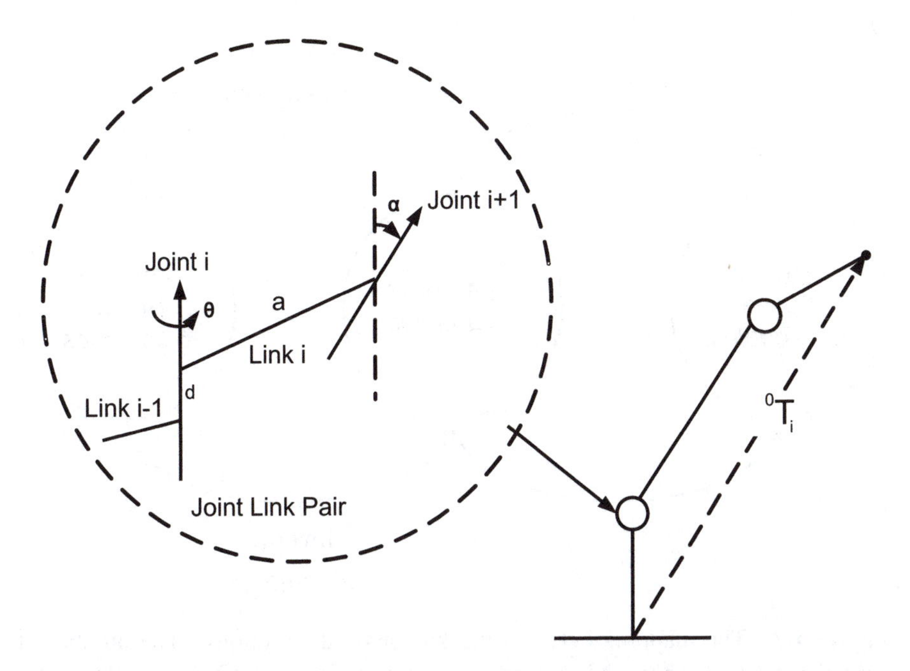

Figure 1.6. Joint-transformation relationships for a robotic manipulator.

(see Figure 1.6). This relationship specifies that the location of the ith coordinate frame with respect to the base coordinate system is the chain product of successive coordinate transformation matrices for each individual joint-link pair, ${}^{i-1}A_i$, which can be expressed as

$$ {}^0T_i = {}^0A_i \, {}^1A_2 \, {}^2A_3 .... {}^{i-1}A_i \tag{1.7} $$

$$ {}^0T_i = \begin{bmatrix} x_i & y_i & z_i & p_i \\ 0 & 0 & 0 & 1 \end{bmatrix} \tag{1.8} $$

Where $[x_i \; y_i \; z_i]$ is the orientation matrix of the ith coordinate with respect to the base coordinate system, and $[p_i]$ is the position vector which points from the origin of the base coordinate system of the ith coordinate frame. Each ${}^{i-1}A_i$ transformation contains information for a single joint-link pair, and it consists of the four parameters: d, the distance between two links along the $i - 1^{th}$ joint axis; θ, the joint angle; a, the distance between two joint axes; and α, the angle between two joint axes.

In any joint-link pair only one parameter can be a variable; θ in a rotary joint, and d for a prismatic or linear joint. The general transformation for a joint-link pair is given by (Paul, 1984)

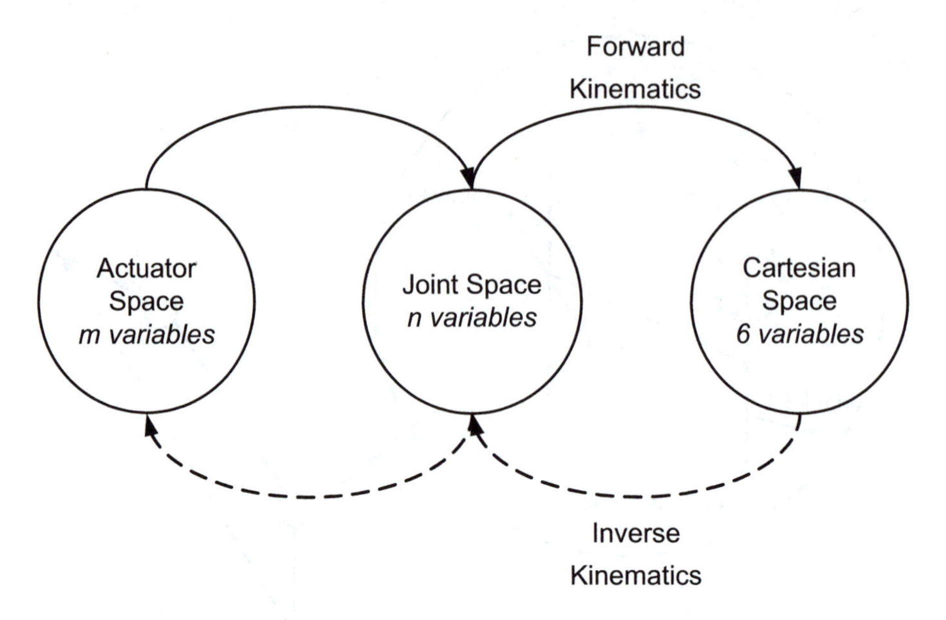

Figure 1.7. The mapping between the kinematic descriptions. The number of variables in Cartesian space is 6, while the number of variables in joint and actuator space is determined by the manipulators's design.

$$^{0}A_i = \begin{bmatrix} \cos\theta_i & -\cos\alpha_i\sin\theta_i & \sin\alpha_i\cos\theta_i & a_i\cos\theta_i \\ \sin\theta_i & \cos\alpha_i\cos\theta_i & -\sin\alpha_i\cos\theta_i & a_i\cos\theta_i \\ 0 & \sin\alpha_i & \cos\alpha_i & d_i \\ 0 & 0 & 0 & 1 \end{bmatrix} \tag{1.9}$$

The solution of the end effector position from the joint variables is termed *forward kinematics*, while the determination of the joint variables from the robot's position in termed *inverse kinematics*. To move the joints to the required position the actuators need to be driven under closed loop control to a required position, within actuator space. The mapping between joint, Cartesian and actuator space space is shown in Figure 1.7. The inverse kinematic is essentially non-linear, as we are given $^{0}T_i$ and are required to find values for $\theta_1.....\theta_n$. If we consider a consider a six-axis robot $^{0}T_6$ has sixteen variables, of which four are trivial, from which we are required to solve for six joints. In principle we have 12 equations with 6 unknowns. However, within the rotational element of the matrix, equation (1.7), only three variables are independent, reducing the number of equations to six. These equations are highly non-linear, transcendental equations, that are difficult to solve. As with any set of non-linear equations their are multiple solutions that need to be considered, the approaches used can either be analytic, or more recently approaches based on neural networks have been investigated.

In order to determine the change of joint position required to change the end effectors' position, use is made on inverse kinematics. Consider the case of a

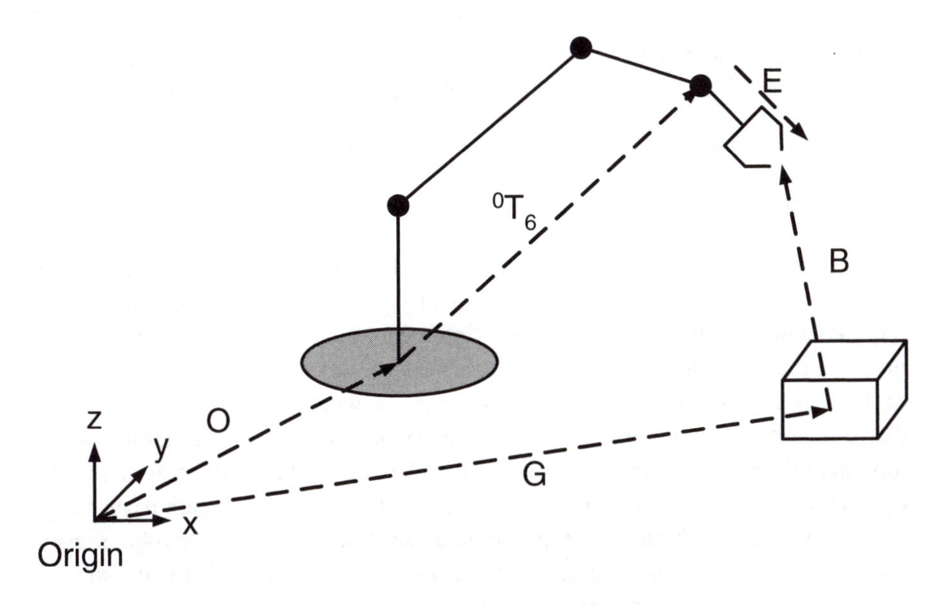

Figure 1.8. The transformations that need to be considered when controlling the position and trajectory of a six axis manipulator.

six-axis manipulator that is required to move an object, where the manipulator is positioned with respect to the base frame by a transform O (see Figure 1.8). The position and orientation of the tool interface of the six-axis manipulator is described by ${}^0T_6$, and the position of the end effector relative to the tool interface is given by E. The object to be moved is positioned at G, relative to the origin, and the location of the end effector relative to the object is B. Hence, it is possible to equate the position of the end effector by two routes, firstly via the manipulator and secondly via the object, giving

$$O\,{}^0T_6E = BG \tag{1.10}$$

and hence

$${}^0T_6 = O^{-1}BGE^{-1} = {}^0A_i\,{}^1A_2\,{}^2A_3 \ldots\ldots {}^5A_6 \tag{1.11}$$

As ${}^0T_6$ is limited to six variables in Cartesian space, the six individual joint positions are determined by solving the resultant six simultaneous equations. However problems will occur when the robot or manipulator is considered to be kinematically redundant. A kinematically redundant manipulator is one that has more than six joints, hence a unique solution is not possible. They are widely found in specialist applications, for example snake-like robots used to inspect the internal structures of nuclear reactors. The inverse kinematics of a redundant manipulator requires the user to specify a number of criteria to solve the ambiguities in the joint positions, for example:

- Maintain the joint positions as close as possible to a specified value, ensuring that the joints do not reach their mechanical limits.
- Minimise the joint torque, potential or kinematic energy of the system.
- Avoid obstacles in the robot's workspace.

In order for a smooth path to be followed, the value of B needs to be updated as a function of time. The robot's positional information is used to generate the required joint position by the inverse kinematic solution of the ${}^0T_6$ matrix. In practice, the algorithms required to obtain these solutions are complicated by the occurrence of multiple solutions and singularities, which are resolved by defining the path, and is solution prior to moving the robot. Usually, it is desirable that the motion of the robot is smooth; hence, the first derivative (that is, the speed) must be continuous. The actual generation of the path required by the application can be expressed as a cubic or higher-order polynomial, see Section 2.4. As the robot moves, the dynamics of the robot changes. If the position loops are individually closed, a poor end-effector response results, with a slow speed and unnecessary vibration. To improve the robot's performance, and increasingly that of CNC machine tools, considerable use of is made of the real-time solution of dynamic equations and adaptive control algorithms, as discussed in Section 10.1.1.

1.3.2 Robotic hands

Dextrous manipulation is an area of robotics where an end effector with co-operating multiple fingers is capable of grasping and manipulating an object. The development of such hands is a significant electromechanical design challenge, as the inclusion of multiple fingers requires a significant numbers of actuators. A dextrous end effector is capable of manipulating an object so that it can be arbitrarily relocated to complete a task. One of the main characteristics of the dextrous manipulation is that it is object and not task centred. It should be noted that *dexterity* and *dextrous* are being used to define attributes to an end effector: a dexterous end effector may not have the ability to undertake a task that a human considers as dextrous. As dextrous manipulation is quintessentially a human activity, a majority of the dextrous robotic end effectors developed to date have significant anthropomorphic characteristics. In view of the importance of this research area a considerable body of research literature on the analysis of the grasp quality and its control is currently available; the review by Okamura et al. (2000) provides an excellent introduction to the field.

As a dextrous end effector needs to replicate some or all the functionality of the human hand, an understanding of human hand functionality is required in the design process. It is recognised that there are five functions attributed to the hand: manipulation, sensation and touch, stabilisation as a means of support, protection, and expression and communication. In robotic systems only the first three need to be considered. The hand can function either dynamically or statically. Its function

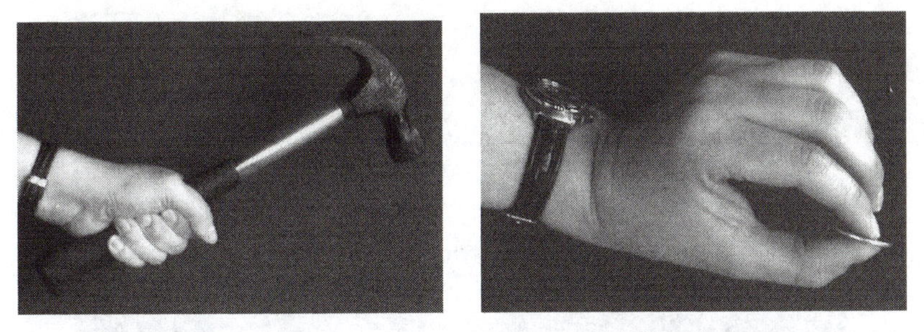

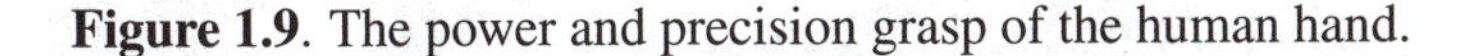

(a) Power grasp (b) Precision grasp

Figure 1.9. The power and precision grasp of the human hand.

as a whole is the sum of many sub-movements; these movements may be used to explore an object, involve actions such as grasping and carrying as well as provide dexterity and maintain stability.

The hand may be used in a multitude of postures and movements, which in most cases involve both the thumb and other digits. There are two basic postures of the human hand: the power grasp and the precision grasp. The power grasp, (Figure 1.9(a)) used where full strength is needed, is where the object is held in a clamp formed by the partly flexed fingers and often a wide area of the palm. The hand conforms to the size and shape of the object. All four fingers are more or less flexed, with each finger accommodating a position so that force can be applied and the force applied to the object to perform a task or resist motion. In a precision grasp (Figure 1.9(b)) there is a greater control of the finger and thumb position than in the power grasp. The precision grasp is carried out between the tip of the thumb and that of one or more of the fingers. The object is held relatively lightly and manipulated between the thumb and related finger or fingers.

The human hand consists of a palm, four fingers and a thumb. The internal structure consists of nineteen major bones, twenty muscles within the hand, a number of tendons from forearm muscles, and a considerable number of ligaments. The muscles in the body of the hand are smaller and less powerful than the forearm muscles and are used more for the precise movements rather than the power grasps. A hand is covered with skin that contains sensors and provides a protective compliant covering.

The classification of movements of the hand in which work is involved can be placed in two main areas: prehensile and non-prehensile. A prehensile movement is a controlled action in which an object is held in a grasp or pinching action partly or wholly in the working envelope of the hand, while a non-prehensile movement is one, which may involve the whole hand, fingers, or a finger but in which no object is grasped or held. The movement may be a pushing one such as pushing an object, or a finger-lifting action such as playing the piano.

The dynamic specification of the hand can be summarised as:

- Typical forces in the range 285–534 N during a power grasp.

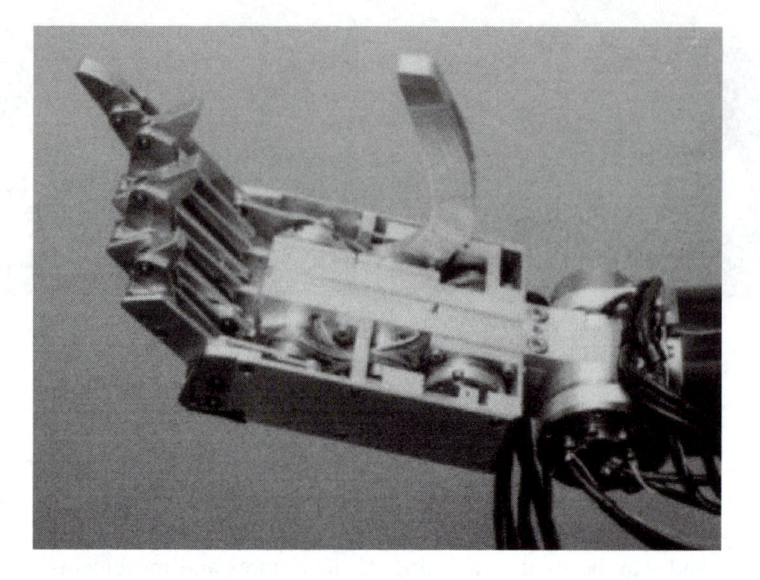

Figure 1.10. The Whole Arm Manipulator's anthropomorphic hand.

- Typical forces in the range 55–133 N during a precision grasp.
- Maximum joint velocity $600°s^{-1}$.
- Maximum repetitive motion frequency, 5 Hz.

End effector technologies

The development of dextrous hands or end effectors has been of considerable importance to the academic robotic research community for many years, and while in no way exhaustive is does however present some of the thinking that has gone into dextrous robotic systems.

- *University of Southampton.* A significant robotic end effector designs was the Whole Arm Manipulator (Crowder, 1991). This manipulator was developed at for insertion into a human sized rubber glove, for use in a conventional glove box. Due to this design requirement, the manipulator has an anthropomorphic end effector with four adaptive fingers and a prehensile thumb. Due to size constraints the degrees of freedom within the hand were limited to three, Figure 3.1.2.

- *Stanford/JPL Hand.* The Stanford/JPL hand (some times termed the Salisbury hand) was designed as a research tool in the control and design of articulated hands. In order to minimise the weight and volume of the hand the motors are located on the forearm of the serving manipulator and use Teflon-coated cables in flexible sleeves to transmit forces to the finger joints.

To reduce coupling and to make the finger systems modular, the designers used four cables for each three degree of freedom finger making each finger identical and independently controllable (Salisbury, 1985).

- *Universities of Southampton and Oxford.* The work at the University of Southampton on prosthetic hands has continued both at Southampton and at Oxford (Kyberd et al., 1998). The mechanics of the hand are very similar to the Whole Arm Manipulator with solid linkages and multiple motors, however the flexibility and power capabilities are closly tailored to prosthetic applications as opposed to industrial handelling.

- *BarrettHand.* One of the most widely cited commercial multifingered dextrous hands is the BarrettHand, this hand combines a high degree of dexterity with robust engineering and is suitable for light engineering applications (Barrett Technology, 2005).

- *UTAH–MIT hand.* The Utah–MIT Dextrous hand (Jacobsen et al., 1986), is an example of an advanced dextrous system. The hand comprises three fingers and an opposed thumb. Each finger consists of a base roll joint, and three consecutive pitch joints. The thumb and fingers have the same basic arrangement, except the thumb has a lower yaw joint in place of the roll joint. The hand is tendon driven from external actuators.

- *Robonaut Hand.* The Robonaut Hand (Ambrose et al., 2000), is one of the first systems being specifically developed for use in outer space: its size and capability is close to that of a suited astronaut's hand. Each Robonaut Hand has a total of fourteen degrees of freedom, consisting of a forearm which houses the motors and drive electronics, a two degree of freedom wrist, and a five finger, twelve degree of freedom hand.

The design of the fingers and their operation are the key to the satisfactory operation of a dextrous hand. It is clear that two constraints exist. The work by Salisbury (1985) indicated that the individual fingers should be multi-jointed, with a minimum of three joints and segments per finger. In addition a power grasp takes place in the lower part of the finger, while during a precision grasp it is the position and forces applied at the fingertip that is of the prime importance. It is normal practice for the precision and power grasp not occur at the same time.

In the design of robotic dextrous end effectors, the main limitation is the actuation technology: it is recognised that an under-actuated approach may be required, where the number of actuators used is less that the actual number of degrees of freedom in the hand. Under-actuation is achieved by linking one or more finger segments or fingers together: this approach was used in Southampton's Whole Arm Manipulator. The location and method of transmission of power is crucial to the successful operation of any end effector, the main being that the end effector size should be compact and consistent with the manipulator.

Actuation

Both fully and under-actuated dextrous artificial hands have been developed using electric, pneumatic or hydraulic actuators. The use of electrically powered actuators have, however, been the most widely used, due to both its convenience and its simplicity compared to the other approaches. The use of electrically powered actuator systems ensures that the joint has good stiffness and bandwidth. One drawback with this approach is the relatively low power to weight/volume ratio which can lead to a bulky solution: however, the developments in magnetic materials and advanced motor design have (and will continue to) reduced this problem. In many designs the actuators are mounted outside the hand with power transmission being achieved by tendons. On the other hand, pneumatic actuators exhibit relatively low actuation bandwidth and stiffness and as a consequence, continuous control is complex. Actuation solutions developed on the basis of pneumatic actuators (if the pump and distribution system are ignored) offer low weight and compact actuators that provide considerable force. Hydraulic actuators can be classified somewhere in between pneumatics and electrically powered actuators. With hydraulics stiffness is good due to the low compressibility of the fluid. While pneumatic actuators can be used with gas pressures up to 5–10 MPa, hydraulic actuators will work with up to 300 MPa. One approach that is being considered at present is the development of artificial muscles. Klute et al. (2002) provide a detailed overview of the biomechanics approach to aspects of muscles and joint actuation. In addition the paper presents details of a range of muscle designs, including those based on pneumatic design which are capable of providing 2000 N of force. This force equates to that provided by the human's triceps. The design consists of a inflatable bladder sheathed double helical weave so that the actuator contracts lengthwise when it expands radially. Other approaches to the the design for artificial mussels have been based on technologies including shape-memory alloy (see Section 9.5), electro-resistive gels, and stepper motors connected to ball screws.

When considering conventional technologies the resultant design may be bulky and therefore the actuators have to be placed somewhere behind the wrist to reduce system inertia. In these systems power is always transmitted to the fingers by using tendons or cables. Tendon transmission systems provide a low inertia and low friction approach for low power systems. As the force transmitted increases considerable problems can be experienced with cable wear, friction and side loads in the pulleys. One of the main difficulties in controlling tendon systems is the that force is unidirectional – a tendon cannot work in compression. The alternative approach to joint actuation is to used a solid link which has a bi-directional force characteristic, thus it can both push and pull a finger segment. The use of a solid link reduced the number of connections to an individual finger segment. The disadvantage of this approach is a slower non-linear dynamic response, and that ball screw or crank arrangement is required close to the point of actuation. Irrespective of the detailed design of the individual fingers, they are required to be mounted on a supporting structure, this is more fully discussed in Pons et al. (1999).

1.3.3 Mobile robotics

In recent years there has been a considerable increase in the types and capabilities of mobile robots, and in general three classes can be identified: UAV (unmanned aerial vehicles), UGV (unmanned ground vehicles) and UUV (unmanned underwater vehicles). In certain cases the design and control theory for a mobile robot has drawn heavily on biological systems, leading to a further class, biologically inspired robotics. An early example of this type of robot was the *Machina Speculatrix* developed by W. Grey Walter (Holland, 2003), which captured a number of principles including simplicity, exploration, attraction, aversion and discernment. Since this original work a considerable number of robots have been developed including both wheeled and legged. The applications for mobile robots are wide-ranging and include:

- *Manufacturings systems.* Mobile robots are widely used to move material around factories. The mobile robot is guided through the factory by the use of underfloor wiring or visual guidelines. In most systems the robots follow a fixed path under the control of the plant's controller, hence they are able to move product on-demand.

- *Security systems.* The use of a mobile robot is considered to be a cost effective approach to patrolling large warehouses or other buildings. Equipped with sensors they are able to detect intruders and fires.

- *Ordinance and explosive disposal.* Large number of mobile robots have been developed to assist with searching and disposal of explosives, one example being the British Army's Wheelbarrow robots that have been extensively used in Northern Ireland. The goal of these robots is to allow the inspection and destruction of a suspect device from a distance without risking the life of a bomb disposal officer.

- *Planetary exploration.* Figure 1.11 shows an artist's impression of one of the two Mars rovers that were landed during January 2004. *Spirit* and *Opportunity* have considerably exceeded their primary objective of exploring Mars for 90 days. At the time of writing, both rovers have been on Mars for over a year and have travelled approximately 3 Km. During this time, sending back to Earth over 15 gigabytes of data, which included over 12,000 images. Of particular interest is that to achieve this performance each rover incorporated 39 d.c. brushed ironless rotor motors (see Section 5.2.1). The motors were of standard designs with a number of minor variation, particularly as the motors have to endure extreme conditions, such as variations in temperature which can range from -120°C to +25°C.

Figure 1.11. An artist's impression of the rover *Spirit* on the surface of Mars. The robotic arm used to position scientific instruments is clearly visible. Image reproduced courtesy of NASA/JPL-Caltech.

1.3.4 Legged robots

While the majority of mobile robots are wheeled, there is increasing interest in legged systems, partly due to increased research activity in the field of biologically inspired robotics. One example is shown in Figure 1.12. Many legged designs have been realised, ranging from military logistic carriers to small replicas of insects. These robots, termed biometric robots, mimic the structure and movement of humans and animals. Of particular interest to the research community is the construction and control of dynamically stable legged robots. In the design of these systems the following constraints exist (Robinson et al., 1999).

- The robot must be self supporting. This puts severe limits on the force/mass and power/mass ratio of the actuators.
- The actuators of the robot must not be damaged during impact steps or falls and must maintain stability following an impact.
- The actuators need to be force controllable because the algorithms used for robot locomotion are force based.

One of the most successful sets of legged robots has been based on a series elastic actuator, which has a spring in series with the transmission and the actuator output (Pratt and Williamson, 1995). The spring reduces the actuators's bandwidth

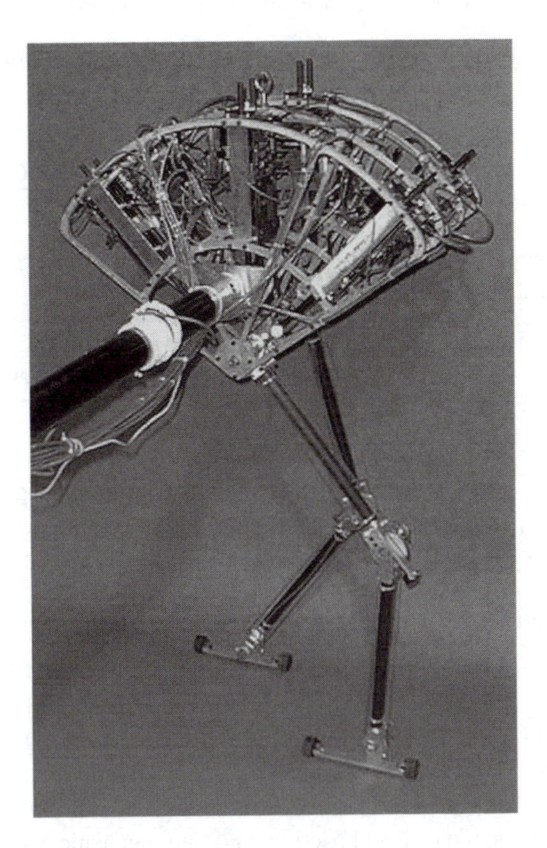

Figure 1.12. Spring Flamingo, a 6 degree-of-freedom planar biped robot, was developed at the MIT Leg Laboratory. Spring Flamingo is capable of human-like walking at speeds of up to 1.25 m s^{-1}. Picture reproduce with permission from Jerry Pratt, Yobotics, Cincinnati, OH.

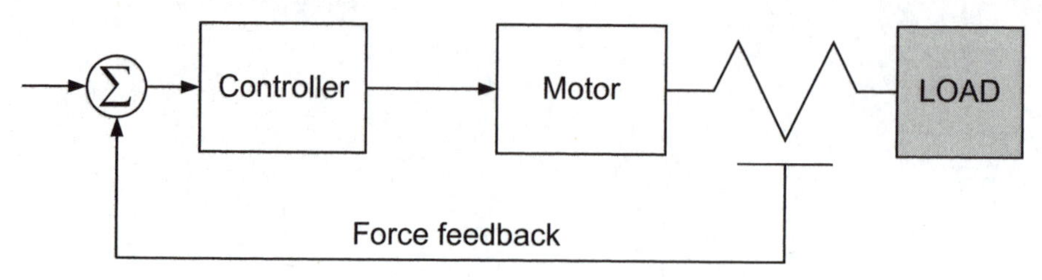

(a) Block diagram of the actuator's control loop

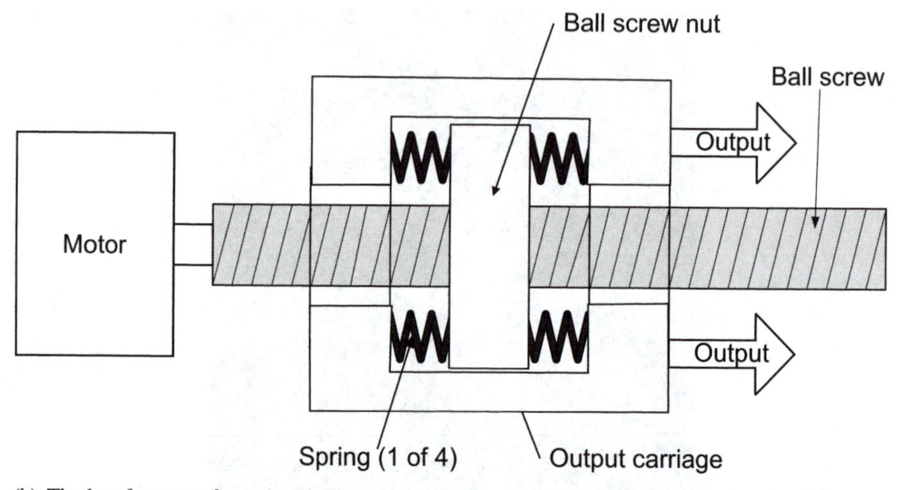

(b) The key features of a series elastic actuator. The output carriage is connected to the ball screw nut solely by the four springs; the required support bearings have been omitted for clarity.

Figure 1.13. The operation of the series elastic actuator.

somewhat, but for a low bandwidth application, such as walking, this is unimportant. In exchange, Series Elastic Actuators are low motion, high force/mass, high power/mass actuators with good force control as well as impact tolerance. In addition they have low impedance and friction, and thus can achieve high quality force control.

Figure 1.13(a) shows the architecture of a series elastic actuator. It should be recognised that the series elastic actuators is similar to any motion actuator with a load sensor and closed loop control system.

The series elastic actuator uses active force sensing and closed loop control to reduce friction and inertia. By measuring the compression of the compliant element, the force on the load can be calculated using Hooke's law. A feedback controller calculates the error between the actual force and the desired force: applying appropriate current to the motor will correct any force errors. The actuator's design introduces compliance between the actuator's output and the load, allowing for greatly increased control gains.

In practice the series elastic actuator consists of two subassemblies: a drive train subassembly and an output carriage subassembly, Figure 1.13(b). When assembled, the output carriage is coupled to the drive train through springs. During operation, the servomotor directly drives the ball screw, the ball nut direction of travel depending on the direction of motor rotation. The rotary motion of the motor is converted to linear motion of a ball nut which pushes on the compression springs that transmit forces to the load. The force on the load is calculated by measuring the compression of the springs using position transducers, such as a linear potentiometer or linear variable differential transformer as discussed in Section 4.3.2.

1.4 Other applications

1.4.1 Automotive applications

In a modern car, small electric motors undertake functions that were formerly considered the domain of mechanical linkages or to increase driver comfort or safety. The conventional brushed d.c. motors, can be found in body and convenience areas, for example windscreen wipers and electric windows. Increasingly brushless motors are also being used in open loop pump drives and air conditioning applications. Figure 1.14 shows a number of electrically operated functions in a modern car. Many top of the range models currently, or will shortly incorporate systems such as intelligent brake-control, throttle-by-wire and steer-by-wire that require a sensor, a control unit and an electric motor. It has been estimated that the electrical load in a car will increase from to around 2.5 kW, with a peak value of over 12 kW. This implies that the electrical system will have to be redesigned from the current 12 V d.c. technology to distribution and utilisation at higher voltages.

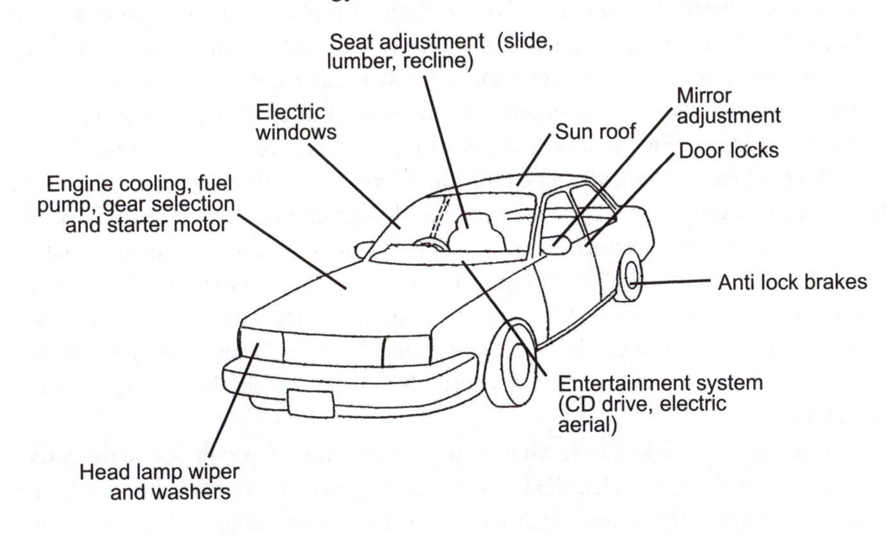

Figure 1.14. Typical electrically operated functions in a modern car.

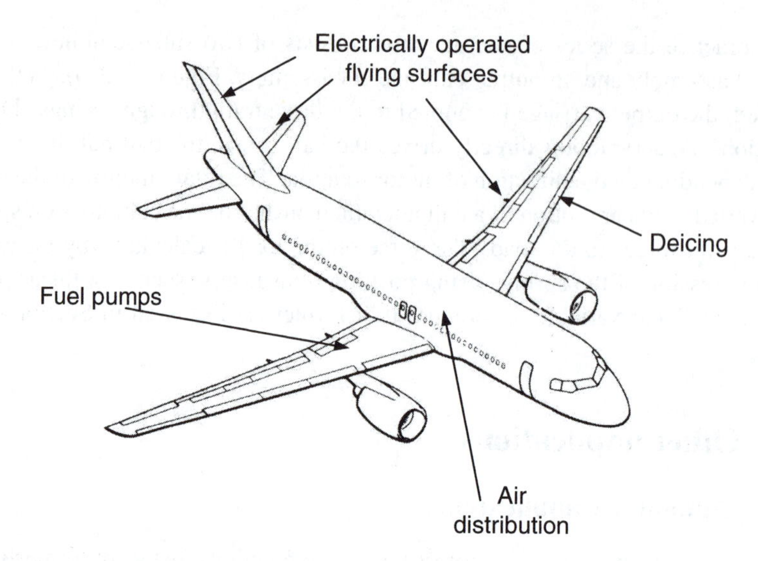

Figure 1.15. The all-electric-aircraft concept, show the possible location of electrically powered actuators, or drives within a future civil aircraft.

One of the possible options is a multivoltage system with some functions remaining at 12 V, and others operating from voltages as high as 48 V, (Kassakian et al., 1996).

1.4.2 Aerospace applications

The flying surfaces of civil aircraft are conventionally powered through three independent and segregated hydraulic systems. In general, these systems are complex to install and costly to maintain. The concept of replacing the hydraulic system with electric actuation, coupled with changes to the electric generation technology and flight control systems, is commonly termed either the *more-electric-aircraft* or the *all-electric-aircraft* depending on the amount of electrical system incorporated, Figure 1.15. Electrically powered flight systems are not new: a number of aircraft developed in the 1950's and 60's incorporated electrically actuated control functions, however they were exceptions to the general design philosophy of the time. Recently there has been increasing interest in electrically powered actuators due to the increased reliability of power electronics, and the continual drive for the reduction of operating cost of the aircraft though weight reduction and increased fuel efficiency. The use of electrically powered flying surfaces in civilian aircraft is still rare, but a number of military aircraft are fitted with electrically powered actuators.

In the more- or all-electric-aircraft the distribution of power for flight actuation will be through the electrical system, as opposed to the currently used bulk hydraulic system. It has been estimated that the all-electric-aircraft could have a weight reduction of over 5000 kg over existing designs, which could be converted

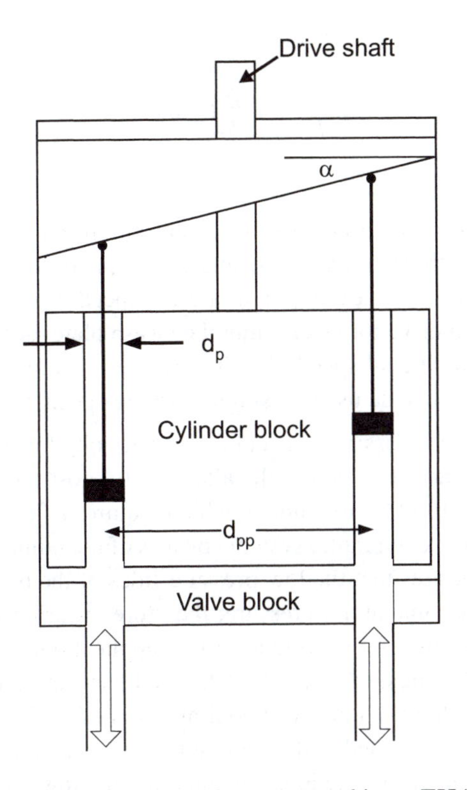

Figure 1.16. The displacement hydraulic pump used in an EHA. Driving the valve cylinder causes the pistons to operate, the amount of stroke is determined either by the rotation speed, or the swash plate angle, α. The clearances between the cylinder block and, the valve block and casing have been exaggerated.

into an increase in range or a reduction in fuel costs. In order to implement power-by-wire, high-performance electrically powered actuators and related systems are required, (Howse, 2003). Electrically powered flight actuators can take one of two principal configurations, the electromechanical actuator with mechanical gearing, and the electrohydrostatic actuator, or EHA, with fluidic gearing, between the motor and the actuated surface.

In an EHA, hydraulic fluid is used to move a conventional hydraulic actuator, the speed and direction of which are controlled by the fluid flow from an electric motor driven hydraulic pump. If a displacement pump (see Figure 1.16) is used, where the piston's diameter is d_p and the pitch diameter is d_{pp}; the flow rate $Q(t)$ as a function of the pump speed, $\omega_p(t)$ can be determined to be

$$Q(t) = D\omega_p(t) \tag{1.12}$$

where the pump constant, D, is given by

$$D = \frac{\pi d_p^2 d_{pp} \tan\alpha}{4} \tag{1.13}$$

Hence the flow rate in a variable-displacement pump unit can controlled by adjustment of the swash plate angle, α, and hence piston displacement . In this approach two motor-drives are required, a fixed speed drive for the pump, and a small variable-speed drive for positioning the swash plate. A different approach is just to control the rotational speed of a displacement pump, ω_p, where α is fixed this design only requires the use of a single variable-speed motor drive.

Figure 1.17 shows a possible concept for an electrohydrostatic actuator suitable for medium power surfaces, such as the ailerons. In most future designs the fixed pump option will be used for the rudder, which requires a far higher power output. In the actuator the basic hydraulic system consists of the pump, actuator, and accumulator. A valve ensures that the low pressure sides of the pump and actuator are maintained at the accumulator's pressure, therefore ensuring that cavitation does not occur in the system. As envisaged in this application, the motor can have a number of special features, in particular a flooded 'air gap', allowing the motor to be cooled by the hydraulic fluid. The cooling oil is taken from the high pressure side of the pump, and returned to the accumulator via an additional valve. The accumulator has a number of functions: maintaining the low pressure in the system to an acceptable value, acting as the hydraulic fluid's thermal radiator, and making up any fluid loss. It is envisaged that the unit is sealed at manufacture, and then the complete actuator considered to be a line replacement unit.

The flow of hydraulic fluid, and hence the actuator's displacement, is determined by the pump's velocity. To obtain the specified required slew rate, the required motor speed of approximately 10,000 rpm will be required, depend on the pump and actuators size. It should be noted that when the actuator is stationary, low speed operation (typically 100 rpm) is normally required, because of the leakage flow across the actuator and pump. The peak pressure differential within a typical system is typically 200 bar.

The motor used in this application can be a sinusoidally wound permanent magnet synchronous motor, the speed controller with vector control to achieve good low speed performance. An outer digital servo loop maintains the demanded actuator position, with a LVDT measuring position. The controller determines the motor, and hence pump velocity. Power conversion is undertaken using a conventional three phase IGBT bridge. In an aircraft application the power will be directly supplied from the aircraft's bus, in the all-electric-aircraft this is expected to be at 270 V dc, as opposed to the current 110 V, 400 Hz ac systems. To prevent excessive bus voltages when the motor drive is regenerating under certain aerodynamic conditions, particularly when the aerodynamic loading back drives the actuator, a bus voltage regulator to dissipate excess power is required, see Section 5.4.

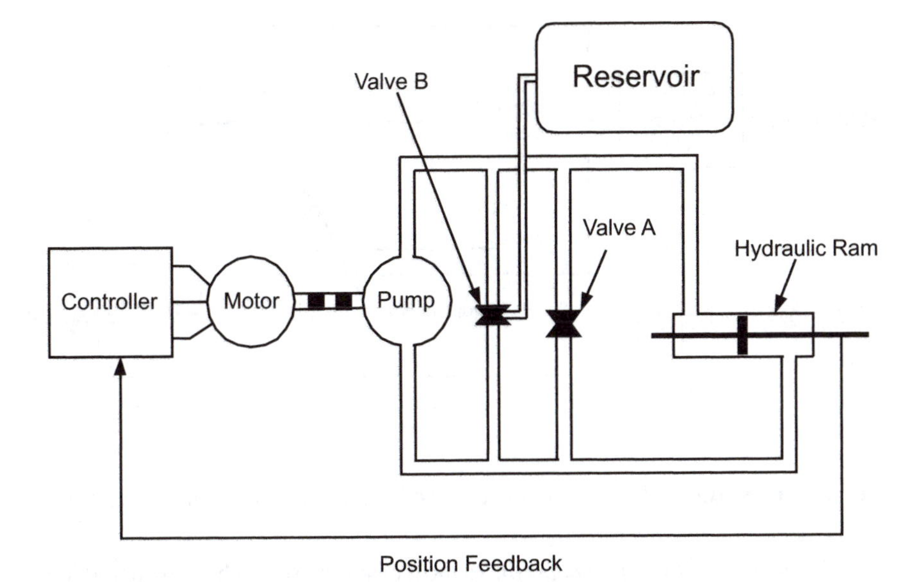

Figure 1.17. Concept of an electrohydrostatic Actuator for use in an aircraft. Value A is a bypass valve that can allow the ram to move under external forces in the case of a failure, while valve B ensures that the pressure of the input side of the pump does not go below that of the reservoir.

1.5 Motion-control systems

In this brief review of the motion requirements of machine tools, robotics and related systems, it is clear that the satisfactory control of the axes, either individually or as a coordinate group, is paramount to the implementation of a successful system. In order to achieve this control, the relationship between the mechanical aspects of the complete system and its actuators needs to be fully understood. Even with the best control system and algorithms available, it will not perform to specification if the load cannot be accelerated or decelerated to the correct speed within the required time and if that speed cannot be held to the required accuracy.

A motion-control system consists of a number of elements (see Figure 1.18) whose characteristics must be carefully determined in order to optimise the performance of the complete system. A motion control system system consists of five elements:

- *The controller,* which implements the main control algorithms (normally either speed or position control) and provides the interface between the motion-control system and the main control system and/or the user.

- *The encoders and transducers,* required to provide feedback of the load's position and speed to the controller.

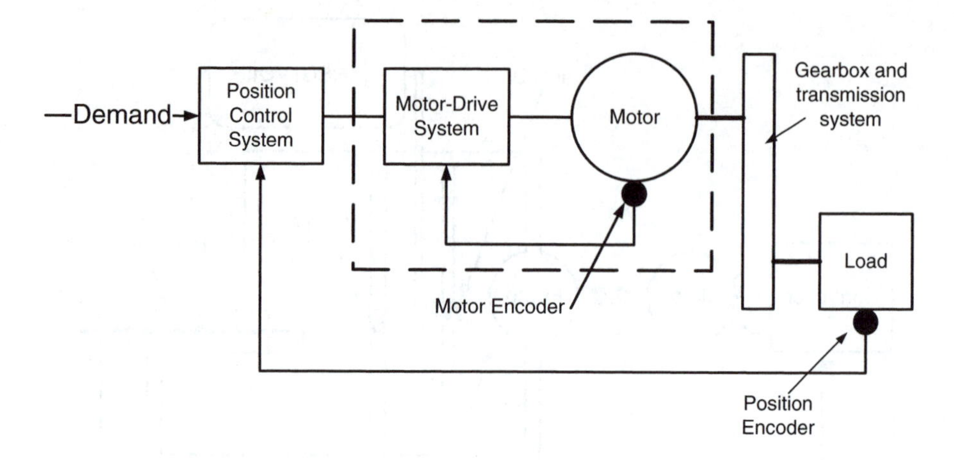

Figure 1.18. Block diagram of an advanced electric motion-control system.

- *The motor controller, and motor.* In most cases, these can be considered to be an integral package, as the operation and characteristics of the motor being totally dependent on its control package.
- *The transmission system.* This takes the motor output and undertakes the required speed changes and, if required, a rotary-to-linear translation.
- *The load.* The driven elements greatly influences the operation of the complete system. It should be noted that a number of parameters, including inertia, external loads, and friction, may vary as a function of time, and need to be fully determined at the start of the design process.

The key to successful implementation of a drive system is full identification of the applications needs and hence its requirements; these are summarised in Table 1.2. In order to select the correct system elements for an application, a number of activities, ranging over all aspects of the application, have to be undertaken. The key stages of the process can be identified as follows:

- *Collection of the data.* The key to satisfactory selection and commissioning of a motor drive system is the collection of all the relevant data before starting the sizing and selection process. The information obtained will mostly relate to the system's operation, but may also include commercial considerations.
- *Sizing of the system.* The size of the various drive components can be determined on the basis of the data collected earlier.
- *Identification of the system to be used.* Once the size of the various elements and the application requirements are known, the identity of the various elements can be indicated. At this stage, the types of motor, feedback transducer, and controller can finalised.

Table 1.2. Requirements to be considered in the selection of a motor-drive system.

Load	Maximum speed Acceleration and deceleration Motion profile Dynamic response External forces
Environmental factors	Safety and risk Electromagnetic compatibility Climatic and humidity ranges Electrical supply specifications and compatibility
Life-cycle costs	Initial costs Operational costs Maintenance costs Disposal costs
System integration	Mechanical fittings Bearing and couplings Cooling Compatibility with existing systems

- *Selection of the components.* Using the acquired knowledge, the selection process can be started. If the items cannot be supplied and need to be re-specified, or the specification of a component is changed, the effect on the complete system must be considered.

- *Verification.* Prior to procuring the components, a complete check must be made to ensure that the system fits together in the space allocated by the other members of the design team.

- *Testing.* Theoretically, if all the above steps have been correctly followed, there should be no problems. But this is not always the case in the real world, commissioning modification may be required. If this is required care must be taken to ensure that the performance of the system is not degraded.

One of the main design decisions that has to be taken is the selection of the correct motor technology. With the rapid development in this field, number of options are available; each option will have benefits and disadvantages. In the consideration of the complete system the motor determines the characteristic of the drive, and it also determines the power converter and control requirements. A wide range of possibilities exist, however only a limited number of combinations will have the broad characteristics which are necessary for machine-tool and robotic applications, namely:

- A high speed-to-torque ratio.

- Four-quadrant capability.
- The ability to produce torque at standstill.
- A high power-to-weight ratio.
- High system reliability.

The following motor-drive systems satisfy these criteria, and are widely used in machine tool, robotic and other high performance applications:

- Brushed, permanent-magnet, d.c. motors with a pulse width modulated or linear drive systems (see Chapter 5);
- Brushless, d.c., permanent-magnet motors, either with trapezoidal or sinusoidal windings (see Chapter 6);
- Vector, or flux-controlled induction motors (see Chapter 7);
- Stepper motors (see Chapter 8).

With the exception of brushed, permanent-magnet d.c. motors, all the other machines are totally dependent on their power controller, and they will be treated as integrated drives. The list above covers most widely used motors, however recent development have allowed the introduction of other motors, ranging from the piezoelectric motor to the switched reluctance motor; these and other motors are briefly discussed in Chapter 9.

1.6 Summary

This chapter has briefly reviewed a number of typical application areas where high performance servo drives are required. It has been clearly demonstrated that the satisfactory performance of the overall system is dependent on all the components in the motor-drive system and its associated controllers; in particular, it is dependent on its ability to provide the required speed and torque performance. The determination of the characteristics that are required is a crucial step in the specification of such systems, and this will be discussed in subsequent chapters.

Chapter 2

Analysing a drive system

To achieve satisfactory operation of any motion-control system, all the components within the system must be carefully selected. If an incorrect selection is made, either in the type or the size of the motor and/or drive for any axis, the performance of the whole system will be compromised. It should be realised that over-sizing a system is as bad as under-sizing; the system may not physically fit and will certainly cost more. In the broadest sense, the selection of a motor-drive can be considered to require the systematic collection of data regarding the axis, and its subsequent analysis.

In Chapter 1 an overview of a number of applications were presented, and their broad application requirements identified. This chapter considers a number of broader issues, including the dynamic of both rotary and linear systems as applied to drive, motion profiles and aspects related to the integration of a drive system into a full application. With the increasing concerns regarding system safety in operation the risks presented to and by a drive are considered, together with possible approaches to their mitigation.

2.1 Rotary systems

2.1.1 Fundamental relationships

In general a motor drives a load through some form of transmission system in a drive system and although the motor always rotates the load or loads may either rotate or undergo a translational motion. The complete package will probably also include a speed-changing system, such as a gearbox or belt drive. It is convenient to represent such systems by an equivalent system (see Figure 2.1); the fundamental relationship that describes such a system is

$$T_m = T_L + I_{tot}\frac{d\omega_m}{dt} + B\omega_m \tag{2.1}$$

where I_{tot} is the system's total moment of inertia, that is, the sum of the inertias of the transmission system and load referred to the motor shaft, and the inertia

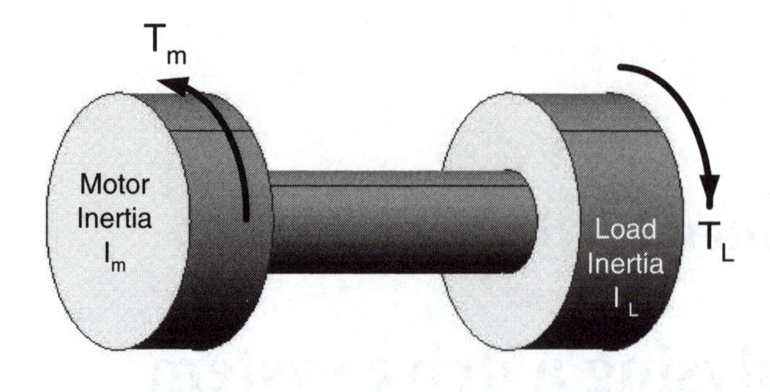

Figure 2.1. The equivalent rotational elements of a motor drive system.

of the motor's rotor (in kg m^2); B is the damping constant (in N rad^{-1} s); ω_m is the angular velocity of the motor shaft (in rad s^{-1}); T_L is the torque required to drive the load referred to the motor shaft (in Nm), including the external load torque, and frictional loads (for example, those caused by the bearings and by system inefficiencies); T_m is the torque developed by the motor (in Nm).

When the torque required to drive the load (that is, $T_L + B\omega_m$) is equal to the supplied torque, the system is in balance and the speed will be constant. The load accelerates or decelerates depending on whether the supplied torque is greater or lower than the required driving torque. Therefore, during acceleration, the motor has to supply not only the output torque but also the torque which is required to accelerate the inertia of the rotating system. In addition, when the angular speed of the load changes, for example from ω_1 to ω_2, there is a change in the system's kinetic energy, E_k, given by

$$\Delta E_k = \frac{I_{tot}(\omega_2^2 - \omega_1^2)}{2} \tag{2.2}$$

I_{tot} is the total moment of inertia that is subjected to the speed change. The direction of the energy flow will depend on whether the load is being accelerated or decelerated. If the application has a high inertia and if it is subjected to rapid changes of speed, the energy flow in the drive must be considered in some detail, since it will place restrictions on the size of the motor and its drive, particularly if excess energy has to be dissipated, as discussed in Section 5.4.

Moment of inertia

In the consideration of a rotary system, the body's *moment of inertia* needs to be considered, which is the rotational analogue of mass for linear motion. For a point mass the moment of inertia is the product of the mass and the square of perpendicular distance to the rotation axis, $I = mr^2$. If a point mass body is considered within a body, Figure 2.2, the following definitions hold:

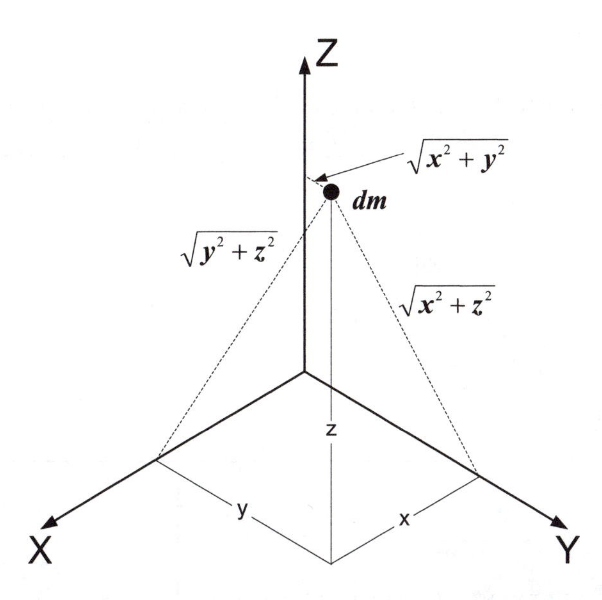

Figure 2.2. Calculation of the moment of inertia for a solid body, the elemental mass, together the values of r for all three axes.

$$I_{xx} = \int (y^2 + z^2)dm \tag{2.3}$$

$$I_{yy} = \int (x^2 + z^2)dm \tag{2.4}$$

$$I_{yy} = \int (x^2 + y^2)dm \tag{2.5}$$

For a number of basic components, the moments of inertia is given in Table 2.1. From this table it is possible to calculate the moment of inertia around one axis, and then compute the moment of inertia, I, around a second parallel axis, using the *parallel axis theorem*, where

$$I = I_G + Md^2 \tag{2.6}$$

where I_G is the moment of inertia of the body, M is the mass and d is the distance between the new axis of rotation and the original axis of rotation.

2.1.2 Torque considerations

The torque which must be overcome in order to permit the load to be accelerated can be considered to have the following components:

- *Friction torque*, T_f, results from relative motion between surfaces, and it is found in bearings, lead screws, gearboxes, slideways, etc. A linear friction model that can be applied to a rotary system is given in Section 2.3.

Table 2.1. Moment of inertia for a number of bodies with uniform density.

Body		I_{xx}	I_{yy}	I_{zz}
Slender bar	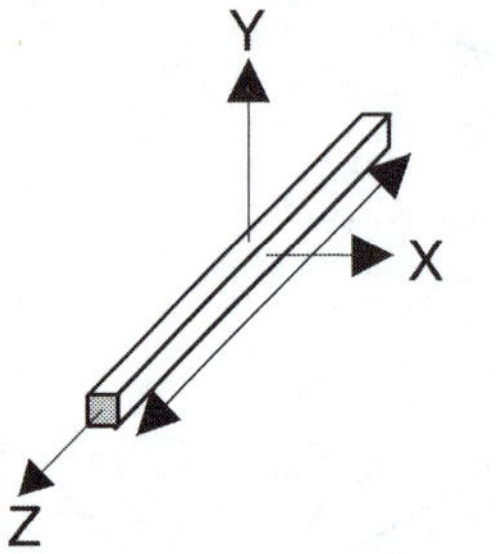	-	$\frac{ml^2}{12}$	$\frac{ml^2}{12}$
Cuboid	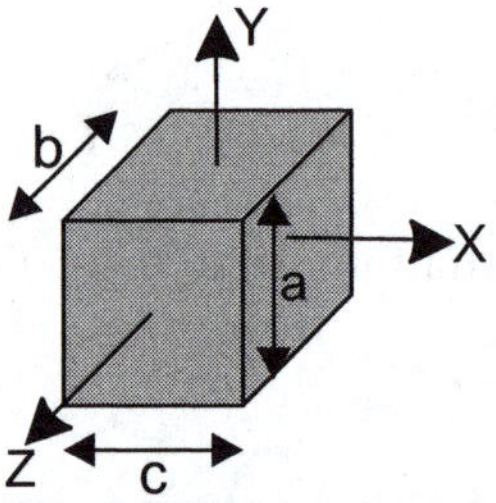	$\frac{m}{12}(a^2+c^2)$	$\frac{m}{12}(a^2+c^2)$	$\frac{m}{12}(b^2+c^2)$
Disc	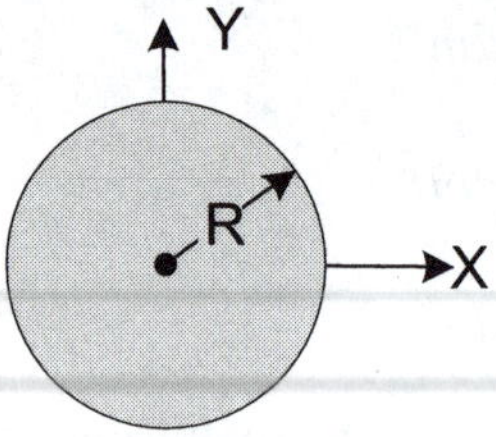	$\frac{mR^2}{4}$	$\frac{mR^2}{4}$	$\frac{mR^2}{2}$
Cylinder	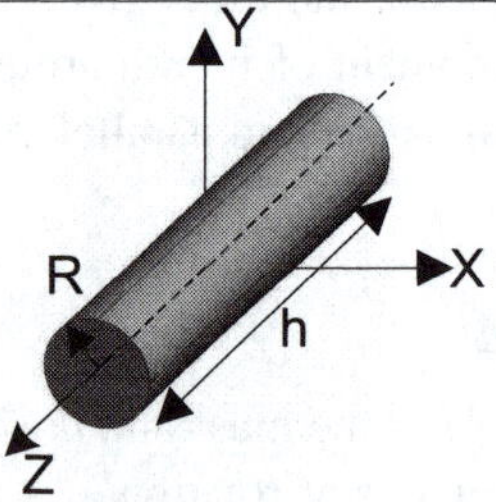	$\frac{m}{12}(3R^2+h^2)$	$\frac{m}{12}(3R^2+h^2)$	$\frac{mR^2}{2}$
Sphere	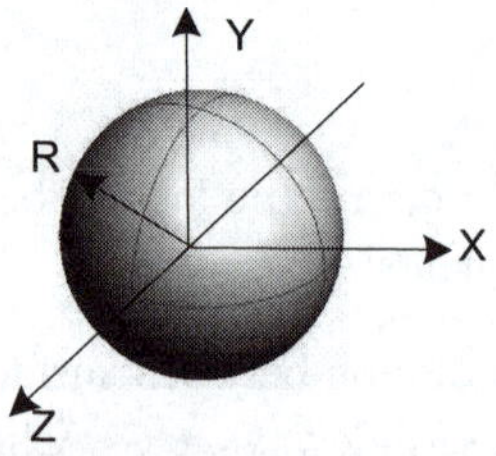	$\frac{2}{5}mR^2$	$\frac{2}{5}mR^2$	$\frac{2}{5}mR^2$

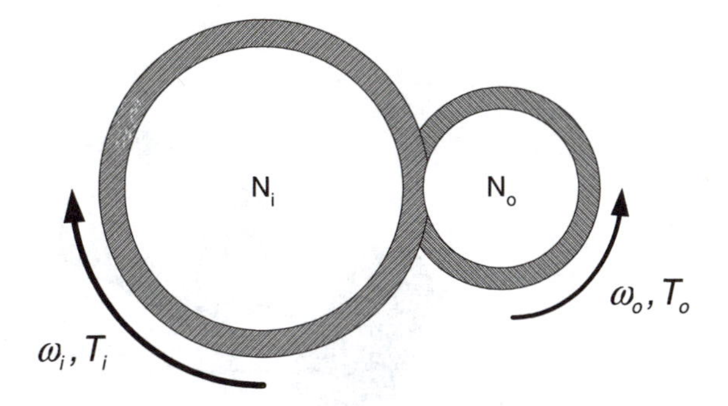

Figure 2.3. The relationship between the input and output of a single stage gear. The gear ratio is calculated from the ratio of teeth on each gearwheel, $N_o : N_i$ or $n : 1$.

- *Windage torque*, T_w is caused by the rotating components setting up air (or other fluid) movement, and is proportional to the square of the speed.
- *Load torque*, T_L, required by the application, the identification of which has been discussed in part within Chapter 1. The load torque is also required to drive the power train, which will be discussed in Chapter 3.

2.1.3 Gear ratios

In a perfect speed-changing system (see Figure 2.3), the input power will be equal to the output power, and the following relationships will apply:

$$T_o = \pm n T_i \tag{2.7a}$$

$$\omega_o = \pm \frac{\omega_i}{n} \tag{2.7b}$$

$$I_{eff} = \frac{I_L}{n^2} \tag{2.7c}$$

The '$\pm$' is determined by the design of the gear train and the number of reduction stages this is discussed more fully in Section 3.1.

If a drive system incorporating a gearbox is considered, Figure 2.4, the dynamics of the system can be written in terms of the load variables, giving

$$T_m - \frac{T_L}{n} = T_{diff} = \alpha_L (I_L + I_m n^2) + \omega_L (B_L + B_m n^2) \tag{2.8}$$

where I_L is the load inertia, I_m is the motor's inertia, B_L is the load's damping term, B_m is the motor's damping term, α_L is the load's acceleration and ω_L is the load's speed. Whether the load accelerates or decelerates depends on the difference between the torque generated by the motor and the load torque reflected through the gear train, T_{diff}. In equation (2.8), the first bracketed term is the effective inertia and the second the effective damping. It should be noted that in determining the

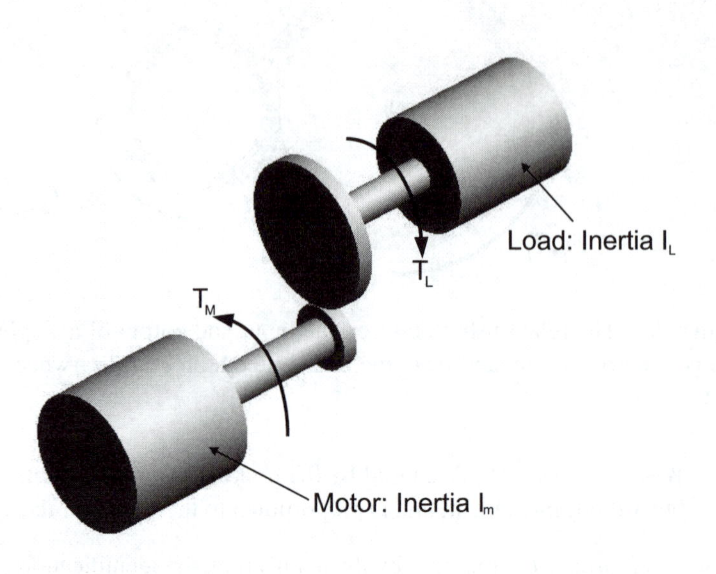

Figure 2.4. A motor connected through gearing to an inertial load.

effective value, all the rotating components need to be considered. Therefore, the inertia of the shafts, couplings, and of the output stage of the gearbox need to be added to the actual load inertia to determine the effective inertia. It should also be noted that if $n \gg 1$ then the motor's inertia will be a significant part of the effective inertia.

As noted in Chapter 1, the drives of robots and machine tools must continually change speed to generate the required motion profile. The selection of the gear ratio and its relationship to the torque generated from the motor must be fully considered. If the load is required to operate at constant speed, or torque, the optimum gear ratio, n^*, can be determined. In practice, a number of cases need to be considered, including acceleration with and without an externally applied load torque and the effects of variable load inertias.

2.1.4 Acceleration without an external load

If a motor is capable of supplying a peak torque of T_{peak} with a suitable drive, the acceleration, α, of a load with an inertia I_L, through a gear train of ratio $n:1$ is given by

$$\alpha = \frac{T_{peak}}{n(I_d + I_L/n^2)} \tag{2.9}$$

The term in parentheses is the total inertia referred to the motor; I_L includes the inertia of the load, and the sum of inertias of the gears, shafts, and couplings rotating at the system's output speed; I_d includes the inertias of the motor's rotor, connecting shafts, gears, and couplings rotating at the motor's output speed. In the case of a belt drive, the inertias of the belt, pulleys, and idlers must be included and referred to the correct side of the speed changer. The optimum gear ratio, n^*, can be determined from equation (2.9), by equating $dT_{peak}/dn = 0$, to give

$$n^* = \sqrt{\frac{I_L}{I_d}} \tag{2.10}$$

Therefore, the inertia on the input side of the gearing has to be equal to the reflected inertia of the output side to give a maximum acceleration of the load of

$$\alpha_{peak} = \frac{T_{peak}}{2I_d n^*} \tag{2.11}$$

The value of α_{peak} is the load acceleration; the acceleration of the motor will be n^* times greater. The acceleration parameters of a motor should be considered during its selection. In practice, this will be limited by the motor's construction, particularly if the motor is brushed and a cooling fan is fitted. Since the acceleration torque is a function of the motor current, the actual acceleration rate will be limited by the current limit on the drive. This needs to be carefully considered when the system is being commissioned.

2.1.5 Acceleration with an applied external load

If an external load, T_L, is applied to an accelerating load (for example, the cutting force in a machine-tool application), the load's acceleration is given by

$$\alpha = \frac{T_{peak} - T_L/n}{n(I_d + I_L/n^2)} \tag{2.12}$$

This value is lower that than that given by equation (2.9) for an identical system.

The optimum gear ratio for an application, where the load is accelerating with a constant applied load, can be determined from this equation, in an identical manner to that described above, giving:

$$n^* = \sqrt{\frac{I_L\alpha - T_L}{I_d\alpha}} \tag{2.13}$$

The peak acceleration for such a system will be,

$$\alpha_{peak} = \frac{T_{peak} - T_L/n^*}{2I_m n^*} \tag{2.14}$$

The use of this value of the optimal gear ratio given by equation (2.10), results in a lower acceleration capability; this must be compensated by an increase in the size of the motor-drive torque rating. In sizing a continuous torque or speed application, the optimal value of the gear ratio will normally be selected by comparing the drives's continuous rating with that of the load. As noted above, the calculation

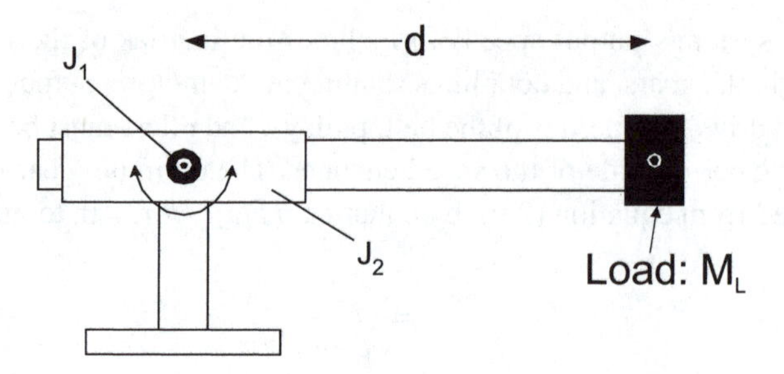

Figure 2.5. The effective load inertia of a rotary joint, J_1, changes as the linear joint, J_2 of polar robot extends or retracts (the YY axis of the joint and the load point out of the page.

of the optimal gear ratio for acceleration is dependent on the drive's peak-torque capability. In most cases, the required ratios obtained will be different, and hence in practice either the acceleration or the constant-speed gear ratio will not be at their optimum value. In most industrial applications, a compromise will have to be made.

2.1.6 Accelerating loads with variable inertias

As has been shown, the optimal gear ratio is a function of the load inertia: if the gear ratio is the optimum value, the power transfer between the motor and load is optimised. However, in a large number of applications, the load inertia is not constant, either due to the addition of extra mass to the load, or a change in load dimension.

Consider polar robot shown in Figure 2.5; the inertia that joint, J_1, has to overcome to accelerate the robot's arm is a function of the square of the distance between the joint's axis and load, as defined by the parallel axis theorem. The parallel axis theorem states that the inertia of the load in Figure 2.5 is given by

$$I_{load} = d^2 M_L + I_{YY} \tag{2.15}$$

where d is the distance from the joint axis to the parallel axis of the load – in this case YY, I_{YY} is the inertia of the load about this axis, and M_L is the mass of the load.

If a constant peak value in the acceleration is required for all conditions, the gear ratio will have to be optimised for the maximum value of the load inertia. At lower values of the inertia, the optimum conditions will not be met, although the load can still be accelerated at the required value.

Example 2.1

Consider the system shown Figure 2.5, where the rotary axis is required to be accelerated at $\alpha_{max} = 10\ rads^{-1}$, irrespective of the load inertia. A motor with inertia $I_m = 2 \times 10^{-3}\ kgm^2$ is connected to the load through a conventional gearbox. As the arm extends the effective load inertia increases from $I_{min} = 0.9\ kgm^2$ to $I_{max} = 1.2\ kgm^2$.

The optimum gear ratio, n^*, can be calculated, using equation 2.10. The gear ratio has limiting values of 6.7 and 31.7, given the range of the inertia. To maintain performance at the maximum inertia the larger gear ratio is selected, hence the required motor torque is:

$$T = 31.7\ \alpha_{max}\left(I_m + \frac{I_{max}}{31.7^2}\right) = 1.3Nm$$

If the lower gear ratio is selected, the motor torque required to maintain the same acceleration is 3 Nm, hence the system is grossly overpowered.

2.2 Linear systems

From the viewpoint of a drive system, a linear system is normally simpler to analyse than a rotary system. In such systems a constant acceleration occurs when a constant force acts on a body of constant mass:

$$\ddot{x} = \frac{F}{m} \tag{2.16}$$

where $\ddot{x}$ is the linear acceleration, F the applied force and m the mass of the object being accelerated. As with the rotary system, a similar relationship to equation (2.1) exists:

$$F_m = F_L + m_{tot}\ddot{x} + B_L\dot{x} \tag{2.17}$$

where m_{tot} is the system's total mass; B_L is the damping constant (in Nm^{-1}s); $\dot{x}$ is the linear velocity (in m s^{-1}); F_L is the force required to drive the load (in N), including the external load forces and frictional loads (for example, those caused by any bearings or other system inefficiencies); F_m is the force (in N) developed by a linear motor or a rotary-to-linear actuator.

Table 2.2. Typical values for the coefficient of friction, μ, between two materials.

Materials	Coefficient of friction
Aluminum and Aluminum	1.05–1.35
Aluminum and Mild steel	0.61
Mild steel and Brass	0.51
Mild steel and Mild steel	0 74
Tool steel on brass	0.24
Tool steel on PTFE	0.05–0.3
Tool steel on stainless steel	0.53
Tool steel on polyethylene	0.65
Tungsten carbide and Mild steel	0.4–0.6

The kinetic energy change for a linear system can be be calculated from

$$\Delta E_k = \frac{m_{tot}(\dot{x}_2^2 - \dot{x}_1^2)}{2} \tag{2.18}$$

for a speed change from $\dot{x}_2$ to $\dot{x}_1$.

2.3 Friction

In the determination of the force required within a drive system it is important to accurately determine the frictional forces this is of particular importance when a retro-fit is being undertaken, when parameters may be difficult to obtain, and the system has undergone significant amounts of wear and tear. Friction occurs when two load-bearing surfaces are in relative motion. The fundamental source of friction is easily appreciated when it is noted that even the smoothest surface consists of microscopic peaks and troughs. Therefore, in practice, only a few points of contact bear the actual load, leading to virtual welding, and hence a force is required to shear these contact points. The force required to overcome the surface friction, F_f, for a normally applied load, N, is given by the standard friction model

$$F_f = \mu N \tag{2.19}$$

where μ is the coefficient of friction; typical values are given in Table 2.2. In order to minimise frictional forces, lubrication or bearings are used, as discussed in Section 3.4.

This basic model is satisfactory for slow-moving, or very large loads. However, in the case of high speed servo application the variation of the Coulomb friction with speed as shown in Figure 2.6(a), may need to be considered. The Coulomb friction at a standstill is higher than its value just above a standstill; this is termed the stiction (or static friction). The static frictional forces is the result of the interlocking of the irregularities of two surfaces that will increase to prevent any relative

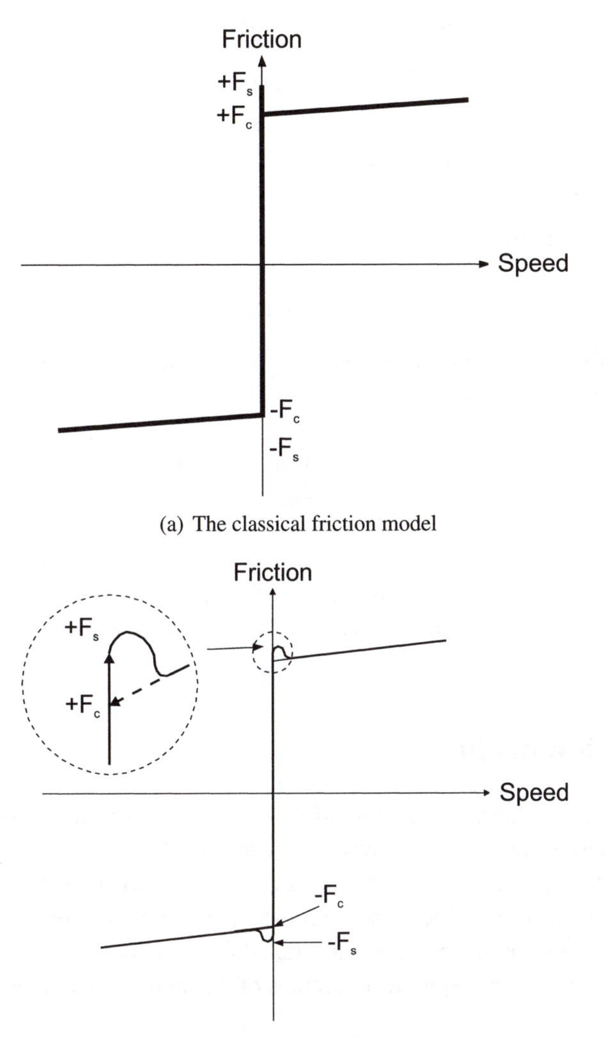

(a) The classical friction model

(b) The general kinetic friction model

Figure 2.6. The friction between two surfaces as a function of speed, using the classical or general kinematic model. F_s is the *breakaway* or *stiction* frictional force. F_c is the coulomb frictional force.

motion. The stiction has to be overcome before the load will move. An additional component to the overall friction is the viscous friction which increases with the speed; if this is combined with Coulomb friction and stiction, the resultant characteristic (known as the *general kinetic friction model*) is shown in Figure 2.6(b) (Papadopoulos and Chasparis, 2002). This curve can be defined as

$$F_f = \begin{cases} F_f(\dot{x}) & \dot{x} \neq 0 \\ F_e & \dot{x} = 0,\ \ddot{x} = 0,\ |F_e| < F_s \\ F_s \mathrm{sgn}(F_e) & \dot{x} = 0,\ \ddot{x} \neq 0,\ |F_e| > F_s \end{cases} \tag{2.20}$$

where the classical friction model is given by

$$F_f(\dot{x}) = F_c \mathrm{sgn}(\dot{x}) + B\dot{x} \tag{2.21}$$

where F_c is the Coulomb friction level and B the viscous friction coefficient. The 'sgn' function is defined as

$$\mathrm{sgn}(\dot{x}) = \begin{cases} +1 & \dot{x} > 0 \\ 0 & \dot{x} = 0 \\ -1 & \dot{x} < 0 \end{cases}$$

In the analysis F_e is the externally applied force, and F_s is the breakaway force, which is defined as the limit between static friction (or stiction) and the kinetic friction.

2.4 Motion profiles

In this section, the methods of computing the trajectory or motion profile that describes the design motion of the system under consideration, are considered. The motion profile refers to the time history of position velocity and acceleration for each degree of freedom. One significant problem is how to specify the problem – the user does not want to write down a detailed function, but rather specify a simple description of the move required. Examples of this approach include:

- In a *point-to-point* system the objective is to move the tool from one predetermined location to another, such as in a numerically-controlled drill, where actual tool path is not important. Once the tool is at the correct location, the machining operation is initiated. The control variables in a point-to-point system are the X- and Y- coordinates of the paths' starting and finishing points (Figure 2.7(a)).

- A *straight-cut* control system is capable of moving the cutting tool parallel to one of the major axes at a fixed rate which is suitable for machining, the control variable being the feed speed of the single axis (Figure 2.7(b)).

- *Contouring* is the most flexible, where the relative motion between the workpiece and the tool is continuously controlled to generate the required geometry. The number of axes controlled at any one time can be as high as five (three linear and two rotational motions); this gives the ability to produce with relative ease, for example, plane surfaces at any orientation, or circular and conical paths. The control variable in a contouring system is the relationship between the speed of all the axes under control. If a smooth curved path is to be generated (Figure 2.7(c)), there will be a constant relationship between the speeds of the X- and Y- axes. To generate a curve, the speeds, and hence the acceleration of the individual axes will vary as a function of the position; this is identical to the function of a robot's controller. The path followed by the cutter in a contouring machine tool is generated by the controller, on the basis of knowledge of the profile required and the size of the cutter. The cutter has to follow a path that will produce the required profile; this requires careful design of the profile to ensure that the cutter will not have to follow corners or radii that would be physically impossible to cut.

The paths described in Figure 2.7 are defined in terms of the machine's X and Y coordinates the actual profile or trajectory of the individual joint axes have to be generated at run time, the rate at which the profile is generated is termed the path-update rate, and for a typical system lies between 60 Hz and 2000 Hz.

In an indexing, or point-to-point, application, either a triangular or a trapezoidal motion profile can be used, the trapezoidal profile being the most energy-efficient route between any two points. If a specific distance must be moved within a specific time, t_m (seconds), the peak speed, and acceleration, can be determined; for a triangular profile in a rotary system, Figure 2.8(a), the required peak speed and acceleration can be determined as

$$N_{max} = \frac{2\theta_m}{t_m} \tag{2.22}$$

$$\alpha = \frac{2N_{max}}{t_m} \tag{2.23}$$

where θ_m is the distance moved in revolutions, N_{max}, the maximum speed required, and α is the acceleration. For a trapezoidal motion profile, and if the time spent on acceleration, deceleration and at constant velocity are equal, Figure 2.8(b), the peak speed and acceleration are given by

$$N_{max} = \frac{3\theta_m}{2t_m} \tag{2.24}$$

$$\alpha = \frac{3N_{max}}{t_m} \tag{2.25}$$

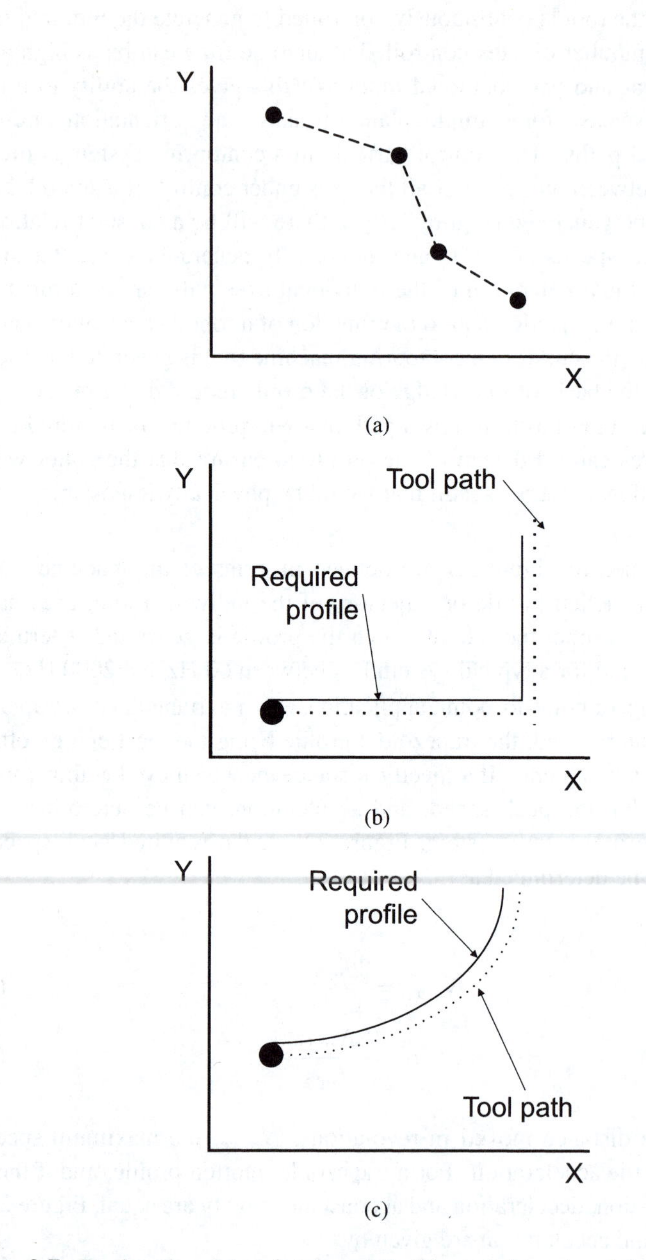

Figure 2.7. Tool paths: (a) point to point, (b) straight cut and (c) contouring. In cases (b) and (c) the tool path is offset by the radius of the cutter or, in the case of a robot application the size of the robot's end effector.

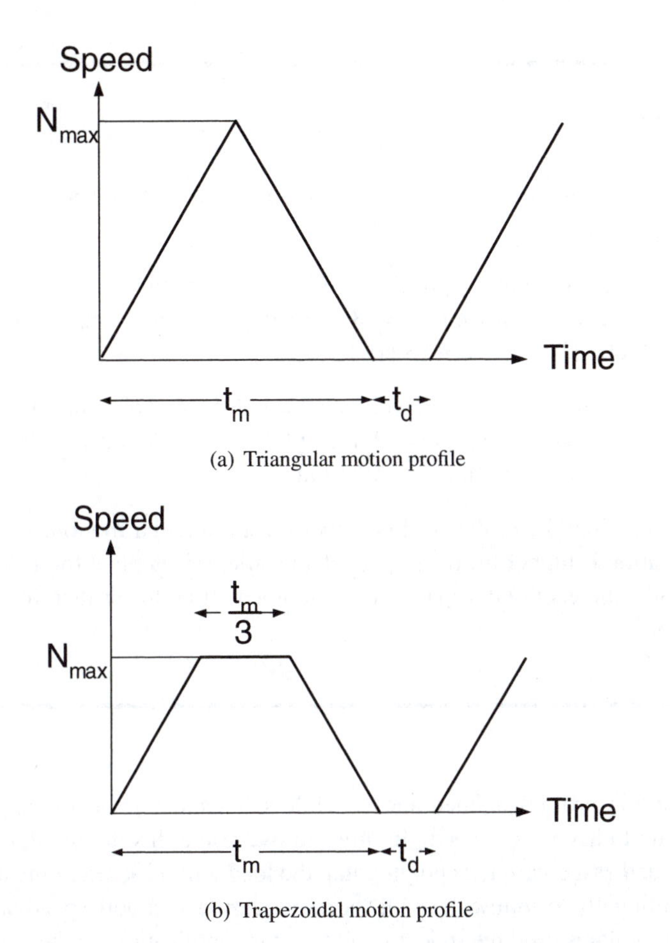

(a) Triangular motion profile

(b) Trapezoidal motion profile

Figure 2.8. Motion profiles, the total distance covered and time is identical in both cases.

In order to determine the power requirements of the drive system, the acceleration/deceleration duty cycle, d, needs to be determined. If the dwell time between each move is t_d, then the duty cycle for a triangular profile is

$$d = \frac{t_m}{t_m + t_d} \tag{2.26}$$

and for the trapezoidal profile

$$d = \frac{2t_m}{3(t_m + t_d)} \tag{2.27}$$

The general form of the above equations can also be used for linear motions, though care must be taken to ensure consistency of the units.

Example 2.2

Compare the triangular and trapezoidal motion profiles, when a disc is required to move $90°$ *in* 1.0 *s.*

1. Using a triangular profile, the peak speed of the table will be $N_{max} = 30$ rev min^{-1} $= 3.142$ rad s^{-1} and the acceleration $\alpha = 3600$ rev min^{-2} $= 6.282$ rad s^{-2}.

2. Using a trapezoidal profile, with one third of the move at constant speed, the peak speed will be $N_{max} = 22.5$ rev min^{-1} $= 2.365$ rad s^{-1} and the acceleration $\alpha = 4050$ rev min^{-2} $= 7.1$ rad s^{-2}.

It can be noted that the peak speed is higher for a triangular motion. Even though the acceleration is higher for the trapezoidal profile, it is applied for a shorter time period, hence the energy dissipated in the motor will be lower than for triangular motion profile.

The motion profiles defined above, while satisfactory for many applications, result in rapid changes of speed. In order to overcome this the motion trajectory can be defined as a continuous polynomial, the load will be accelerating and decelerating continually to follow the path specified, giving a smooth speed profile. If a cubic polynomial is used the trajectory for a rotary application can be expressed as

$$\theta(t) = a_0 + a_1 t + a_2 t^2 + a_3 t^3 \tag{2.28}$$

The generation of the polynomial's coefficients can be calculated from defined parameters, typically the positions and speeds at the start and end of the move. This will allow the joint's velocity and accelerations to be determined as a function of time.

If a path is required that moves a load from θ_1 to θ_2, and the speeds at both ends of of the motion path are zero, it is possible to determine the speeds and acceleration required are defined by the following equations:

$$\theta(t) = \theta_1 + \frac{3}{t_m^2}(\theta_2 - \theta_1)t^2 + \frac{2}{t_m^3}(\theta_2 - \theta_1)t^3 \tag{2.29}$$

$$\dot{\theta}(t) = \frac{6}{t_m^2}(\theta_2 - \theta_1)t + \frac{6}{t_m^3}(\theta_2 - \theta_1)t^2 \tag{2.30}$$

$$\ddot{\theta}(t) = \frac{6}{t_m^2}(\theta_2 - \theta_1) + \frac{12}{t_m^3}(\theta_2 - \theta_1)t \tag{2.31}$$

where t_m is the time required to complete the move. As in the case of the triangular and trapizoidal profile, a polynomial profile can be applied to linear motions, in which case equation (2.28) will be expressed as

$$x(t) = a_0 + a_1 t + a_2 t^2 + a_3 t^3 \tag{2.32}$$

Example 2.3

Determine the polynomial profile for the following application. If a joint is at rest at $\theta = 15°$, and is required to move to $\theta = 75°$ in 3 s, using the profile defined by equation (2.28).

In making the single smooth motion, four constraints are evident, the starting and finishing positions are known and the initial and final velocities are zero, allowing the following coefficients to be determined:

$$\begin{aligned} a_0 &= 15.0 \\ a_1 &= 0 \\ a_2 &= 20.0 \\ a_3 &= -4.44 \end{aligned}$$

when substituted in the equation (2.28), we obtain:

$$\begin{aligned} \theta &= 15.0 + 20.0t^2 - 4.44t^3 \\ \dot{\theta} &= 40.0t + 13.33t^2 \\ \ddot{\theta} &= 40.0 - 26.66t \end{aligned}$$

Figure 2.9 shows the position, velocity and acceleration functions for the required profile. It should be noted that the velocity profile of a movement where the distance moved as a function of time is a cubic polynomial, is a parabola, and the acceleration is linear.

In many case the path is required to pass through an number of intermediate or via points without changing speed. The via points can be determined as a set of positions. In order to determine the path, equation (2.28) is applied for each

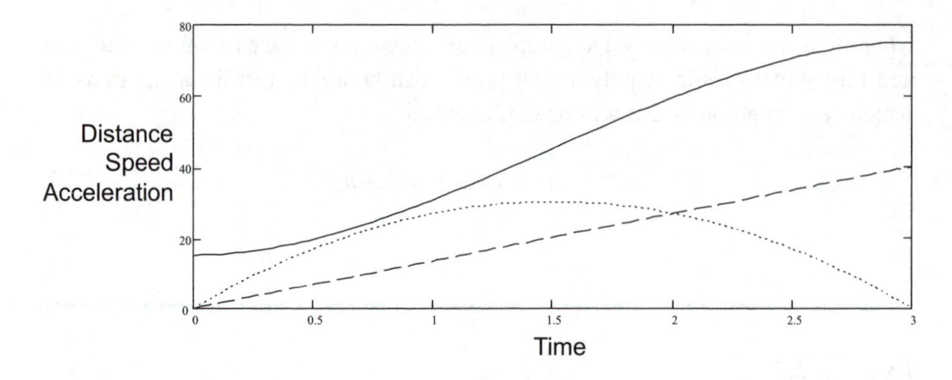

Figure 2.9. Position, velocity and acceleration profiles for a single cubic segment where $\theta_1 = 15°$, $\theta_2 = 75°$, the move time is given by $t_m = 3$ s, at the start and end positions the system is at rest. The distance moved is shown as a solid line, the speed as dotted and the acceleration as dashed.

segment, however the velocity at the start and/or end points of cubic segments are non-zero, and can be specified by a one of the following approaches:

- The user specifies the velocities at the via point.
- The system computes a velocity based on a function of the joint position.
- The system computes the velocity at the via point to maintain constant acceleration through the via point.

For certain applications higher order polynomials can be used to specify paths, for example for a quintic polynomial, equation (2.33) can be used. However this needs all three parameters at the start and finish to be specified, which gives a linear set of six equations that require solving, to determine the coefficients of the profile.

$$\theta(t) = a_0 + a_1 t + a_2 t^2 + a_3 t^3 + a_4 t^4 + a_5 t^5 \tag{2.33}$$

2.5 Assessment of a motor-drive system

The first step to the successful sizing and selection of a system is the collating of the information about the system and its application. Apart from the electrical and mechanical aspects this must also include details of the operating environment; for if these details are not considered at an early stage, the system which is selected may not be suitable for the application.

2.5.1 Mechanical compatibility

The mechanical requirements of the motor must be identified at an early stage in the sizing and selection procedure. Items that are frequently overlooked include any restrictions in dimensions and orientation resulting from the mechanical design. If such restrictions can be identified at an early stage unsatisfactory performance of the equipment after it is installed may be prevented. In particular, if the motor is mounted in the vertical position, special shimming or bearing preloads may be required. Apart from ensuring that all the mechanical parts fit together, the problems of assembly and maintenance should be considered at this stage; there is no more frustrating (or costly) problem than a junction box that cannot be reached, or having to dismantle half a machine to replace a motor or position encoder.

The correct identification and determination of the load and any externally applied forces often is the most critical aspect of sizing a motor-drive combination. The worst-case masses, forces, and speeds need to be accurately determined if the motor drive is to be correctly sized. Even if the exact mass of a component is unknown, it can be determined from the component's volume and density.

In an application where the linear-motion axis is horizontal, these values can normally be determined without significant problems. However, when the axis of motion is vertical, the total weight of the load, if it is not counterbalanced, will appear as a constant load on the motor, which must be taken into account. When the load is counterbalanced, the design must be carefully analysed. Two possible counterbalancing schemes can be used, either using a second mass, or a pneumatic system. If the former strategy is used, Figure 2.10(a), the effective load inertia must include the counterbalance when the load's operating requirement is calculated. If a pneumatic system is used, Figure 2.10(b), to support the load, the percentage of the load which is supported must be determined, and the motor must again handle the unsupported load as a constant-thrust loading. Systems of this type should be carefully designed to ensure there is no possibility of damage because of a failure of the system. This may require the fitting of brakes, overspeed detectors, and, in the case of the hydraulic or pneumatic systems, pressure switches.

The area that most often prevents successful motor-drive sizing is poor estimation of the frictional forces. Modern antifriction devices such as air bearings, hydrostatic ways, linear-roller or ball-thrust bearings have done much to minimise this problem, as discussed in Section 3.4. In most cases the frictional forces can be safely estimated for horizontal motion using equation (2.19). Again in a vertical application, the forces on the slideways need to be carefully resolved. If the drive is being used as a retrofit on an older design of machine tool which incorporates dovetail slides or similar bearings, the actual coefficients of friction can be many times the nominal value for the given material-to-material contact. In practice, the actual friction should be carefully determined by practical measurement or an adequate safety margin should be used in the sizing of the motor-drive in order to reflect the degree of uncertainty which is present.

2.5.2 Electromagnetic compatibility

Electromagnetic interference (EMI) can affect all types of electrical and electronic equipment to varying degrees; such interference has increased in importance be-

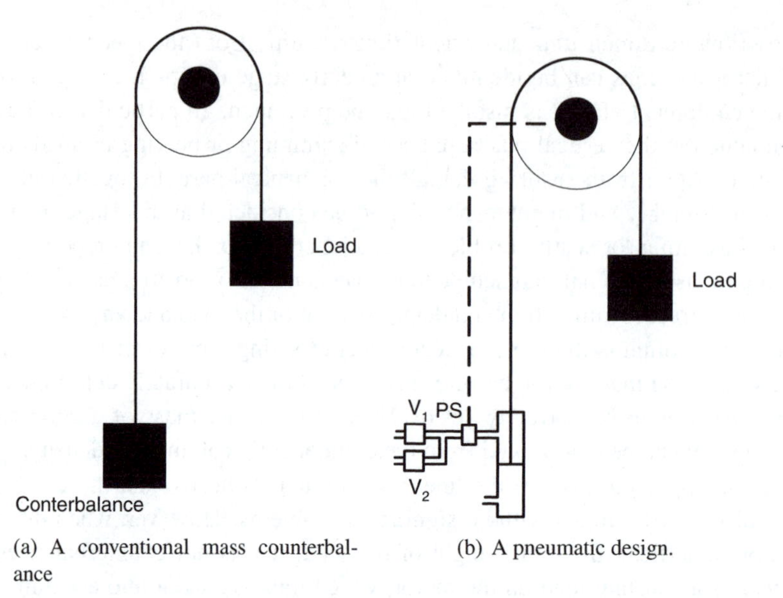

(a) A conventional mass counterbalance

(b) A pneumatic design.

Figure 2.10. The use of a counterbalance to reduce the motor's continuous loading in a vertical drive application. In the pneumatic design the valves V_1 and V_2 ensure that the pressure difference in the cylinder remains constant; if the air pressure is lost a low pressure switch activates a brake.

cause of recent European legislation, which recognises the importance of removing these potential problem areas at the design stage. In an assessment of equipment for compatibility, the emissions and the susceptibility of the equipment must be considered over a very broad frequency range, normally d.c. (0 Hz) to 110 MHz. In consequence, electromagnetic compatibility has a considerable influence over the design and application of a system.

The most obvious sources of electromagnetic radiation are the power converters which are used in motor controllers and any associated switch gear. With the increasing use of microprocessor-based controllers, any interference can have serious safety implications. In general, the main route of interference into or out of a piece of equipment is via the cabling. A cable which is longer than one metre should be considered to be a problem area. The use of the grounding and shielding of cables must be a high priority, as must careful design of the panel layout. While these measures cost little to implement, failure to do so will be costly. Figure 2.11 shows the precautions which should be taken in a typical drive system.

Radio-frequency interference (RFI) is electromagnetic emission in the range 150 kHz to 110 MHz which are of sufficient strength to be capable of interfering with any form of telecommunications; the main sources of RFI are power electronics, relays, and motor commutators. Recent legislation has placed strict limits on RFI emissions, and these limits should be met before a piece of equipment can be placed in service.

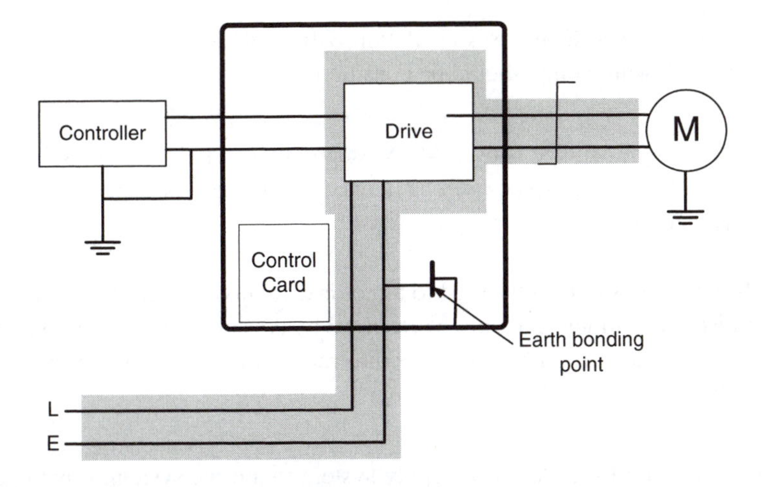

Figure 2.11. A possible wiring layout for a drive system in a normal industrial environment. The motor cables are twisted to reduce interference, the cabinet is provided with a bonding point. While the controller is separately earthed, care needs to be taken to prevent earth loops. In order to prevent problems, any sensitive electronics should be placed a distance away from various elements as shown by the shaded areas.

Due to problems of interference between systems, a range of standards (including the British Standards BS 800, BS 6667 and BS EN 60529) specify the amount of RFI which is permitted in the supply. However, the European Community has produced a directive, EEC82/499, that covers the maximum RFI levels in the supply and the output; this has been in force throughout Europe since 1992. The suppression of RFI requires considerable thought and can only be achieved by the use of filters. The screening of cables will not affect the RFI being transmitted around a system. It should be noted that directive EEC82/499 states that screened cables are not an acceptable form of RFI suppression. The only acceptable form of suppression is by the use of in-line filters in all the supply lines, including the earth return. The design of suitable filtering should be carried out in partnership with the system user, its supplier, and the electricity supply authority.

2.5.3 Wiring considerations

While the installation of motor power cables needs to be considered for any EMI and RFI effects, other factors must also be considered if the separation of the motor and drive is excessive. The pulses from a pulse width modulated drive will act as an impulse on the motor cables, and reflection will occur due to the mismatch in impedance between the motor and the cable. If the length of the cable is excessive, the reflected pulse will combine with other pulses to increase the motor voltage to over twice the nominal supply voltage. This reflected-wave theory is well understood in terms of transmission-line theory. The net effect of this high-voltage spike is the possibility that the motor's insulation will break down. In the selection of a motor, the following points should be considered:

- If the supply voltage is over 440 V (such as the supply of 460–480 V in the USA), the voltage spikes will be in excess of those experienced in European applications.

- If the drive is to be retrofitted to a motor with unknown insulation specifications, this problem can only be resolved by consultation with the supplier of the original motor, and it may require replacement of the motor by a motor with enhanced insulation capabilities.

In the design of the electrical supply system to a drive system, it is important to ensure that the system is fully and correctly earthed. A good earthing system is required:

- to ensure safety of operators and other personnel by limiting touch voltages to safe values, by provide a low resistance path for fault current so that the circuit protective devices operate rapidly to disconnect the supply. The resistance of the earth path must be low enough so that the potential rise on the earth terminal and any metalwork connected to it is not hazardous.

- to limit EMI and RFI as discussed in Section 2.5.2, by providing a noise-free ground .

- to ensure correct operation of the electricity supply network and ensure good power quality.

The actual design of a complete earthing system is complex and reference should be made to the relevant national standards, within the UK reference should be made to BS7671:2001 (*Requirements for electrical installations. IEE Wiring Regulations. Sixteenth edition*) and BS7430:1998 (*Code of practice for earthing*).

In the construction of a drive system, bonding is applied to all accessible metalwork – whether associated with the electrical installation (known as exposed-metalwork) or not (extraneous-metalwork) – is connected the system earth. The bonding must be installed so that the removal of a bond for maintenance of equipment does not break the connection to any other bond.

As noted above the provision of a good earth is fundamental to the prevention of EMI and RFI problems. It is common practice to use a single point or star earthing system to avoid the problems of common mode impedance coupling. However, care needs to exercised when shielded cables are used, as loops may inadvertently be formed, which will provide a path for any noise current.

2.5.4 Supply considerations

While the quality of public-utility supplies in Western Europe is normally controlled within tight specifications, considerable voltage fluctuations may have to be accommodated in a particular application. In cases were the drive system is used on sites with local generation (for example, on oil rigs and ships), considerable care needs to be taken in the specification of the voltage limits. Since the peak speed of a motor is dependent on the supply voltage, consideration needs to be given to what happens during a period of low voltage. As a guideline, drives are normally sized so that they can run at peak speed at eighty per cent of the nominal supply voltage. If a system is fed from a vulnerable supply, considerable care will have to be taken to ensure that the drive, its controller, and the load are all protected from damage; this problem is particularly acute with the introduction of microprocessor systems, which may lock-up or reset without warning if they are not properly configured, leading to a possibly catastrophic situation.

In practice, the supply voltage can deviate from a perfect sinewave due to the following disturbances.

- **Overvoltage**. The voltage magnitude is substantially higher than its nominal value for a significant number of cycles. This can be caused sudden decreases in the system load, thus causing the supply to rise rapidly.

- **Undervoltage or brownout**. The voltage is substantially lower than its nominal value for a significant number of cycles. Undervoltages can be caused by a sudden increase in load, for example a machine tool or induction motor starting.

- **Blackout or outage**. The supply collapses to zero for a period of time that can range from a few cycles to an extended period of time.

- **Voltage spikes**. These are superimposed on the normal supply waveform, and are non-repetitive. A spike can be either differential-mode or a common-mode. Occasional large voltage spikes can be caused by rapid switching of power factor correction capacitors, power lines or motors in the vicinity.

- **Chopped voltage waveform**. This refers to a repetitive chopping of the waveform and associated ringing. Chopping of the voltage can be caused by ac-to-dc line frequency thyristor converters, Figure 2.12(a).

- **Harmonics**. A distorted voltage waveform contains harmonic voltage components at harmonic frequencies (typically low order multiples of the line frequency). These harmonics exist on a sustained basis. Harmonics can be produced by a variety of sources including magnetic saturation of transformers or harmonic currents injected by power electronic loads, Figure 2.12(b).

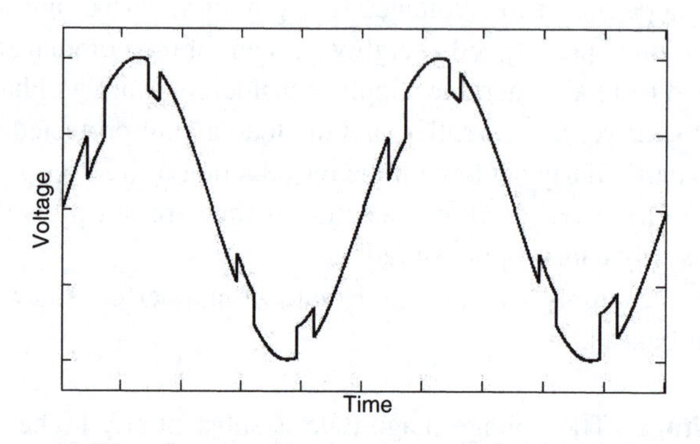

(a) Distortion caused by chopping applied to the supply.

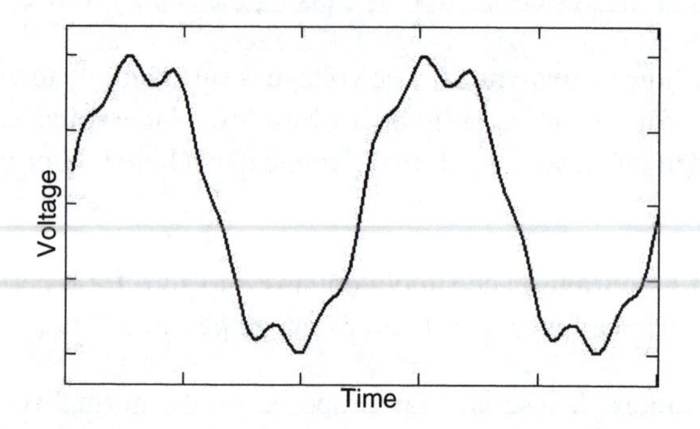

(b) Distortion caused by the addition of externally generated harmonics to the supply

Figure 2.12. Distortion in the supply input to a drive system due to externally generated harmonics or chopping.

- **Electromagnetic interference**. This refers to high-frequency noise, which may be conducted on the power line or radiated from its source, see Section 2.5.2.

The effect of power line disturbances on drive systems depends on a number of variables including the type and magnitude of the disturbance, the type of equipment and how well it is designed and constructed, and finally the power conditioning equipment fitted to the system or the individual drive.

Sustained under- and over- voltages will cause equipment to trip out, which is both highly undesirable and with a high degree of risk in certain applications. Large voltage spikes may cause a hardware (particularly in power semiconductors) failure in the equipment. Manufacturers of critical equipment often provide a certain degree of protection by including surge arrestors or snubbers in their designs. However, spikes of very large magnitude in combination with a higher frequency may result in a stress-related hardware failure, even if normal protection standards are maintained. A chopped voltage and harmonics have the potential to interfere with a drive system if it is not designed to be immune from such effects.

Power conditioning

Power conditioning provides an effective way of suppressing some or all of the electrical disturbances other than the power outages and frequency. Typical methods of providing power conditioning include:

- metal-oxide varistors, which provide protection against line-mode voltage spikes,
- electromagnetic interference filters, which help to prevent the effect of the chopped waveforms on the equipment as well as to prevent the equipment from conducting high-frequency noise into supply,
- isolation transformers with electrostatic shields, which not only provide galvanic isolation, but also provide protection against voltage spikes,
- ferroresonant transformers, which provide voltage regulation as will as line spike filtering.

Interface with the utility supply

All power electronic converters (including those used to protect critical loads) can add to the supply line disturbances by distorting the supply waveform. To illustrate the problems due to current harmonies in the input current of a power electronic load, consider the block diagram of Figure 2.13. Due to the finite internal impedance of the supply source, the voltage waveform at the point of common coupling to other loads will become distorted, which may cause additional malfunctions. In addition to the waveform distortion, other problems due to the harmonic

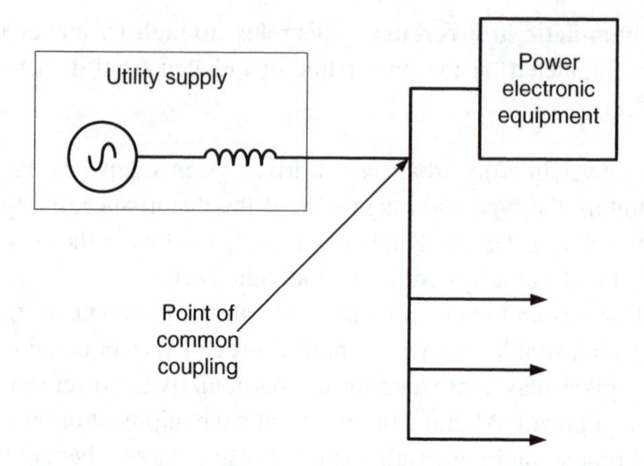

Figure 2.13. The utility interface, showing the *point of common coupling*, where the supply distortion caused by each individual load is combined, due to the finite impedance of the supply – here represented by a simple inductance.

currents include: additional heating and over-voltages (due to resonance conditions) in the utility distribution and equipment, errors in metering and malfunction of utility relays and interference with communications and control signal.

One approach to minimise this impact is to filter the harmonic currents and the EMI produced by the power electronic loads. An alternative, in spite of a small increase in the initial cost, is to design the power electronic equipment such that the harmonic currents and the EMI are prevented or minimised from being generated in the first place.

Standards

In view of the increased amount of power electronic equipment connected to the utility systems, various national and international agencies have been considering limits to the amount of harmonic current injection to maintain good power supply quality. As a consequence a number of standards have been developed, including:

- EN 60555–1:1987, *The Limitation of Disturbances in Electricity Supply Networks caused by Domestic and Similar Appliances Equipped with Electronic Device*, and EN 61000-3-2:1995 *Electromagnetic compatibility (EMC) – Part 3–2: Limits – Limits for harmonic current emissions (equipment input current up to and including 16A per phase)*. Both these European Standards are prepared by CENELEC (European Committee for Electrotechnical Standardisation).

- IEC Norm 555-3, prepared by the International Electrotechnical Commission.

- *IEEE Recommended Practices and Requirements for Harmonic Control in Electrical Power Systems*, IEEE Standard 519-1992.

The CENELEC and IEC standards specify the limits on harmonics within the supply, while the IEEE standards contain recommended practices and requirement tor harmonic control in electric power system as well as specifying requirements on the user as well as on the utility.

2.5.5 Protection from the environment

A significant proportion of drive systems have to operate in relatively poor environments. The first line in this protection is the provision of a suitable protective enclosure. Two basic classes exist for non-hazardous areas and hazardous areas. An enclosure for non-hazardous areas is classified by the use of an IP code number specified in IEC publication 60529, which indicates the degree of protection from ingress of solid objects including personnel contact, dust, and liquid. In the United States the National Electrical Manufacturers Association (NEMA) classification should be referred to.

Table 2.3. The IP rating system allows designers to specify a motor, or an enclosure, or other components, with a specified degree of protection from dust, water, and impact. The two numerals can be used to specify the protection afforded to a component. In a number of cases a third numeral can be attached this defines the protection against impact.

	First Number *Protection against solid objects*	**Second Number** *Protection against liquids*
0	No protection	No protection
1	Objects up to 50.00 mm	Protection against vertically falling drops of water
2	Objects up to 12.00 mm	Direct sprays up 15° from the vertical
3	Objects up to 2.50 mm	Direct sprays up 60° from the vertical
4	Objects up to 1.00 mm	Water sprayed in all direction; limited ingress is permitted
5	Protection against dust; limited ingress is permitted, but no harmful deposits	Protection from low pressure jets of water in all directions; limited ingress is permitted
6	Totally protected against dust	
7		Protection from immersion in water up to a depth between 15 cm and 1 m
8		Immersion under pressure

A brief definition of the IP classifications are given in Table 2.3. If a drive has to operate in a hazardous environment, where an explosive gas/air mixture is present, the formal United Kingdom definition is contained in BS 6345; careful consideration has to be given to the design of the enclosure and all external connections. It is recommended that the designers of systems for this type of environment consult the relevant specialist agencies.

These general application problems can never be solved by one specific formula; rather, the requirements of the various equipment must be recognised, and an optimum system should be selected by careful attention to detail. For example a system protected to IP54 is dust protected, and also protected against splashing water.

The NEMA system takes a different approach, by classifying individual cubicles or systems for a specific application, for example, a NEMA-3 system is defined as being for outdoor use and providing a degree of protection against windblown dust, rain and sleet, and will be undamaged by the formation of ice on the enclosure – this equates to IP64 protection.

2.5.6 Drive hazards and risk

It is a legal requirement, placed on both the supplier and user, that the equipment should be designed, manufactured, installed, operated, and maintained to avoid dangerous situations. Within the United Kingdom these requirements are embodied in the relevant Acts of Parliament, and they are enforced by the Health and Safety Executive, which issues a range of notes for guidance for the designers of equipment. Regulations in other countries will be covered by national legislation, and this needs to be considered during the design process. In understanding risks it is worth considering the concepts of hazards, risk and danger – and how they can be determined and designed out of a system.

A **hazard** is any condition with the potential to cause an accident, and the exposure to such a hazard is known as the corresponding **danger**. As part of the design process an estimate the of damage that may result if an accident occurs, together with the likelihood that such damage will occur, is termed the **risk** associated with the hazard.

Principles of risk management

Some hazards are inherent within a design; for example, the spindle of a lathe is hazardous by its very nature. Other hazards are contingent upon some set of conditions, such as improper maintenance, unsafe design, or inadequate operating instructions. Several distinct types of hazards can be associated with machine tools and similar systems:

- Entrapment and entanglement hazards, where part or all of a person's body or clothes may be pinched or crushed as parts move together, including gears and rollers.

- Contact hazards, where a person can come into contact with hot surfaces, sharp edges, or live electric components.
- Impact hazard, where a person strikes the machine or a part of the machine strikes the person.
- Ejection hazards, where material or a loose component is thrown from the machine.
- Noise and vibration hazards, which can cause loss of hearing, a loss of tactile sense, or fatigue. In addition, an unexpected sound may cause a person to respond in a startled manner.
- Sudden release of stored energy from mechanical springs, capacitors, or pressurised gas containers.
- Environmental and biological hazards associated with a design, its manufacture, operation, repair, and disposal.

Within any form of risk assessment, the first step is to identify the hazards, namely those with the potential for causing harm. It should be noted that some physical hazards might be present for the complete life cycle of the system whilst others may exist only during the installation, or during maintenance. The second step is to identify the possible accidents or failure modes associated with each hazard, or combinations of hazards, that could lead to the release of the hazard potential and then to determine the times in the life cycle at which such events could occur. To be successful in finding the majority of these events requires the use of a systematic approach, such as a hazard and operability study.

Accidents, however, do not just happen and the third step is to study the possible range of triggering mechanisms, or conditions, which can give rise to each failure or accident. For some events a combination or sequence of triggering conditions will be needed, in other cases only one. The underlying causes, or the conditions which initiate the trigger, often relate back to earlier phases of the project, for example to the design or planning stages. Risk assessment is the estimation of the probabilities or likelihoods that the necessary sequence of triggering events will occur for each particular hazard potential to be released, and an estimation of the consequences of each accident or failure. The latter may involve fatalities, serious injuries, long term health problems, environmental pollution and financial losses. Risk management is an extension of risk assessment and typically it involves the steps described above, together with the introduction of preventative measures. The measures may be designed to reduce or eliminate the hazards themselves, the triggering conditions, or on the magnitude of the potential consequences.

A risk assessment methodology

This section describes the development of a practical risk assessment methodology, as part of risk management of engineering systems; in particular how the process

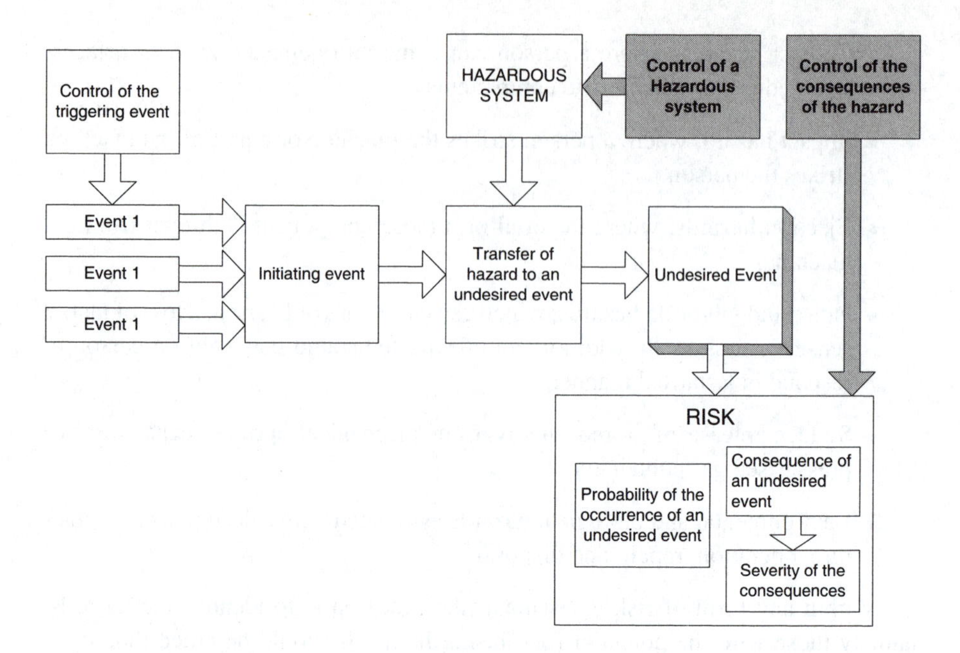

Figure 2.14. Risk management model showing the path from the triggering event to the undesired event and the subsequent risk.

is undertaken. It is clear that the risk assessment methodology should satisfy a number of basic requirements, as shown in Figure 2.14. The approach should be capable of:

- Identifying significant hazards at various stages in the equipment's life.
- Identifying the failure mechanisms that could lead to a release of each hazard's potential and the associated triggering conditions.
- Assessing the nature and severity of the consequences of each type of physical failure and other undesired events.
- Enabling estimates to be made of the likelihood of each type of physical failure and other undesired events.
- Assessing the resulting risks.
- Determining the control measures that could reduce the likelihood of undesired events and mitigate their consequences.

The following five step methodology for dealing effectively with hazards has been found to be effective:

1. *Review existing standards.* These will include those provided by the British Standards Institute (BSI), Institution of Electrical Engineers, American National Standards Institute, Underwriters Laboratory, and Institution of Electronic and Electrical Engineers. This review will determine if standards and requirements exist for the product or system being considered.

2. *Identify known hazards.* Studying recognised standards should make it possible to identify the hazards usually associated with a system.

3. *Identify unknown hazards.* These hazards include those identified in standards that must be eliminated. The design team must follow a systematic approach to identify these undiscovered hazards lurking within the design and in its use or misuse by the operator. Several techniques can be used to identify the unknown hazards, including hazard and operability studies (HAZOP), hazards analysis (HAZAN), fault tree analysis (FTA), and failure modes and effects analysis (FMEA).

4. *Determine the characteristics of hazards.* This stage attempts to determine the frequency, relative severity, and charcterictics of each hazard. By doing so, the designer can focus initially upon those hazards that can result in the most damage and/or those that have the greatest risk associated with them.

5. *Elimination and reduction of the hazard.* Following identification of the hazard, they can be ranked in order of severity and occurrence; the designer can concentrate on their elimination.

Hazards analysis

HAZAN seeks to identify the most effective way in which to reduce the threat of hazards within a design by estimating the frequency and severity of each threat and developing an appropriate response to these threats. Although there are some similarities between HAZAN and HAZOP (e.g., both focus upon hazards, and both try to anticipate the consequences of such hazards), nevertheless there are clear distinctions between the two methods. In particular, HAZOP is qualitative in nature, in contrast to HAZAN, which is quantitative. The stages of HAZAN in the form of three brief questions:

- How frequently will the hazard occur?
- How large is the possible consequences of the hazard?
- What action is to be taken to eliminate or reduce the hazard?

HAZAN is based upon probabilistic analysis in estimating the frequency with which some threat to safety may occur, together with the severity of its consequences. Through such analysis, engineers can focus their initial efforts toward reducing those hazards with the highest probabilities of occurrence and/or the most severe consequences.

Failure modes and effects analysis When using failure modes and effects analysis (FMEA) to troubleshoot a design, one begins by focusing upon each basic component one at a time and tries to determine every way in which that component might fail. All components of a design should be included in the analysis, including such elements as warning labels, operation manuals, and packaging. One then tracks the possible consequences of such failures and develops appropriate corrective actions. As part of a FMEA exercise an analysis of all the system's components are produced. A format can be used through which all components or parts can be listed, together with the following information:

- Failure models, identifying all ways in which the part can fail to perform its intended function should be identified.
- Failure causes, identifies the underlying reasons leading to a particular failure.
- Identifying how that a particular failure mode has occurred.
- Details of the protective measures that have been incorporated to prevent any failure.
- A weighted value of the severity, occurrence and detection of the event.

Example 2.4

Consider the risks associated with the tachogenerator within a motor drive system.

An illustration of the FEMA format, which takes a bottom up approach is shown in Table 2.4.

The rating is a subjective measures of the consequence of an undesirable event upon the operators, company and the system itself. Depending on the scale used the resolution can be company specific. In this example the scale runs from 1 to 7, with 2 being major repairs being required.

Table 2.4. FMEA risk assessment for a tachogenerator as fitted to a motor drive system. P is the probability, S the seriousness of the fault, D the likelihood that the fault will reach the customer and $R = P \times S \times D$) is the priority measure. P, S and D are measure on a scale of 1 to 5.

Failure Mode	Cause	Hazard	P	S	D	R	Corrective action
Plug failure	Using as a step	Overspeed	1	4	1	4	Safety cover and warning label
Incorrect wiring	Assembly fault	Overspeed	2	4	1	8	QA documentation
Broken coupling	Metal fatigue due to misalignment	Overspeed	2	4	2	16	QA and inspection
Wiring failure	Fatigue caused by vibration	Overspeed	1	4	2	8	Design and use of cable restraints

Risk assessment

Risk assessment is the second stage of the risk management methodology. All undesired events can be grouped into one of two categories, termed here as physical undesired events and operational undesired events. A physical undesired event involve's some degree of physical failure, for example, as a result of wear or corrosion of part of a subsystem during use. The latter may, or may not, lead on to an operational undesired event. An operational undesired event is defined as an event leading to death or injury, or a near miss, in which there is no physical failure of any part of the equipment being assessed. The next step requires a determination of the likelihood of each significant undesired event and the severity of its consequences. Success is dependent upon the comprehensive identification of possible undesired events and knowing how these can be related back to the initiating events, which caused them. The process involves determining the likelihood that the initiating event will be detected, before serious damage can occur; determining the corresponding likelihood of recovery from, or correction of, the initiating event; assessing the likelihood that the initiating event will escalate to give rise to an undesired event; and finally, determining the consequences of the undesired event, and their severity. Since data on the frequencies of these types of event are unlikely to be available in most situations, the various likelihoods are obtained by expert judgement using specially selected teams of experts, for example as convened in the process industries for HAZOP studies or historical data. As part of the process the consequences of all significant undesired events also need to be assessed. The severity of the consequences can be expressed in financial terms for the physical damage that may occur and in terms of injury/harm to people for the operational undesired events.

Risk assessment of software Many engineering systems now incorporate computer based control systems, which must be incorporated into any risk assessment process. It has been estimated that for every million lines of code, there are 20,000 bugs. Of these, 90% are found during the testing phase, of the remaining 2000, 200 are found in the first year of use. The problem is such that the remainder will remain dormant until a set of trigger conditions occur. The risks of using software with any system can be minimised by techniques such as protective redundancy or N modular redundancy. N modular redundancy depends on the assumption that a programme can be completely, constantly and unambiguously specified, and that the independently developed programmes will fail independently. To fully implement this a number of versions of the programme must be developed using different languages or compilers, and running of hardware supplied by different manufactures.

Preventative measures When the most significant risks have been determined, the next stage of the methodology requires that the underlying causes should be targeted for control, as shown in Figure 2.14. In general each possible triggering

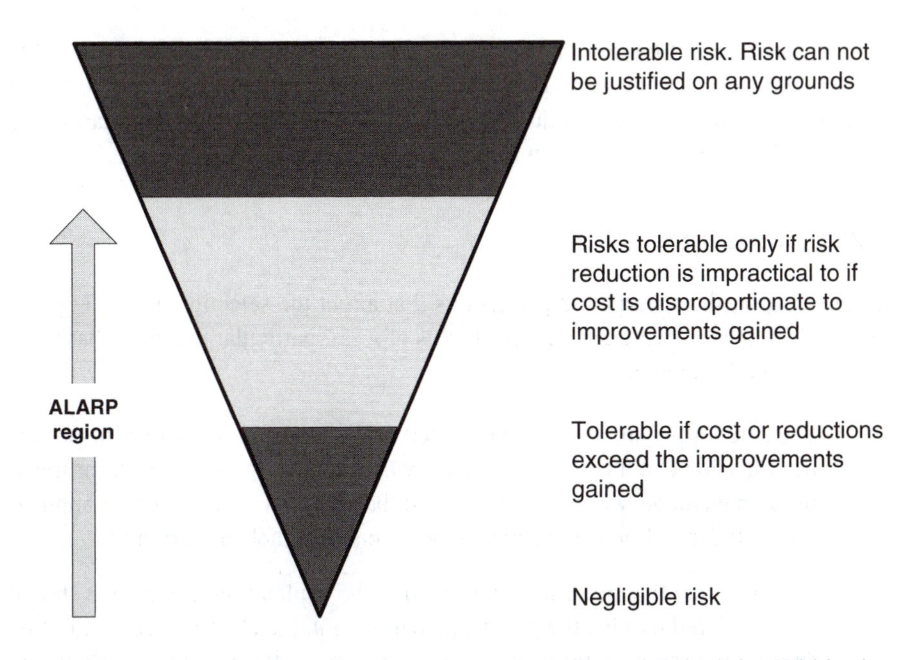

Figure 2.15. A diagrammatic representation of the ALARP principle. Within the ALARP region the costs of removing a risk have to be justified against the benefits.

condition leading to an undesired event can be attributed to a stage of design, assembly, or maintenance and this indicates whether it is the manufacturer or user of the system who is failing in their responsibilities. Finally, remedial action should be specified, indicating how the undesired events and their effects can be controlled. This involves a hierarchy of control measures, with these being applied not only to the cause of the initiating event and triggering conditions, but also to the hazards and the consequences of failure. For example, one might use interlock systems, and other safety features to protect or distance the operator from the hazard, thereby reducing her or his exposure to the danger. Finally, if a hazard cannot be eliminated or further reduced, one should prepare an appropriate set of warnings and instructions for the operator so that he or she can take precautions to avoid the danger.

ALARP Within the scope of considering preventative assessment the concept of As Low As Reasonably Practical, or ALARP, needs to be discussed. The ALARP principle is fundamental to the regulation of health and safety in the UK and requires that risks should be weighed against the costs of implementing the control measures. These measures must then be taken to reduce or eliminate the risks unless the cost of doing so is obviously unreasonable compared with the risk. The principles are summarised in Figure 2.15. Within the ALARP process, a value has to be put on the hazards, its consequence and control. As can be appreciated this can be a very problematic area, particularly when human life has to be assigned a monetary value. In addition the level of risk needs to be quantified, and while this in engineering terms can be rigorously achieved, the public's perceived level of risk is far more difficult to quantify. This can be illustrated by comparing the public perception of risk between public and private transport.

2.6 Summary

In this chapter the criteria and parameters that affect the selection and sizing of a motor-drive system were considered. In this process, particular emphasis has to be placed on the following:

- The load torque and the speed reflected back to the motor through the power-transmission chain have to be correctly determined. A range of power transmission elements were considered, and the selection for a particular application will depend on the speed and the precision which are required.
- The drive and the motor must be suited to the application. Points that should be considered include the physical environment, the electrical compatibility, and the emission requirements. In certain cases, the optimum system will need to be modified in because of the customer's commercial requirements.
- Drives by their very nature can pose a significant hazard to the user or its immediate environment, hence in all cases a suitable risk analysis should be undertaken.

Chapter 3

Power transmission and sizing

While the previous chapters have considered the analysis of a proposed motor-drive system and obtaining the application requirements, it must be recognised that the system comprises a large number of mechanical component. Each of these components, for example couplings, gearboxes and lead screws, will have their own inertias and frictional forces, which all need to be considered as part of the sizing process. This chapter considers power transmission components found in applications, and discusses their impact on overall system performance, and concludes with the process required to determine the detailed specifications of the motor and the drive.

The design parameters of the mechanical transmission system of the actuator must be identified at the earliest possible stage. However, it must be realised that the system will, in all probability, be subjected to detailed design changes as development proceeds. It should also be appreciated that the selection of a motor and its associated drive, together with their integration into a mechanical system, is by necessity an iterative process; any solution is a compromise. For this reason, this chapter can only give a broad outline of the procedures to be followed; the detail is determined by the engineer's insight into the problem, particularly for constraints of a non-engineering nature, such as a company's or a customer's policy, which may dictate that only a certain range of components or suppliers can be used.

In general, once the overall application, and the speed and torque (or in the case of a linear motor, speed and force) requirements of the total system have been clearly identified, various broad combinations of motors and drives can be reviewed. The principles governing the sizing of a motor drive are largely independent of the type of motor being considered. In brief, adequate sizing involves determining the motor's speed range, and determining the continuous and intermittent peak torque or force which are required to allow the overall system to perform to its specification. Once these factors have been determined, an iterative process using the manufacturer's specifications and data sheets will lead to as close an optimum solution as is possible.

3.1 Gearboxes

As discussed in Section 2.1.3 a conventional gear train is made up of two or more gears. There will be a change in the angular velocity and torque between an input and output shaft; the fundamental speed relationship is given by

$$n = \pm\frac{\omega_i}{\omega_o} = \pm\frac{N_o}{N_i} \tag{3.1}$$

where N_i and ω_i are the number of teeth on, and the angular velocity of, the input gear, and N_o and ω_o are the number of teeth on, and the angular velocity of, the output gear. In equation (3.1) a negative sign is used when two external gears are meshing, Figure 3.1(a), or a positive sign indicates that system where an internal gear is meshing with an internal gear, Figure 3.1(b).

In the case where an idler gear is included, the gear ratio can be calculated in an identical fashion, hence for an external gear train, Figures 3.1(c) and 3.1(d),

$$n = \frac{\omega_{in}}{\omega_{out}} = \left(-\frac{N_2}{N_1}\right)\left(-\frac{N_3}{N_2}\right) = \frac{N_3}{N_1} \tag{3.2}$$

The direction of the output shaft is reversed for an internal gear train, Figure 3.1(d). In practice the actual gear train can consist of either a spur, or helical gear wheels. A spur gear (see Figure 3.2(a)) is normally employed within conventional gear trains, and has the advantage of producing minimal axial forces which reduce problems connected with motion of the gear bearings. Helical gears (see Figure 3.2(b)) are widely used in robotic systems since they give a higher contact ratio than spur gears for the same ratio; the penalty is axial gear load. The limiting factors in gear transmission are the stiffness of the gear teeth, which can be maximised by selecting the largest-diameter gear wheel which is practical for the application, and backlash or lost motion between individual gears. The net result of these problems is a loss in accuracy through the gear train, which can have an adverse affect on the overall accuracy of a controlled axis.

In many applications conventional gear trains can be replaced by complete gearboxes (in particular those of a planetary, harmonic, or cycloid design) to produce compact drives with high reduction ratios.

3.1.1 Planetary gearbox

A Planetary gearbox is co-axial and is particularly suitable for high torque, low speed applications. It is extremely price-competitive against other gear systems and offers high efficiency with minimum dimensions. For similar output torques the planetary gear system is the most compact gearbox on the market. The internal details of a planetary gearbox are shown in Figure 3.3; a typical planetary gear box consists of the following:

- A sun gear, which may or may not be fixed.

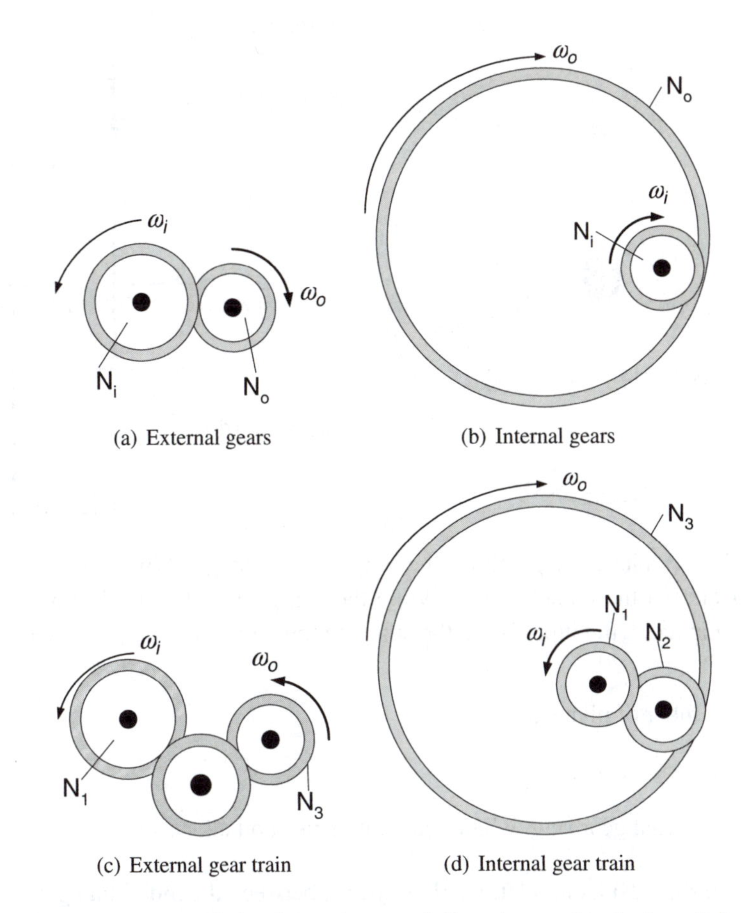

(a) External gears (b) Internal gears

(c) External gear train (d) Internal gear train

Figure 3.1. Examples of the dependency of direction and velocity of the output shaft on the type of gearing.

(a) Spur gears. (b) Helical gears.

Figure 3.2. Conventional gears.

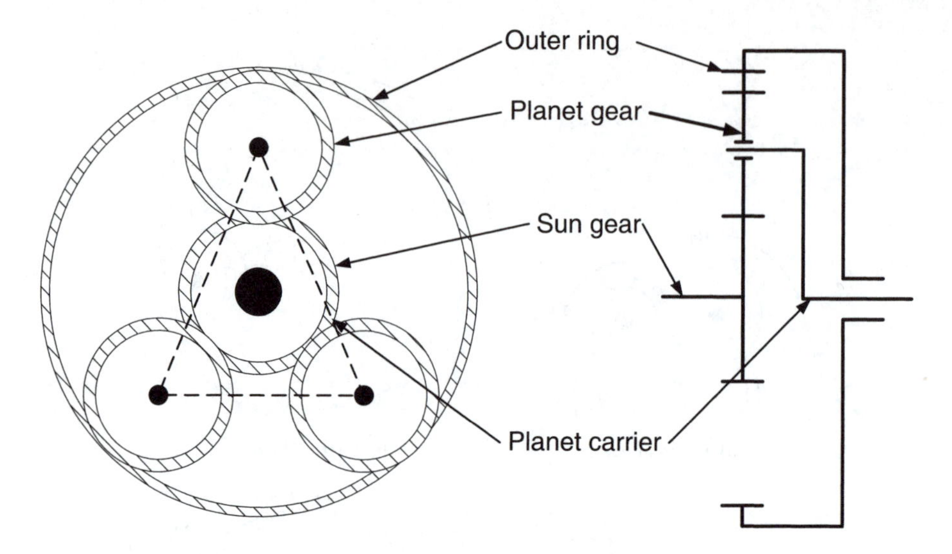

Figure 3.3. A planetary gearbox; the output from the gearbox is from the three planet gears via the planet carrier, while the sun gear is driven. In this case the outer ring is fixed, the input is via the sun, and the output via the planet carrier.

- A number of planetary gears.
- Planet gear carrier.
- An internal gear ring, which may not be used on all systems.

This design results in relatively low speeds between the individual gear wheels and this results in a highly efficient design. One particular advantage is that the gearbox has no bending moments generated by the transmitted torque; consequently, the stiffness is considerably higher than in comparable configuration. Also, they can be assembled coaxially with the motor, leading to a more compact overall design. The relationship for a planetary gearbox can be shown to be (Waldron and Kinzel, 1999)

$$\frac{\omega_{sun} - \omega_{carrier}}{\omega_{ring} - \omega_{carrier}} = -\frac{N_{ring}}{N_{sun}} \tag{3.3}$$

where ω_{sun}, $\omega_{carrier}$ and ω_{ring} are the angular speeds of the sun gear, planet carrier and ring with reference to ground. N_{ring} and N_{sun} are the number of teeth on the sun and ring respectively. Given any two angular velocities, the third can be calculated – normally the ring if fixed hence $\omega_{ring} = 0$. In addition it is important to define the direction of rotation; normally clockwise is positive, and counter-clockwise is negative.

Example 3.1

A planetary gearbox has 200 teeth on its ring, and 40 teeth on its sun gear. The input to the sun gear is 100 rev min^{-1} clockwise. Determine the output speed if the ring is fixed, or rotating at 5 rev min^{-1} either clockwise or counterclockwise.

Rearranging equation (3.3), gives

$$\omega_{carrier} = \frac{N_{sun}\omega_{sun} - N_{ring}\omega_{ring}}{N_{sun} - N_{ring}}$$

- When the ring is rotated at 5 rev min^{-1} clockwise, the output speed is 18.75 rev min^{-1} conterclockwise.
- When the ring is fixed, the output speed is 25 rev min^{-1} conterclockwise.
- When the ring is rotated at 5 rev min^{-1} counterclockwise, the output speed is 31.25 rev min^{-1} counterclockwise.

This simple example demonstrates that the output speed can be modified by changing the angular velocity of the ring, and that the direction of the ring adds or subtracts angular velocity to the output.

3.1.2 Harmonic gearbox

A harmonic gearbox will provide a very high gear ratio with minimal backlash within a compact unit. As shown in Figure 3.4(a), a harmonic drive is made up of three main parts, the circular spline, the wave generator, and the flexible flexspline. The design of these components depends on the type of gearbox, in this example the flexispline forms a cup. The operation of an harmonic gearbox can be appreciated by considering the circular spline to be fixed, with the teeth of the flexspline to engage on the circular spline. The key to the operation is the difference of two teeth (see Figure 3.4(b)) between the flexspline and the circular spline. The bearings on the elliptical-wave generator support the flexspline, while the wave generator causes it to flex. Only a small percentage of the flexispline's teeth are engaged at the ends of the oval shape assumed by the flexspline while it is rotating, so there is freedom for the flexspline to rotate by the equivalent of two teeth relative to the circular spline during rotation of the wave generator. Because of the large number of teeth which are in mesh at any one time, harmonic drives have a high torque capability; in addition the backlash is very small, being typically less that 30" of arc.

In practice, any two of the three components that make up the gearbox can be used as the input to, and the output from, the gearbox, giving the designer considerable flexibility. The robotic hand shown in Figure incorporates three harmonic gearboxes of a pancake design where the flexispline is a cylinder equal in width to the wave generator.

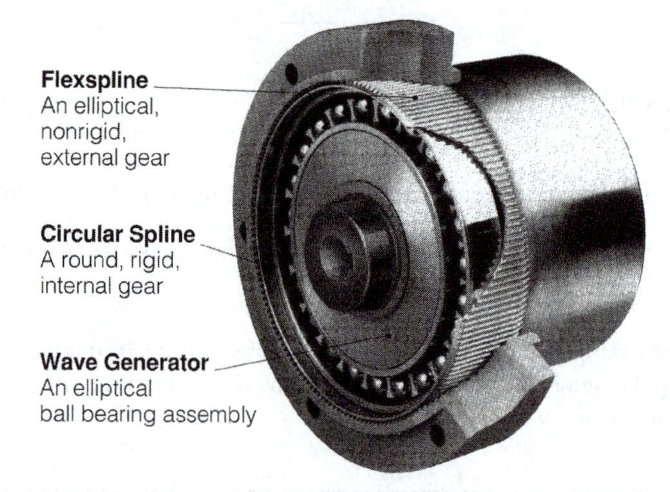

(a) Components of a harmonic gearbox

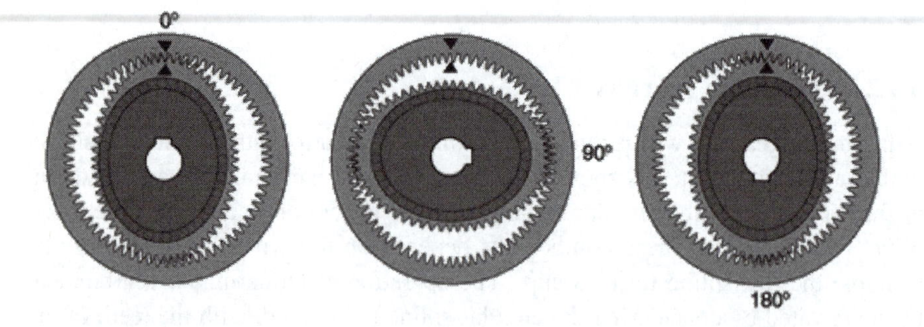

(b) Operation of a harmonic gear box, for each 360° rotation of the wave generator the flexspline moves 2 teeth. The deflection of the flexspline has been exaggerated.

Figure 3.4. Construction and operation of an HDC harmonic gear box. Reproduced with permission from Harmonic Drive Technologies, Nabtesco Inc, Peabody, MA.

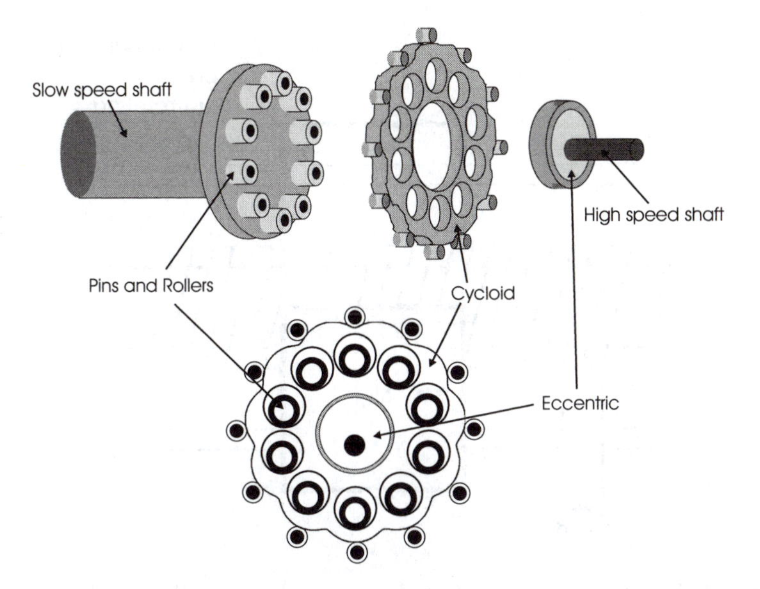

Figure 3.5. A schematic diagram of a cycloid speed reducer. The relationship between the eccentric, cylcoid and slow-speed output shaft is clearly visible. It should be noted that in the diagram only one cycloid disc is shown, commercial systems typically nave a number of discs, to improve power handelling.

3.1.3 Cycloid gearbox

The cycloid gearbox is of a co-axial design and offers high reduction ratios in a single stage, and is noted for its high stiffness and low backlash. The gearbox is suitable for heavy duty applications, since it has a very high shock load capability of up to 500%. Commercially cycloid gearboxes are available in a range of sizes with ratios between 6:1 and 120:1 and with a power transmission capability of up to approximately 100 kW. The gearbox design, which is both highly reliable and efficient, undertakes the speed conversion by using rolling actions, with the power being transmitted by cycloid discs driven by an eccentric bearing.

The significant features of this type of gearbox are shown in Figure 3.5. The gear box consists of four main components:

- A high speed shaft with an eccentric bearing.
- Cycloid disc(s).
- Ring gear housing with pins and rollers.
- Slow speed shaft with pins and rollers.

As the eccentric rotates, it rolls the cycloid disc around the inner circumference of the ring gear housing. The resultant action is similar to that of a disc

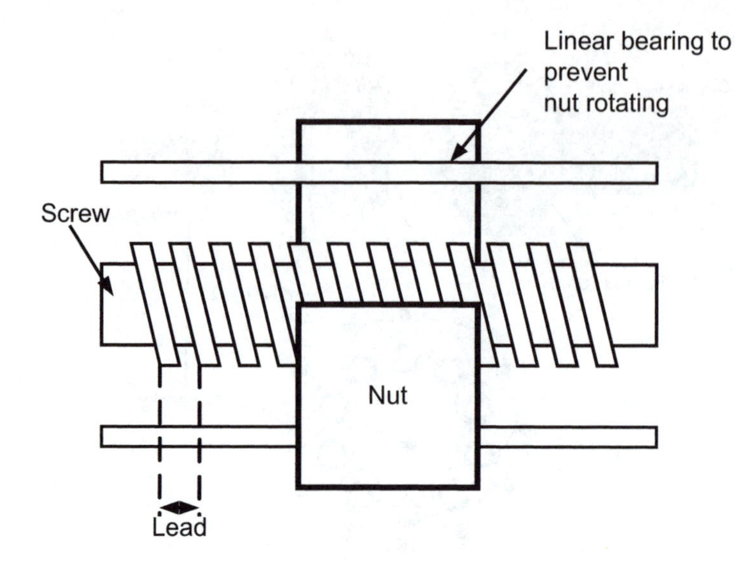

Figure 3.6. The construction of a lead screw. The screw illustrated is single start with an ACME thread.

rolling around the inside of a ring. As the cycloid disc travels clockwise around the gear ring, the disc turns counterclockwise on its axis. The teeth of the cycloid discs engage successively with the pins on the fixed gear ring, thus providing the reduction in angular velocity. The cycloid disc drives the low speed output shaft. The reduction ratio is determined by the number of 'teeth' on the cycloid disc, which has one less 'tooth' than there are rollers on the gear ring. The number of teeth on the cycloid disc equals the reduction ratio, as one revolution of the high speed shaft, causes the cycloid disc to move in the opposite direction by one 'tooth'.

3.2 Lead and ball screws

The general arrangement of a lead screw is shown in Figure 3.6. As the screw is rotated, the nut, which is constrained from rotating, moves along the thread. The linear speed of the load is determined by the rotational speed of the screw and the screw's lead. The distance moved by one turn of the lead screw is termed the lead: this should not be confused with the pitch, which is the distance between the threads. In the case of a single start thread, the lead is equal to the pitch; however the pitch is smaller than the lead on a multi-start thread. In a lead screw there is direct contact between the screw and the nut, and this leads to relatively high friction and hence an inefficient drive. For precision applications, ball screws are used due to their low friction and hence their good dynamic response. A ball screw is identical in principle to a lead screw, but the power is transmitted to the nut via ball bearings located in the thread on the nut (see Figure 3.7).

The relationship between the rotational and linear speed for both the lead and

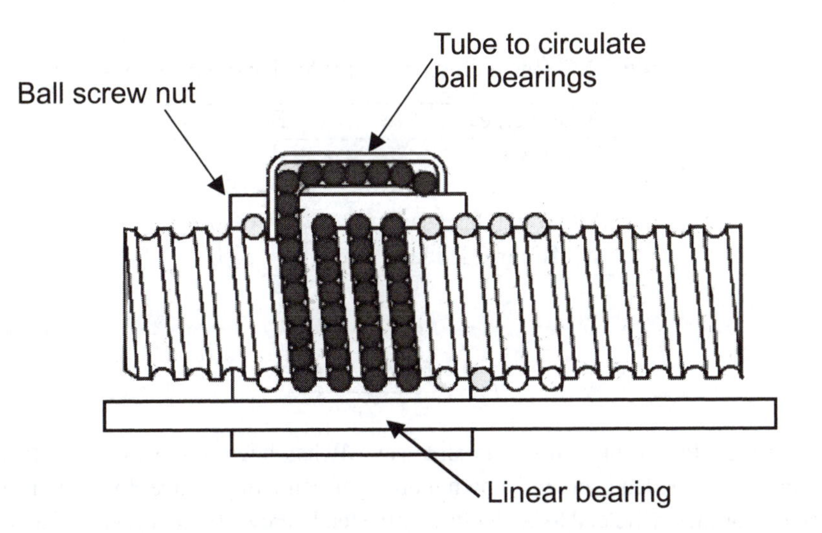

Figure 3.7. The cross section of a high performance ball screw, the circulating balls are clearly visible.

ball screw is given by:

$$N_L = \frac{V_L}{L} \tag{3.4}$$

where N_L is the rotational speed in rev min^{-1}, V_L is the linear speed in m min^{-1} and L is the lead (in metres). The inertia of the complete system is the sum of the screw inertia J_s and the reflected inertia of the load J_L

$$I_{tot} = I_s + I_L \tag{3.5}$$

where

$$J_s = \frac{M_s r^2}{2} \tag{3.6}$$

$$J_L = M_L \left[\frac{L}{2\pi}\right]^2 \tag{3.7}$$

where M_L is the load's mass in kg, M_s is the screw's mass in kg and r is the radius of the lead screw (in metres). In addition, the static forces, both frictional and the forces required by the load, need to be converted to a torque at the lead screw's input. The torque caused by external forces, F_L, will result in a torque requirement of

$$T_L = \frac{LF_L}{2\pi} \tag{3.8}$$

and a possible torque resulting from slideway friction of

Table 3.1. Typical efficiencies for lead and ball screws

System type	Efficiency
Ball screw	0.95
Lead screw	0.90
Rolled-ball lead screw	0.80
ACME threaded lead screw	0.40

$$T_f = \frac{LM_L g \cos\theta \, \mu}{2\pi} \tag{3.9}$$

where θ is the inclination of the slideway. It has been assumed so far that the efficiency of the lead screw is one hundred per cent. In practice, losses will occur and the torques will need to be divided by the lead-screw efficiency, ϵ, see Table 3.1, hence

$$T_{required} = \frac{(T_f + T_L)}{\epsilon} \tag{3.10}$$

A number of linear digital actuators are based on stepper-motor technology, as discussed in Chapter 8, where the rotor has been modified to form the nut of the lead screw. Energisation of the windings will cause the lead screw to move a defined distance, which is typically in the range 0.025–0.1 mm depending on the step angle and the lead of the lead screw. For a motor with a step angle of θ radians, fitted to a lead screw of lead L, the incremental linear step, S, is given by

$$S = \frac{\theta L}{2\pi} \tag{3.11}$$

Example 3.2

Determine the speed and torque requirements for the following lead screw application:

- *The length (L_s) of a lead screw is 1 m, its radius (R_s) is 20 mm and is manufactured from steel (ρ = 7850 kg m^{-3}). The lead (L) is 6 mm rev^{-1}. The efficiency (ϵ) of the lead screw is 0.85.*

- *The total linear mass (M_L) to be moved is 150 kg. The coefficient of friction (μ)between the mass and its slipway is 0.5. A 50 N linear force (F_L) is being applied to the mass.*

- *The maximum speed of the load (V_L) has to be 6 m min^{-1} and the time (t) the system is required to reach this speed in 1 s.*

The mass of the lead screw and its inertia are calculated first:

$$M_s = \rho \pi R_s^2 L_s = 9.97\,\text{kg and } J_s = \frac{M_s R_s^2}{2} = 1.97 \times 10^{-3}\,\text{kg m}^{-2}$$

The total inertia can be calculated by adding the reflected inertia from the load to the lead screw's inertia:

$$J_{tot} = J_s + M_L \left(\frac{L}{2\pi}\right) = 2.11\,\text{kg m}^{-2}$$

The torque required to drive the load against the external and frictional forces, allowing for the efficiency of the lead screw, is given by

$$T_{ext} = \frac{1}{\epsilon}\left(\frac{LF_L}{2\pi} + \frac{LM_L g\mu}{2\pi}\right) = 0.79\,\text{Nm}$$

The input speed required is given by

$$N_L = \frac{V_L}{L} = 1000\,\text{rev min}^{-1} = 104.7\,\text{rad s}^{-1}$$

and the input torque to accelerate the system is given by

$$T_{in} = \frac{N_L}{t} J_{tot} + T_{ext} = 1\,\text{Nm}$$

3.3 Belt drives

The use of a toothed belt or a chain drive is an effective method of power transmission between the motor and the load, while still retaining synchronism between the motor and the load (see Figure 3.8). The use of belts, manufactured in rubber or plastic, offers a potential cost saving over other methods of transmission. Typical applications that incorporate belt drives include printers, ticketing machines, robotics and scanners. In the selection of the a belt drive, careful consideration has to be given to ensuring that positional accuracy is not compromised by selection of an incorrect component. A belt drive can be used in one of two ways, either as a linear drive system (for example, positioning a printer head) or as speed changer.

In a linear drive application, the rotational input speed is given by

Figure 3.8. Synchronous belts and pulleys suitable for servo-drive applications.

$$N_i = \frac{V_L}{\pi D} \tag{3.12}$$

where D is the diameter of the driving pulley (in metres), and V_L is the required linear speed (in m s^{-1}). The inertia of the transmission system, J_{tot}, must include the contributions from all the rotational elements, including the idler pulleys, any rotating load, and the belt:

$$I_{tot} = I_p + I_L \tag{3.13}$$

where I_p is the sum of the inertias of all the rotating elements. The load and belt inertia is given by

$$I_L = \frac{MD^2}{4} \tag{3.14}$$

where the mass, M, is the sum of the linear load (if present) and the transmission-belt masses. An external linear force applied to the belt will result in a torque at the input drive shaft of

$$T_{in} = \frac{DF}{2} \tag{3.15}$$

In a linear application, the frictional force, F_f, must be carefully determined as it will result in an additional torque

$$T_f = \frac{DF_f}{2} \tag{3.16}$$

If a belt drive is used as a speed changer, the output speed is a ratio of the pulley diameters

$$n = \frac{D}{d} \tag{3.17}$$

and the input torque which is required to drive the load torque, T_i, is given by

$$T_i = \frac{T_{out}}{n} \tag{3.18}$$

The inertia seen at the input to the belt drive is the sum of the inertias of the pulleys, the belt, the idlers, and the load, taking into account the effects of the gearing ratio; that is

$$I_{tot} = I_{p1} + I_{belt} + \frac{I_L + I_{p2}}{n^2} \tag{3.19}$$

Where the inertia of the belt can be calculated from equation (3.14) and J_{p2} is the inertia of the driven pulley modified by the gear ratio. The drive torque which is required can then be computed; the losses can be taken into account by using equation (3.10).

The main selection criteria for a belt or chain is the distance, or pitch, between the belt's teeth (this must be identical to the value for the pulleys) and the drive characteristics. The belt pitch and the sizes of the pulleys will directly determine the number of teeth which are in mesh at a particular time, and hence the power that can be transmitted. The power that has to be transmitted can be determined by the input torque and speed. The greater the number of teeth in mesh, the greater is the power that can be transmitted; the number of teeth in mesh on the smaller pulley, which is the system's limiting value, and can be determined from

$$\text{Teeth in mesh} = \left[\pi - 2\sin^{-1}\frac{(D-d)}{C}\right] \times \frac{\text{Teeth on the small pulley}}{2\pi} \tag{3.20}$$

The selection of the correct belt requires detailed knowledge of the belt material, together with the load and drive characteristics. In the manufacturer's data sheets, belts and chains are normally classified by their power-transmission capabilities. In order to calculate the effect that the load and the drive have on the belt, use is made of an application factor, which is determined by the load and/or drive. Typical values of the application factors are given in Table 3.2, which are used to determine the belt's power rating, P_{belt}, using

$$P_{belt} = \text{Power requirements} \times \text{application factor} \tag{3.21}$$

Table 3.2. Typical application factors for belt drives

Load	Drive characteristic		
	Smooth running	Slight shocks	Moderate shocks
Smooth	1.0	1.1	1.3
Moderate shocks	1.4	1.5	1.7
Heavy shocks	1.8	1.9	2.0

Example 3.3

Determine the speed and torque requirements for the following belt drive:

- *A belt drive is required to position a 100 g load. The drive consists of two aluminium pullies ($\rho = 2770$ kgm^{-3}), 50 mm in diameter and 12 mm thick driving a belt weighting 20 g. The efficiency (ϵ) of the drive is 0.95.*
- *The maximum speed of the load (V_L) is 2 m min^{-1} and the acceleration time (t) is 0.1 s.*

Firstly calculate the moment of inertia of the pulley

$$M_p = \rho\pi R_p^2 t_p = 0.065 \text{ kg hence } I_p = \frac{M_p R_p^2}{2} = 2 \times 10^{-5} \text{ kg m}^2$$

The reflected inertia of the belt and load is given by

$$I_L = \frac{MD^2}{4} = 7.5 \times 10^{-5} \text{ kg m}^2$$

The total driven inertia can now be calculated

$$I_{tot} = 2J_p + I_L = 11.5 \times 10^{-5} \text{ kg m}^2$$

The required peak input speed is

$$N_i = \frac{V_L}{piD} = 763 \text{ rev min}^{-1}$$

and hence the the torque torque requirement can be determined

$$T_{in} = \frac{1}{\epsilon}\left(\frac{IN_i}{t}\right) = 0.098 \text{ Nm}$$

3.4 Bearings

In the case of a rotating shaft, the most widely used method of support is by using one or a number bearing. A considerable number of different types of bearing are commonly available. The system selected is a function of the loads and speeds experienced by the system; for very high speed application air or magnetic bearings are used instead of the conventional metal-on-metal, rolling contacts. When considering the dynamics of a system, the friction and inertia of individual bearings, though small, must need to be into account.

3.4.1 Conventional bearings

The bearing arrangement of a rotating component, e.g. a shaft, generally requires two bearings to support and locate the component radially and axially relative to the stationary part of the machine. Depending on the application, load, running accuracy and cost the following approaches can be considered:

- Locating and non-locating bearing arrangements.
- Adjusted bearing arrangements.
- Floating bearing arrangements.

Locating and non-locating bearing arrangements

The locating bearing at one end of the shaft provides radial support and at the same time locates the shaft axially in both directions. It must, therefore, be fixed in position both on the shaft and in the housing. Suitable bearings are radial bearings which can accommodate combined loads, e.g. deep groove ball bearings. The second bearing then provides axial location in both directions but must be mounted with radial freedom (i.e. have a clearance fit) in its housing. The deep groove ball bearing and a cylindrical roller bearing, shown in Figure 3.9(a), illustrate this concept.

Adjusted bearing arrangements

In an adjusted bearing arrangements the shaft is axially located in one direction by the one bearing and in the opposite direction by the other bearing. This type of arrangement is referred to as *cross located* and is generally used on short shafts. Suitable bearings include all types of radial bearings that can accommodate axial loads in at least one direction, for example the taper roller bearings shown in Figure 3.9(b).

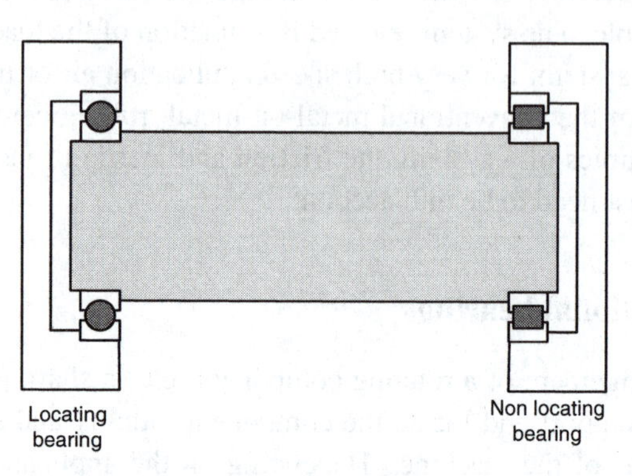

(a) Locating and non-locating bearing arrangement

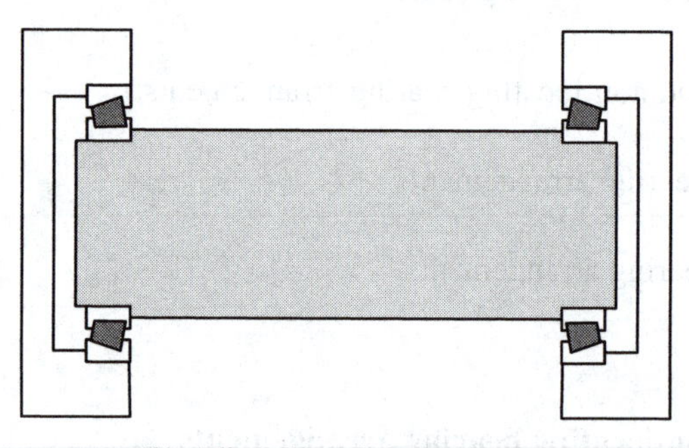

(b) Adjusted bearing arrangement

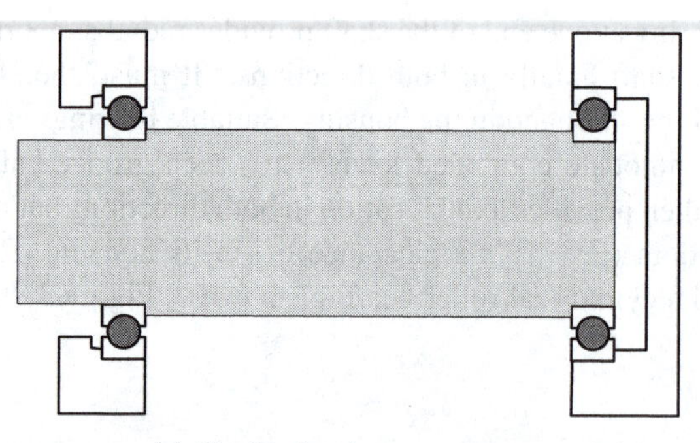

(c) Floating bearing arrangements

Figure 3.9. Three approaches to supporting a rotating shaft.

Table 3.3. Typical coefficients of friction for roller bearings.

Bearing types	Coefficient of friction , μ_b
Deep grove	0.0015–0.003
Self-aligning	0.001–0.003
Needle	0.002
Cylindrical, thrust	0.004

Floating bearing arrangements

Floating bearing arrangements are also cross located and are suitable where demands regarding axial location are moderate or where other components on the shaft serve to locate it axially. Deep groove ball bearings will satisfy this arrangement, Figure 3.9(c).

Bearing friction

Friction within a bearing is made up of the rolling and sliding friction in the rolling contacts, in the contact areas between rolling elements and cage, as well as in the guiding surfaces for the rolling elements or the cage, the properties of the lubricant and the sliding friction of contact seals when applicable.

The friction in these bearing is either caused by the metal-to-metal contact of the balls or rollers on the bearing cage, or by the presence of lubrication within the bearing. The manufacturer will be able to supply complete data, but, as an indication, the friction torque, T_b, for a roller bearing can be determined using the following generally accepted relationship

$$T_b = 0.5 B_l d \mu_b \tag{3.22}$$

where d is the shaft diameter and B_l is the bearing load computed from the radial load, F_r and the axial load, F_a in the bearings, given by

$$B_l = \sqrt{F_r^2 + F_a^2} \tag{3.23}$$

The value of the coefficient of friction for the bearing, μ_b, will be supplied by the manufacturer; some typical values are given in Table 3.3.

The friction due to the lubrication depends on the amount of the lubricant, its viscosity, and on the speed of the shaft. At low speeds the friction is small, but it increases as the speed increases. If a high-viscosity grease is used rather than an oil, the lubrication friction will be higher and this can, in extreme cases, give rise to overheating problems. The contribution of the lubricant to the total bearing friction can be computed using standard equations.

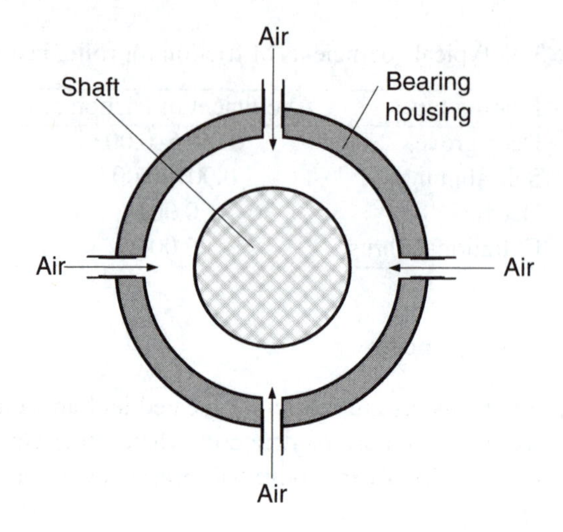

Figure 3.10. Cross section of an air bearing: the dimension of the airgap have been greatly exaggerated.

3.4.2 Air bearings

Air bearings can either be of an aerostatic or an aerodynamic design. In practice aerodynamic bearings are used in turbomachinery, where speeds of up to 36 000 rev min^{-1} in high temperature environments are typically found. In an aerostatic air bearing, Figure 3.10, the two bearing surfaces are separated by a thin film of pressurised air. The compressed air is supplied by a number of nozzles in the bearing housing. The distance between the bearing surfaces is about 5 to 30 μm. As the object is supported by a thin layer of air, the friction between the shaft and its housing can be considered to be virtually zero.

The use of an air bearing gives the system designer a number of advantages including:

- High rotational accuracy typically greater than 5×10^{-8} m is achievable and will remain constant over time as there is no wear due to the absence of contact between the rotating shaft and the housing.
- Low frictional drag, allow high rotational speeds; shaft speed of up to 200 000 rev min^{-1} with suitable bearings can be achieved.
- Unlimited life due to the absence of metal to metal contact, provided that the air supply is clean.
- High stiffness which is enhanced at speed due to a lift effect.

In machine tool applications,the lack of vibration and high rotational accuracy of an air bearing will allow surface finishes of up to 0.012 microns to be achieved.

Figure 3.11. A Radial magnetic bearing, manufactured by SKF Magnetic Bearings, Calgary, Canada.

3.4.3 Magnetic bearings

In a magnetic bearing the rotating shaft is supporting in a powerful magnetic field, and as with the air bearing gives a number of significant advantages:

- No contact, hence no wear, between the rotating and stationary parts. As particle generation due to wear is eliminated, magnetic bearings are suited to clean room applications.
- Operating through a wide temperature range, typically −250°C to 220°C: for this reason magnetic bearings are widely used in superconducting machines.
- A non-magnetic sheath between the stationary and rotating parts allows operation in corrosive environments.
- The bearing can be submerged in process fluid under pressure or operated in a vacuum without the need for seals.
- The frictional drag on the shaft is minimal, allowing exceptionally high speeds.

To maintain clearance, the shaft's position is under closed loop control by controlling the strength of the magnetic field, hence a magnetic bearing requires the following components:

- The bearing, consisting of a stator and rotor to apply electromagnetic forces to levitate the shaft.
- A five axis position measurement system.
- Controller and associated control algorithms to control the bearing's stator current to maintain the shaft at a pre-defined position.

The magnetic bearing stator has a similar construction to a brushless d.c. motor and consists of a stack of laminations wound to form a series of north and south poles. The current is supplied to each winding will produce an attractive force that levitates the shaft inside the bearing. The controller controls the current applied to the coils by monitoring the position signal from the positioning sensors in order to keep the shaft at the desired position through out the operating range of the machine. Usually there is 0.5 mm to 2 mm air gap between the rotor and stator depending on the application. A magnetic bearing is shown in Figure 3.11.

In addition to operation as a bearing, the magnetic field can be used to influence the motion of the shaft and therefore have the inherent capability to precisely control the position of the shaft to within microns and additionally to virtually eliminate vibrations.

3.5 Couplings

The purpose of a coupling is to connect two shafts, end-to-end, to transmit power. Depending on the application speed and power requirements a wide range of couplings are commercially available, and this section summarises the couplings commonly found in servo type applications.

A flexible coupling is capable of compensating for minor amounts of misalignment and random movement between the two shafts. Such compensation is vital because perfect alignment of two shafts is extremely difficult and rarely attained. The coupling will, to varying degrees, minimise the effect of misaligned shafts. If not properly compensated a minor shaft misalignment can result in unnecessary wear and premature replacement of other system components.

In certain cases, flexible couplings are selected for other functions. One significant application is to provide a break point between driving and driven shafts that will act as a *mechanical fuse* if a severe torque overload occurs. This assures that the coupling will fail before something more costly breaks elsewhere along the drive train. Another application is to use the coupling to dampen the torsional vibration that occurs naturally in the driving and/or driven system.

Currently there are a large number of flexible couplings due to the wide range of applications. However, in general flexible couplings fall into one of two broad categories, *elastomeric* or *metallic*. The key advantages and limitations of the designs are briefly summarised in Tables 3.4 and 3.5 to allow the user to select the match the correct coupling to the application.

Elastomeric couplings use a non-metallic element within the coupling, through which the power is transmitted, Figure 3.12(a). The element is manufactured from a compliant medium (for example rubber or plastic) and can be in compression or shear. Compression flexible couplings designs, include those based on jaw, pin and bushing, and doughnut designs while shear couplings include tyre and sleeve moulded elements.

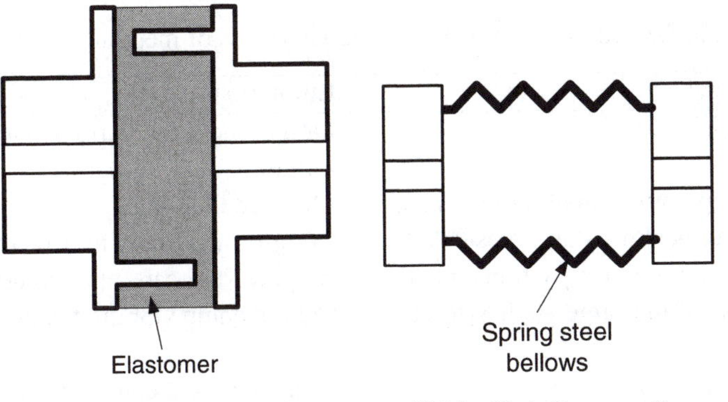

(a) Flexible elastomer coupling. (b) Metallic bellows coupling.

Figure 3.12. Cross sections of commonly used couplings.

Table 3.4. Summary of the key characteristics of elastomeric couplings.

Advantages	Limitations
No lubrication required	Difficult to balance as an assembly
Good vibrational damping and shock absorption	Not torsionally stiff
Field replaceable elastomers elements	Larger than a metallic coupling of the same torque capacity
Capable of accommodating more misalignment than a metallic bellow coupling	Poor overload torque capacity

In practice there are two basic failure modes for elastomeric couplings. Firstly break down can be due to fatigue from cyclic loading when hysteresis that results in internal heat build up if the elastomer exceeds its design limits. This type of failure can occur from either misalignment or torque beyond its capacity. Secondly the compliant component can break down from environmental factors such as high ambient temperatures, ultraviolet light or chemical contamination. It should be noted that all elastomers have a limited shelf life and will in practice require replacement as part of maintenance programme, even if these failure conditions a not present.

Metallic couplings transmit the torque through designs where loose fitting parts are allowed to roll or slide against one another (for example in designs based on gear, grid, chain) or through the flexing/bending of a membrane (typically designed as a disc, diaphragm, beam, or bellows), Figure 3.12(b). Those with moving parts generally are less expensive, but need to be lubricated and maintained. Their pri-

Table 3.5. Summary of the key characteristics of metallic couplings.

Advantages	Limitations
Torsionally stiff	Fatigue or wear plays a major role in failure
High temperature capability	May need lubrication
Good chemical resistance possible	Complex assembly may be required
Low cost per unit torque transmitted	Require very careful alignment
High speed and large shaft size capability	Cannot damp vibration or absorb shock
Zero backlash	High electrical conductivity

mary cause of failure in a flexible metallic couplings is wear, so overloads generally shorten the couplings life through increased wear rather than sudden failure.

3.6 Shafts

A linear rotating shaft supported on bearings can be considered to be the simplest element in a drive system: their static and dynamic characteristics need to be considered. While it is relatively easy, in principle, to size a shaft, it can pose a number of challenges to the designer if the shaft is particularly long or difficult to support. In most systems the effects of transient behaviour can be neglected for the purpose of selecting the components of the mechanical drive train, as the electrical time constants are lower than the mechanical time constant, and therefore they can be considered independently. While such effects are not commonly found, they must be considered if a large-inertia load has to be driven by a relatively long shaft, where excitation generated either by the load (for example, by compressors) or by the drive's power electronics needs to be considered.

3.6.1 Static behaviour of shafts

In any shaft, torque is transmitted by the distribution of shear stress over its cross-section, where the following relationship, commonly termed the *Torsion Formula*, holds

$$\frac{T}{I_o} = \frac{G\theta}{L} = \frac{\tau}{r} \tag{3.24}$$

where T is the applied torque, I_o is the polar moment of area, G is the shear modulus of the material, θ is the angle of twist, L is the length of the shaft, τ is the shear stress and r the radius of the shaft.

In addition we can use the torsion equation to determine the stiffness of a circular shaft

$$K = \frac{T}{\theta} = \frac{G\pi r^4}{2L} \tag{3.25}$$

where the polar moment of area of a circular shaft is given by

$$I_o = \frac{\pi r^4}{2} \tag{3.26}$$

Example 3.4

Determine the diameter of a steel shaft required to transmit 3000 Nm, without exceed the shear stress of 50 MNm^{-1} *or a twist of 0.1 rad m*$^{-1}$*. The shear modulus for steel is approximately 80 GNm*$^{-2}$*.*

Using equation (3.24) and equation (3.26) it is possible to calculate the minimum radius for both the stress and twist conditions

$$r_{stress} = \frac{\tau_{max} I_o}{T} = \sqrt[3]{\frac{2T}{\pi \tau_{max}}} = 33.7mm$$

$$r_{twist} = \sqrt[4]{\frac{2TL}{G\theta\pi}} = 22.1mm$$

To satisfy both constraints the shaft should not have a radius of less than 33.7 mm.

3.6.2 Transient behaviour of shafts

In most systems the effects of transient behaviour can be neglected for the purpose of selecting the components of the mechanical drive train, because, in practice, the electrical time constants are normally smaller than the mechanical time constant. However, it is worth examining the effects of torque pulsations on a shaft within a system. These can be generated either by the load (such as a compressor) or by the drive's power electronics. While these problems are not commonly found, they must be considered if a large inertia load has to be driven by a relatively long shaft.

The effect can be understood by considering Figure 3.13; as the torque is transmitted to the load, the shaft will twist and carry the load. The twist at the motor end, θ_m will be greater than the twist at the load end, θ_L, because of the flexibility of the shaft; the transmitted torque will be proportional to this difference. If K is

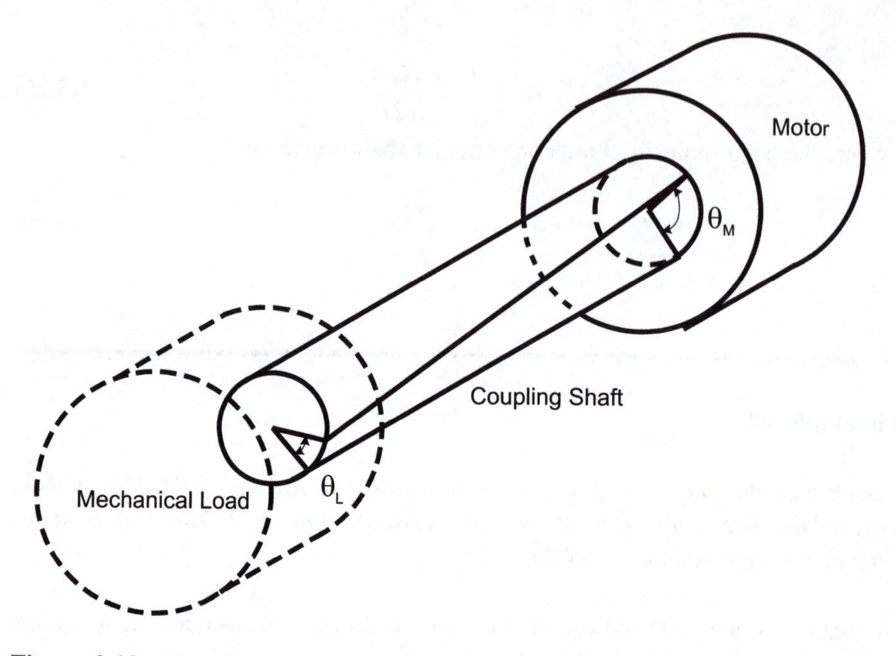

Figure 3.13. The effect of coupling a motor to a high-inertia load via a flexible shaft.

the shaft stiffness (Nm rad^{-1}), and B is the damping constant (Nm rad $^{-1}$s) then for the motor end of the shaft

$$T_m = I_m s^2 \theta_m + Bs(\theta_m - \theta_L) + K(\theta_m - \theta_L) \tag{3.27}$$

and at the load end the torque will turn the load in the same direction as the motor, hence

$$Bs(\theta_m - \theta_L) + K(\theta_m - \theta_L) = I_m s^2 \theta_m + T_L \tag{3.28}$$

where s is the differential operator, d/dt. If these equations are solved it can be shown that the undamped natural frequency of the system is given by

$$\omega_o = \sqrt{\frac{K}{I_m} + \frac{K}{I_L}} \tag{3.29}$$

$$\omega_o = \sqrt{1 - \zeta^2} \tag{3.30}$$

$$\zeta^2 = \sqrt{\frac{B}{2\sqrt{K}}\left(\frac{1}{I_m} + \frac{1}{I_L}\right)} \tag{3.31}$$

and the damped oscillation frequency is given by

$$\omega_n = \sqrt{1 + \frac{B^2}{4K}\left(\frac{1}{I_m} + \frac{1}{I_L}\right)} \tag{3.32}$$

In order to produce a stable system, the damped oscillation frequency must be significantly different to any torque pulsation frequencies produced by the system.

3.7 Linear drives

For many high performance linear applications, including robotic or similar high performance applications, the use of leadscrews, timing belts or rack and pinions driven by rotary motors, are not acceptable due to constrains imposed by backlash and limited acceleration. The use of a linear three phase brushless motor (Section 6.3) or the Piezoelectric motor (Section 9.3), provides a highly satisfactory solution to many motion control problems. If the required application requires only a small high-speed displacement, the voice coil (Section 9.1) can be considered.

The following advantages are apparent when a linear actuator is compared to conventional system based on a driving a belt or leadscrew:

- When compared to a belt and pulley system, a linear motor removes the problems associated with the compliance in the belt. The compliance will causes vibration when the load comes to rest, and this limits the speed and acceleration of a belt drive. It should be noted that a high performance belt drive can have a repeatability error in excess of 50 μm.

- As there are no moving parts are in contact, a linear motor has significant advantages over ball and leadscrew drives due to the removal of errors causes by wear on the nut and screw and e to friction, which is common if the drive has a high duty cycle. Even with the use of a high performance ballscrew the wear may become significant for certain application over time.

- As the length of a leadscrew or ballscrew is increased, so its maximum operating speed is limited, due to the flexibility of the shaft leading to vibration, particularly if a resonant frequency is hit – this is magnified as the length of the shaft increases. While the speed of the shaft can be decreased, by increasing the pitch, the system's resolution is compromised.

While the linear motor does provide a suitable solution for many applications, it is not inherently suitable for vertical operation, largely due to the problems associated with providing a fail-safe brake. In addition it is more difficult to seal against environmental problems compared with a rotary system, leading to restrictions when the environment is particularly hostile, for example when there is excessive abrasive dust or liquid present. Even with these issues, linear motors are widely used is many applications, including high speed robotics and other high performance positioning systems.

3.8 Review of motor-drive sizing

This chapter has so far discussed the power transmission elements of a drive system, while Chapter 2 has looked at issues related to the determination of a drive's requirements. This concluding section provides an overview of how this information is brought together, and the size of the motor and its associated drive are identified. The objective of the sizing procedure is to determine the required output speed and torque of the motor and hence to allow a required system to be selected. The process is normally started once the mechanical transmission system has been fully identified and quantified.

The main constraints that have to be considered during the sizing procedure when a conventional motor is being used can be summarised as follows:

- The peak torque required by the application must be less than both the peak stall torque of the motor and the motor's peak torque using the selected drive.
- The root-mean-square (r.m.s.) torque required by the application must be less than both the continuous torque rating of the motor and the continuous torque which can be delivered by the motor with the specified drive system.
- The maximum speed required by the application must be no greater than approximately eighty per cent of the maximum no-load speed of the motor drive combination; this allows for voltage fluctuations in the supply to the drive system.
- The motor's speed-torque characteristics must not be violated; in addition with a direct current (d.c.) brushed motor, the commutation characteristics of the motor must not be exceeded.

It should be noted that if a linear motor is used in an application the same set of constraints need to be considered, however force is considered to be the main driver as opposed to torque in the sizing process.

The operating regimes of the motor and its associated controller must be considered; two types of duty can be identified. The main determining factor is a comparison of the time spent accelerating and decelerating the load with the time spent at constant speed. In a continuous duty application the time spent accelerating and decelerating is not critical to the application, hence the maximum required torque (the external-load torque plus the drive-train's friction torque) needs to be provided on a continuous basis; the peak torque and the r.m.s. torque requirements are not significantly different to that of the continuous torque. The motor and the controller are therefore selected primarily by considering the maximum-speed and continuous-torque requirements.

An intermittent-duty application is defined as an application where the acceleration and deceleration of the load form a significant part of the motor's duty cycle. In this case the total system inertia, including the motor inertia, must be

Table 3.6. Typical d.c brushed motor motor data. All motors are rated for a maximum speed of 5000 rev min^{-1} at terminal voltage of 95 V.

Type	Continuous stall torque: Nm	Peak torque: Nm	Moment of inertia: kg m ^{2}	Voltage constant: V s rad $^{-1}$	Current constant: Nm A^{-1}
M1	0.5	2.0	1.7×10^{-4}	0.18	0.18
M2	0.7	4.0	2.8×10^{-4}	0.18	0.18
M3	1.2	8.0	6.0×10^{-4}	0.18	0.18

Table 3.7. Typical current data (in amps) supplied by manufacturers, for drives capable of driving d.c. brushed motor. All the drives are capable of supplied the 95 V required for the motors detailed in Table 3.6.

Type	Continuous current	Peak current
D1	5	10
D2	10	20
D3	14	20

considered when the acceleration torque is being determined. Thus, the acceleration torque plus the friction torque, and any continuous load torque present during acceleration, must be exceeded by the peak-torque capability of the motor-drive package. Additionally, the drive's continuous torque capability must exceed the required r.m.s. torque resulting from the worst-case positioning move.

The difference between these two application regimes can be illustrated by considering a lathe, shown in Figure 1.2. The spindle drive of a lathe can be considered to be a continuous-duty application since it runs at a constant speed under a constant load, while the axis drives are intermittent-duty applications because the acceleration and deceleration required to follow the tool path are critical selection factors.

The confirmation of suitable motor-drive combinations can be undertaken by the inspection of the supplier's motor-drive performance data, which provides information on the maximum no-load speed and on the continuous torque capability, together with the torque sensitivity of various motor frame sizes and windings. Tables 3.6 and 3.7 contain information extracted from typical manufacturer's data sheets relating to d.c brushed motors and drives, for a more detailed discussion see Chapter 5. In the sizing process it is normal to initially consider only a small number of the key electrical and mechanical parameters. If significant problems with motor and drive selection are experienced, a detailed discussion with the suppliers will normally resolve the problem.

As discussed above, two operating regimes can be identified: the following key features can be summarised as

- In a *continuous duty application* the acceleration and deceleration requirements are not considered critical; the motor and the controller can be satisfactorily selected by considering the maximum-speed and continuous-torque requirements.

- An *intermittent-duty application* is defined as an application where the acceleration and deceleration of the load form a significant part of the motor's duty cycle, and need to be considered during the sizing process.

3.8.1 Continuous duty

For continuous duty, where the acceleration performance is not of critical importance, the following approach can be used:

- Knowledge of the required speed range of the load, and an initial estimation of the gear ratios required, will permit the peak motor speed to be estimated. In order to prevent the motor from not reaching its required speed, due to fluctuations of the supply voltage, the maximum required speed should be increased by a factor of 1.2. It should be noted that this factor is satisfactory for most industrial applications, but it may be refined for special applications, for example, when the system has to operate from a restricted supply as is found in aircraft and offshore-oil platforms.

- Using the drive and the motor manufacturers' data sheet, it will be possible to locate a range of motors that meets the speed requirement when the drive operates at the specified supply voltage. If the speed range is not achievable, the gear ratio should be revised.

- From the motor's data, it will normally be possible to locate a motor-drive that meets the torque requirement; this will also allow the current rating of the drive to be determined. A check should then be undertaken to ensure that the selected system can accelerate the load to its required speed in an acceptable time.

Example 3.5

Determine the motor's speed and torque requirement for the system detailed below, and hence identify a suitable motor and associated drive:

- *The maximum load speed 300 rev min*$^{-1}$*, a non-optimal gearbox with a ratio of 10:1 has been selected. The gearbox's moment of inertia referred to its input shaft is* 3×10^{-4} *kgm*^{2}*.*

- *The load's moment of inertia has been determined to* 5×10^{-2} *kgm*^{2}
- *The maximum load torque is 8 Nm.*

Based on this information the minimum motor speed required can be determined including an allowance for voltage fluctuations

$$\text{Motor speed} = 300 \times \text{gear ratio} \times 1.2 = 3600 \text{ rev min}^{-1}$$

$$\text{Continuous torque} = \frac{8}{10} = 0.8 \text{ Nm}$$

Consideration of the motor data given in Table 3.6 indicates that motor M3 is capable of meeting the requirements. The required speed is below the peak motor speed of 5000 rev min $^{-1}$ $\pm$ 10%, and the required torque is below the motor's continuous torque rating. At the continuously torque demand the motor requires 4.5A, hence the most suitable drive from those detailed in Table 3.7, will be drive D1.

To ensure that the motor-drive combination is acceptable, the acceleration can be determined for the drives peak output of 10 A. At this current the torque generated by the motor is 1.8 Nm, well within the motor rating. Using equation (2.12), and noting that the gearbox's moment of inertia is added to that of the motor to give:

$$\alpha = \frac{T_{peak} - T_L/n}{n(I_d + I_L/n^2)} = \frac{1.8 - 8/10}{10(9 \times 10^{-4} + 5 \times 10^{-2}/10^2)} = 71.4 \text{ rad s}^{-1}$$

Hence the load will be accelerated to a peak speed of 300 rev min^{-1}, within 0.5 seconds, which is satisfactory. In practice the acceleration rate would be controlled, so that the system, it particular the gear teeth, would not experience significant shock loads.

3.8.2 Intermittent duty

When the acceleration performance is all important, the motor inertia must be considered, and the torque which is necessary to accelerate the total inertia must be determined early in the sizing process. A suitable algorithm is as follows:

- Using the application requirements and the required speed profile determine the required speeds and acceleration.
- Estimate the minimum motor torque for the application using equation 2.1.

- Select a motor-drive combination with a peak torque capability of at least 1.5 to 2 times the minimum motor-torque requirement to ensure a sufficient torque capability.
- Recalculate the acceleration torque required, this time including the inertia of the motor which has been selected.
- The peak torque of the motor-drive combination must exceed, by a safe margin of at least fifteen per cent, the sum of the estimated friction torque and the acceleration torque and any continuous torque loading which is present during acceleration. If this is not achievable, a different motor or gear ratio will be required.
- The motor's root-mean-square (r.m.s.) torque requirement can then be calculated as a weighted time average, using;

$$T_{rms} \leqslant T_{cm} + \sqrt{T_f^2 + dT_a^2} \tag{3.33}$$

  where T_{cm} is the continuous motor-torque requirement, T_f is the friction torque at the motor, T_a is the acceleration torque, and d is the duty cycle.
- The selected motor-drive combination is evaluated for maximum speed and continuous torque capabilities as in Section 3.8.1.
- If no motor of a given size can meet all the constraints, then a different, usually larger, frame must be considered, and the procedure must be repeated.

In practice, it is usual for one or two iterations to be undertaken in order find an acceptable motor-drive combination. The approximate r.m.s.-torque equation used above not only simplifies computation, but it also allows an easy examination of the effects of varying the acceleration/deceleration duty cycle. For example, the effects of changes in the dwell time on the value of r.m.s. torque can be immediately identified. Should no cost-effective motor-drive be identified, the effects of varying the speed-reduction ratio and inertias can easily be studied by trying alternative values and sizing the reconfigured system.

Sometimes, repeated selections of motors and drives will not yield a satisfactory result; in particular, no combination is able to simultaneously deliver the speed and the continuous torque which is required by the application, or to simultaneously deliver the peak torque and the r.m.s. torque required. In certain cases motor-drive combinations can be identified, but the size or cost of the equipment may appear to be too high for the application, and changes will again be required.

Example 3.6

Identify a suitable motor and its associated drive for the application detailed below:

- *The load is a rotary disc which has a moment of inertia of* 1.34 *kgm*^{2}. *The estimated frictional torque referred to the table's drive input is 5 Nm, and the external load torque is 8 Nm.*
- *The table is driven through a 20:1 gear box, which has a moment of inertia of* 3×10^{-4} *kgm*^{2} *referred to its input shaft.*
- *The table is required to index 90.0° (θ) in one second (t_m), and then dwell for a further two seconds. A polynomial speed profile is required.*

The selection process starts with the determination of the peak load speed and acceleration, using equation 2.28, the maximum speed occurs at $t = 0.5$ s and maximum acceleration occurs at $t = 0$ s.

$$\dot{\theta}(0.5) = \frac{6\theta t}{t_m^2} + \frac{6\theta t^2}{t_m^3} = 7.1 \text{ rad s}^{-1}$$
$$\ddot{\theta}(0) = \frac{6\theta}{t_m^2} = 9.4 \text{ rad s}^{-2}$$

The peak torque can now be calculated, at the input to the table. The torque is determined by the peak acceleration, and the load and friction torques, giving

$$T_{peak} = 8 + 5 + 1.34 \times 9.4 = 25.6 \text{ Nm}$$

This equates to 1.28 Nm peak torque from the motor. Using the motors defined in Table 3.6, it appears that motor M1 is a suitable candidate as it is capable of supplying over four times the required torque. If the motor's moment of inertia is now included in the calculation, the peak torque requirement is

$$T_{peak} = \frac{25.6}{20} + (3 \times 10^{-4} + 1.7 \times 10^{-4}) \times (9.4 \times 20) = 1.37 \text{ Nm}$$

which is well within the capabilities of the motor and drive D1, detailed in Table 3.7. The required peak current is 7 amps. The r.m.s. torque can now be calculated using equation (3.33):

$$T_{rms} \leqslant T_{cm} + \sqrt{T_f^2 + dT_a^2} = 0.67 \text{ Nm}$$

This figure is in excess of the continuous torque rating of the motor M1, and in certain applications could lead to the motor overheating. In addition, while the current is below the peak rating it it is greater than the continuous rating: in practice this could result in the drive cutting-out due to motor overheating. Thus a case can be made to change the motor and drive – in practice this decision would be made after careful consideration of the application.

If motor M2 is selected, and the above calculations are repeated, the motor's torque requirement becomes 1.4Nm, and the r.m.s torque becomes 0.69Nm. While marginal, M2 can be used along as the friction or load torque do not increase, if the drive is also changed to D2 there is no possibility of any overheating problems in the system.

As a final check the motor's peak speed is determined to be 1350 rev min^{-1}; this is well within the specification of the selected motor and drive.

This short example illustrates how a motor and drive can be selected, however the final decision needs a full understanding of the drive and its application. If the application only requires a few indexing moves, the selection of motor M1 could be justified however if a considerable number of indexes are required, motor M2 could be the better selection. This example has only considered the information given above; in practice the final decision will be influenced on the technical requirements of the complete process, and commercial requirements. While this example has been undertaken for a d.c. brushed motor, the same procedure is followed for any other type of drive – the only differences being the interpretation of the motor and drive specifications.

3.8.3 Inability to meet both the speed and the torque requirements

In the selection of motors, the limitations of both the motor and the drive forces a trade-off between the speed and the torque capabilities. Thus, it is usually advantageous to examine whether some alteration in the mechanical elements may improve the overall cost effectiveness of the application. Usually the speed-reduction ratios used in the application are the simplest mechanical parameter which can be investigated. If the speed required of the motor is high, but the torque seems manageable, a reduction in the gear ratio may solve the problem. If the torque required seems high but additional speed is obtainable, then the gear ratio should be increased. The goal is to use the smallest motor-drive combination that exceeds both the speed and torque requirement by a minimum of ten to twenty per cent. Sometimes the simple changing of a gear or pulley size may enable a suitable system to be selected.

A further problem may be the inability to select a drive that meets both the peak- and the continuous- torque requirements. This is particularly common in intermittent-motion applications. Often, the peak torque is achievable but the drive is unable to supply the continuous current required by the motor. As has been shown earlier, while optimum power transfer occurs when the motor's rotor and the reflected load inertia are equal this may not give the optimum performance for an intermittent drive, hence the gear ratios in the system need to be modified and the sizing process repeated.

Example 3.7

Consider Example 3.6 above, and consider the impact of performance due to a change in the reduction ratio.

In Example 3.6 the speed and torque requirements, using motor M1, were calculated to be 1.35 Nm and 1350 rev min^{-1}. The speed requirement is well within the motor specifications. If the gear ratio was changed to 40:1, the motor's peak speed requirement increases to 2700 rev min^{-1} and the r.m.s. torque drops to 0.35 Nm, and a peak torque of 0.818 Nm. These figures are well within the specification of motor M1 and drive D1.

This example illustrates a different approach to resolving the sizing problem encountered earlier. The change in gear ratio can easily be achieved at the design state, and in all possibility be cheaper that going to a larger motor and drive system.

3.8.4 Linear motor sizing

So far in this section we have considered the sizing of conventional rotary motors. We will now consider the sizing of a linear motor. Due to the simplicity of a linear drive, the process is straightforward compared to combining a leadscrew, ballscrew or belt drive with a conventional motor. As with all other sizing exercises, the initial process is to identify the key parameters, before undertaking the detailed sizing process. A suitable algorithm is as follows:

- Using the application requirements and the required speed profile determine the required speed and acceleration.
- Estimate the minimum motor force for the application using equation 2.2.
- Select a motor-drive combination with a peak force capability of at least 1.5 to 2 times the minimum force requirement to ensure a sufficient capability.

- Recalculate the acceleration force required, this time including the mass of the moving part of the selected motor.
- The peak force of the motor-drive combination must exceed, by a safe margin of at least fifteen per cent, the sum of the estimated friction force and the acceleration force and any continuous force which is present during acceleration. If this is not achievable, a different motor will be required.
- The motor's root-mean-square (r.m.s.) torque requirement can then be calculated as a weighted time average, in addition this will allow the motor's temperature to be estimated.

Example 3.8

Determine the size of the linear motor, and drive required to move a mass of $M_L = 40\ kg$*, a distance of* $d = 750\ mm$ *in time of* $t_m = 400\ ms$*.*

- *The system has a dwell time of* $t_d = 300\ ms$*, before the cycle repeats.*
- *Assume that the speed profile is triangular, and equal times are spent accelerating, decelerating and at constant speed.*
- *Assume the frictional force,* $F_f = 3N$*.*
- *The motors's parameters are: force constant is* $K_F = 40NA^{-1}$*, back emf constant* $K_{emf} = 50\ Vm^{-1}s$*, winding resistance,* $Rt_w = 2\ \Omega$ *and thermal resistance of the coil assembly,* $Rt_{c-a} = 0.15\ °CW^{-1}$*.*

The acceleration and peak speed can be determined using the process determined in Section 2.4, hence

$$\dot{x} = \frac{3d}{2t_m} = 2.4 \text{ m s}^{-1}$$

and

$$\ddot{x} = \frac{3\,\dot{x}}{t_m} = 28.8 \text{ m s}^{-2}$$

The acceleration force required is given by

$$F_a = M_L\ddot{x} + F_f = 1155 \text{ N}$$

This now allows the calculation of the root mean square force requirement

$$F_{rms} = \sqrt{\frac{2t_m F_a^2 + t_m F_f^2}{t_m + t_d}} = 635.5 \text{ N}$$

The drives current and voltage requirements can therefore be calculated

$$V_{drive} = \dot{x} K_{emf} + I_{peak} R_w = 178 \text{ V}$$

$$I_{peak} = \frac{F_a}{K_F} = 28.8 \text{ A}$$

$$I_{continuous} = \frac{F_{rms}}{K_F} = 15.9 \text{ A}$$

As linear motors are normally restricted to temperature rises of less that 100°C, the temperature rise over ambient needs to be calculated

$$T_{rise} = I_{continuous}^2 R_w Rt_{c-a} = 76°\text{C}$$

3.9 Summary

This chapter has reviewed the characteristics of the main mechanical power transmission components commonly used in the construction of a drive system, together with their impact on the selection of the overall drive package. The chapter concluded by discussing the approach to sizing drives. One of the key points to be noted is that the motor-drive package must be able to supply torques and speed which ensure that the required motion profile can be followed. To assist with the determination of the required values, a sizing procedure was presented. It should be remembered over-sizing a drive is as under-sizing.

Chapter 4

Velocity and position transducers

Within a closed-loop control system, feedback is used to minimise the difference between the demanded and actual output. In a motion-control system, the controlled variable is either the velocity or the position. The overall performance of a motion-control system will depend, to a large extent, on the type and quality of the transducer which is used to generate the feedback signal. It should be noted that velocity- or position-measuring transducers need not be used; other process variables (for example, the temperature and the chemical composition) can be used to determine the speed or position of a drive within a manufacturing process. However, as this book is concerned with robotic and machine-tool applications, the primary concentration will be on velocity and position transducers. In order to appreciate the benefits and limitations of the available systems, the performance of measurement systems in general must be considered.

4.1 The performance of measurement systems

The performance of a measurement system is dependent on both the static and dynamic characteristics of the transducers selected. In the case of motion-control systems where the measured quantities are rapidly changing, the dynamic relationships between the input and the output of the measurement system have to be considered, particularly when discrete sampling is involved. In contrast, the measured parameter may change only slowly in some applications; hence the static performance only needs to be considered during the selection process. The key characteristics of a transducer are as follows.

- **Accuracy** is a measure of how the output of the transducer relates to the true value at the input. In any specification of accuracy, the value needs to be qualified by a statement of which errors are being considered and the conditions under which they occur.
- **Dead band** is the largest change in input to which the transducer will fail to respond; this is normally caused by mechanical effects such as friction,

backlash, or hysteresis.

- **Drift** is the variation in the transducer's output which is not caused by a change in the input; typically, it is caused by thermal effects on the transducer or on its conditioning system.

- **Linearity** is a measure of the consistency of the input/output ratio over the useful range of the transducer.

- **Repeatability** is a measure of the closeness with which a group of output values agree for a constant input, under a given set of environmental conditions.

- **Resolution** is the smallest change in the input that can be detected with certainty by the transducer.

- **Sensitivity** is the ratio of the change in the output to a given change in the input. This is sometimes referred to as the gain or the scale factor.

A clear understanding is required of the interaction between accuracy, repeatability and resolution as applied to a measurement system. It is possible to have measurement systems with either high or low accuracy and repeatability; the measurements compared to the target position are shown in Figure 4.1. A motor drive system needs to incorporate a position measurement system with both high accuracy and repeatability to ensure that the target point is measured. If the system has low resolution, Figure 4.2, the uncertainty regarding each measure point increases.

All measurement systems suffer from inherent inaccuracies; and estimation of the uncertainty requires knowledge of the form that the error takes. In general, an error can be classified either as a random or a systematic error. Random errors arise from chance or random causes, and they must be considered using statistical methods. Systematic errors are errors which shift all the readings in one direction; for example, a shift in the zero point will cause all the readings to acquire a constant displacement from the true value.

4.1.1 Random errors

If a large set of data is taken from a transducer under identical conditions, and if the errors generated by the measurement system are random, the distribution of values about the mean will be Gaussian, Figure 4.3. In this form of distribution, sixty eight per cent of the readings lie within ± 1 standard deviation of the mean and ninety five per cent lie within ± 2 standard deviations. In general, if a sample of n readings are taken with values $x_1, x_2 \ldots x_n$, the mean $\bar{x}$ is given by

$$\bar{x} = \frac{1}{n} \sum_{i=1}^{n} n_i \tag{4.1}$$

and the standard deviation, s, by

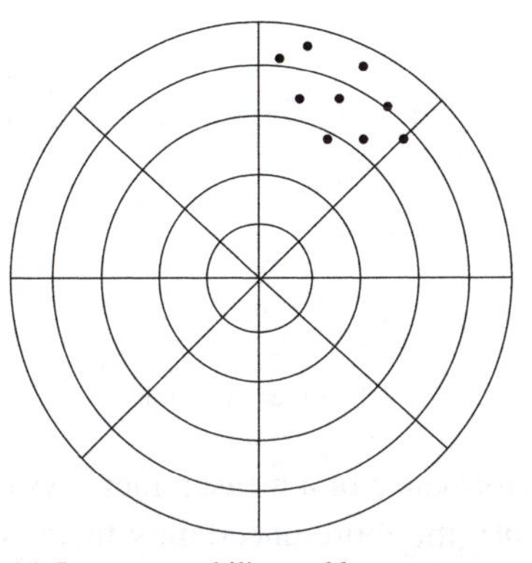

(a) Low repeatability and low accuracy.

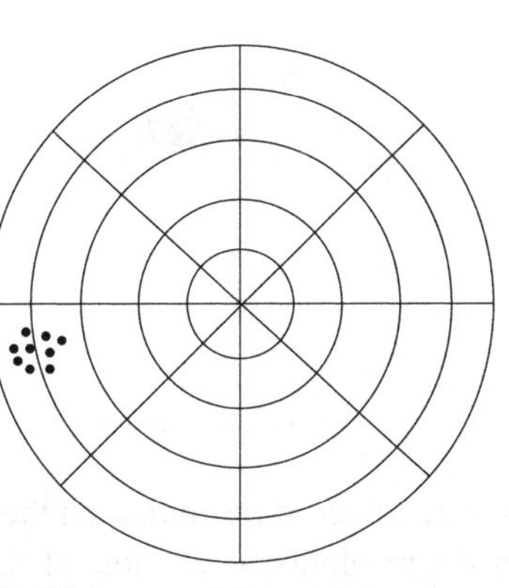

(b) High repeatability and low accuracy.

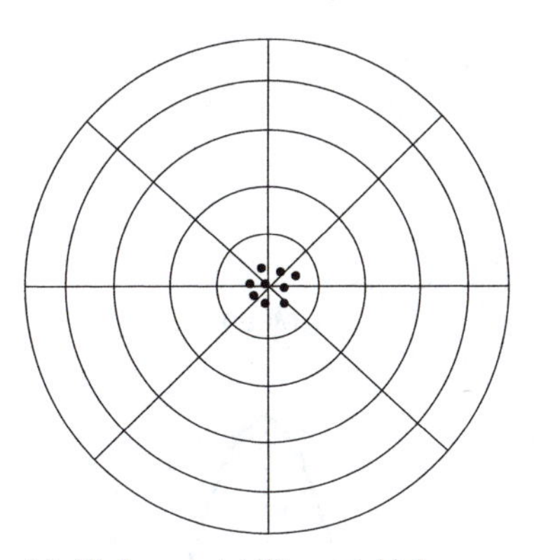

(c) High repeatability and high accuracy.

Figure 4.1. Effect of accuracy and repeatability on the performance of a measurement system. The dots represent the individual measurements. Only when the system has both high accuracy and repeatability can the measurement error with respect to the target point be minimised.

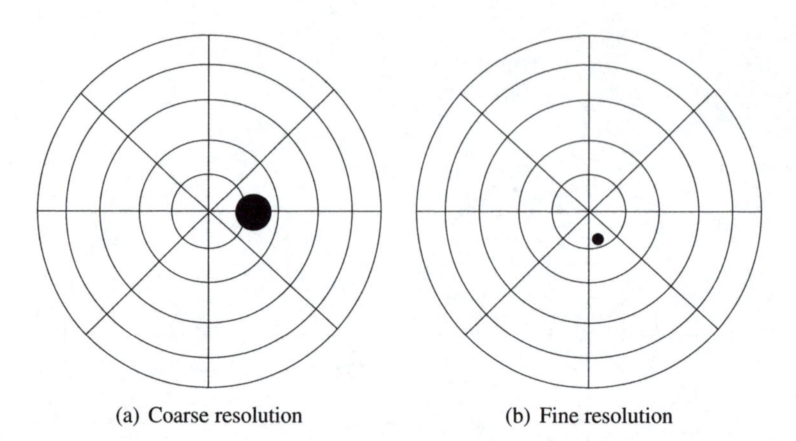

Figure 4.2. Effect of resolution on the performance of a measurement system: the coarser the resolution (i.e. area of the dot), the more uncertainty there is in the measurement.

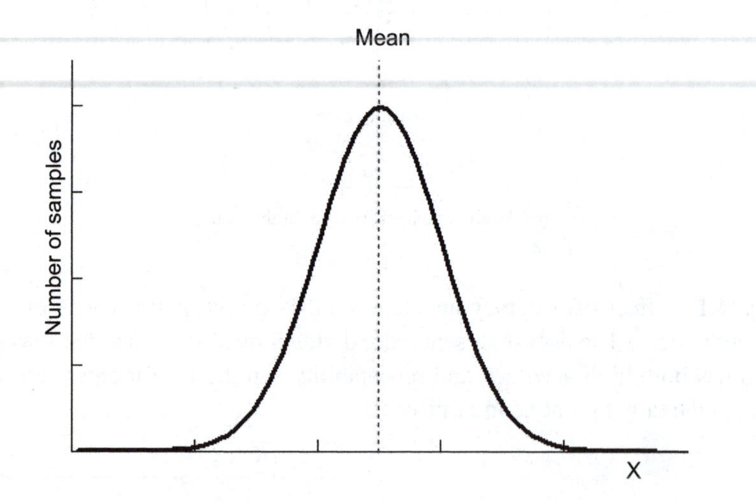

Figure 4.3. A Gaussian data distribution.

$$s = \sqrt{\frac{\sum_{i=1}^{n}(x_i - \bar{x})^2}{n-1}} \qquad (4.2)$$

The mean value which is obtained is dependent on the number of samples taken and on the spread; the true mean value can never be determined since this would require an infinite number of samples. However, by the use of the standard error of the mean, s_m, the probability of how close the mean of a set of data is to the true mean of the system can be evaluated. The standard error is given by

$$s_m = \frac{s}{\sqrt{n-1}} \qquad (4.3)$$

It is possible, using probability theory, to state that with a Gaussian distribution the probability of an individual reading, x_i, being within $\pm s_m$ of the true value is sixty eight per cent and that the probability of being within $\pm 2s_m$ is ninety five per cent.

4.1.2 Systematic errors

It can be seen from equation (4.3) that by taking a large number of samples, the random errors can be reduced to a very low value. However, when a systematic error occurs all the measurements are shifted in one direction by an equal amount. Figure 4.4 shows the spread of readings caused by both types of errors. The terms *accurate* and *precise* are used to cover both these situations; a measurement is accurate if the systematic error is small, and it is precise if the random error is small. A prime example of a systematic error is a zero offset, that is, when a instrument or a measured value does not return to zero when the parameter being measured is zero. This can be introduced by the transducer itself, or, more probably, by any conditioning electronics being used. Systematic errors are cumulative, so if a measurement, M, is a function of x, y, z, such that

$$M = f(x, y, z) \qquad (4.4)$$

then the maximum value of the systematic error, ΔM, will be

$$\Delta M = \delta x^2 + \delta y^2 + \delta z^2 \qquad (4.5)$$

where δx, δy and δz are the respective errors in x, y, and z. However, this approach can be considered to be rather pessimistic, because the systematic errors may not all operate in the same direction, and therefore they can either increase or decrease the reading. It is useful, therefore, to quote the systematic error in the form

$$\Delta M = \sqrt{\delta x^2 + \delta y^2 + \delta z^2} \qquad (4.6)$$

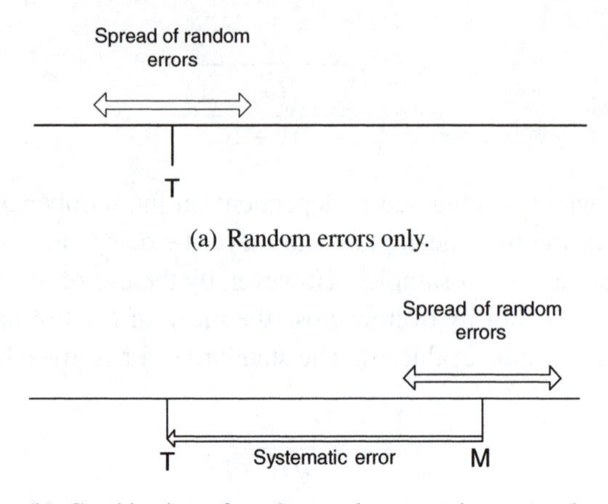

(a) Random errors only.

(b) Combination of random and systematic errors, the spread caused v the random error has been shifted by the systematic error,

Figure 4.4. The effects of systematic and random errors on measurements where T is the true value and M is the mean value of the data.

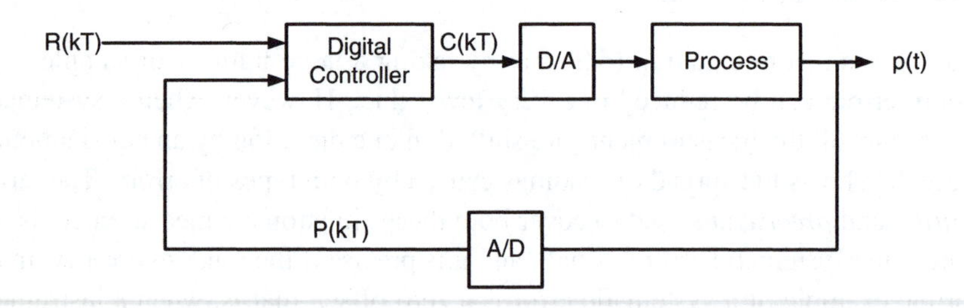

Figure 4.5. A block diagram of a digital-control system, showing the location of the analogue to digital (A/D) and the digital to analogue (D/A) converters.

4.1.3 Digital-system errors

There is an increasing reliance on digital-control techniques in drive systems. Digital controllers require the transducer's output to be sampled and digitised. The actual process of sampling will introduce a number of errors of its own. Consider Figure 4.5, where a reference signal, $R(kT)$, a feedback signal, $P(kT)$, and the resultant computed value, $C(kT)$, are discrete signals, in contrast to the output, $p(t)$, which is a continuous function of time. If the sampling period, T, is small compared with the system's time constant, the system can be considered to be continuous; however, if the sampling time is close to the system's time constant, the effects of digital sampling must be considered. A more detailed discussion of digital controllers is to be found in Section 10.1.1.

A sampler can be considered to be a switch that closes for a period of time every T seconds; with an ideal sampler for an input $p(t)$, the output will be

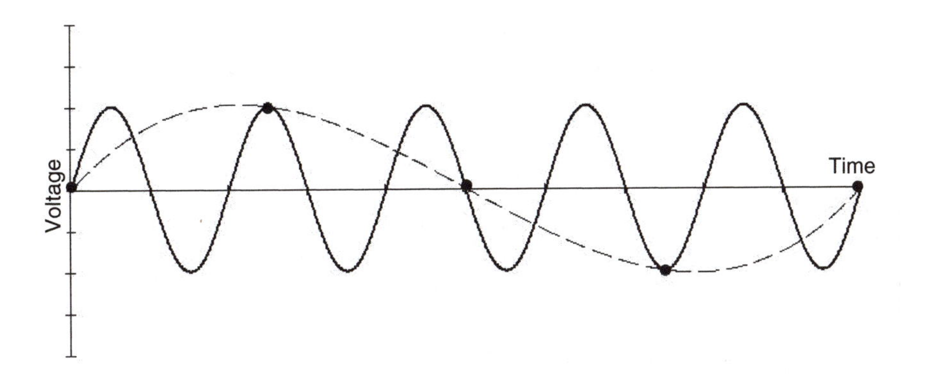

Figure 4.6. Aliasing caused by a sampling frequency. The sampling point are shown as dots, the sampling frequency is below frequency of the waveform being sampled. The reconstituted waveform is shown as the dotted line.

$$p^*(t) = p(nT)\delta(t - nT) \tag{4.7}$$

where δ is the Dirac delta function (Nise, 1995). The input signal can be accurately followed if the sampling time is small compared to the rate of change of the signal; this ensures that the transients are not missed. In order to obtain an accurate picture of the signal being sampled, the sampling frequency must be selected with care. The sampling frequency is largely determined by the loop time of the control system; a high sample rate will place restrictions on the complexity of the algorithms that must be employed. If the highest frequency present in the signal to be sampled is f_p, then the minimum sampling rate is $2f_p$ as defined by Shannon's sampling theorem. The effect of a sampling frequency which is considerable less than the frequency of a signal is shown in Figure 4.6. It can be seen that the reconstituted signal is at a far lower frequency than the original waveform; this signal is referred to as the alias of the original signal. It is impossible to determine whether the sampled data is from the original signal or its alias. A frequently made mistake is the selection of a sampling rate at twice the frequency of interest, without considering the effect of noise, particularly interference from the mains supply. The solution, to this problem is to apply an anti-alias filter which blocks frequencies higher than those of interest.

4.1.4 Analogue-digital and digital-analogue conversion errors

Conversion of an analogue signal to a digital value involves a process of quantisation. In an analogue-to-digital (A/D) converter, the change from one state to the next will occur at a discrete point (the intermediate values are not considered, Figure 4.7). The difference between any two digital values is known as the quan-

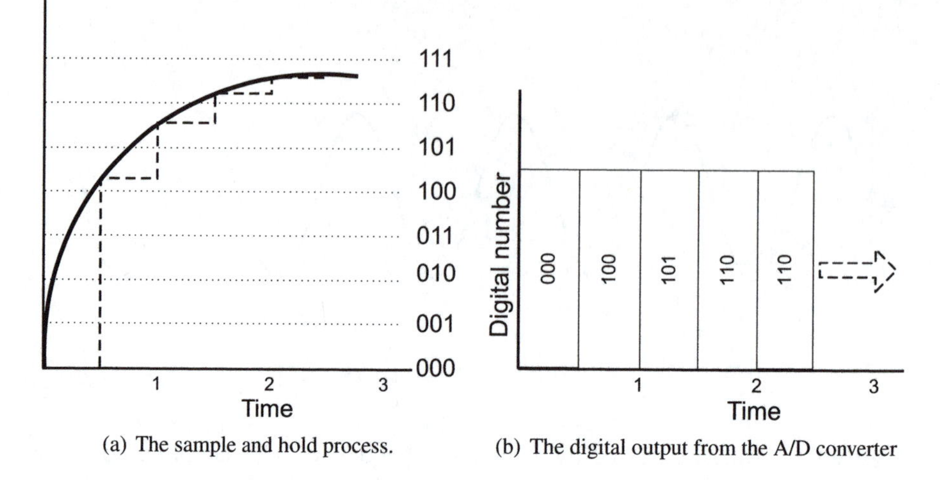

Figure 4.7. The analogue to digital conversion process. The voltage being converted is the solid line in (a), the input to the ADC is the dotted line, showing the change of the sampled value.

tisation size, V_q, and it is commonly termed the resolution of the converter. For an n-bit system the steps due to quantisation step V_q, and the subsequent error E_q are equal to

$$V_q = \frac{\text{Full scale input}}{2^n} \tag{4.8}$$

$$E_q = \frac{1}{2}\frac{\text{Full scale input}}{2^n} = \frac{\text{Full scale input}}{2^{n+1}} \tag{4.9}$$

The resolution is equal to the input voltage, V_q, which will change the state of the least-significant bit (LSB).

Transitions occur from one digital number to the next at integral multiples of the LSB, giving a maximum uncertainty of one bit within the system. The resolution can only be decreased by increasing the number of bits within the converter. A range of techniques are used for analogue to digital conversion, including high-speed-flash (or parallel) converters, integrating, and successive-approximation converters. It is not common to construct a discrete system; one of the commonly available proprietary devices is usually used in the selection of a suitable device, and consideration must be given to the device's conversion time, resolution, and gain. A variant of the successive approximation converter is the tracking converter that forms an integral part of a resolver's decoder; this is discussed later in this chapter.

Digital-to-analogue (D/A) converters are used to provide analogue signals from a digital systems. One of the problems with a D/A converter is that glitches occur as the digital signal (that is, the switches) change state, requiring a finite settling

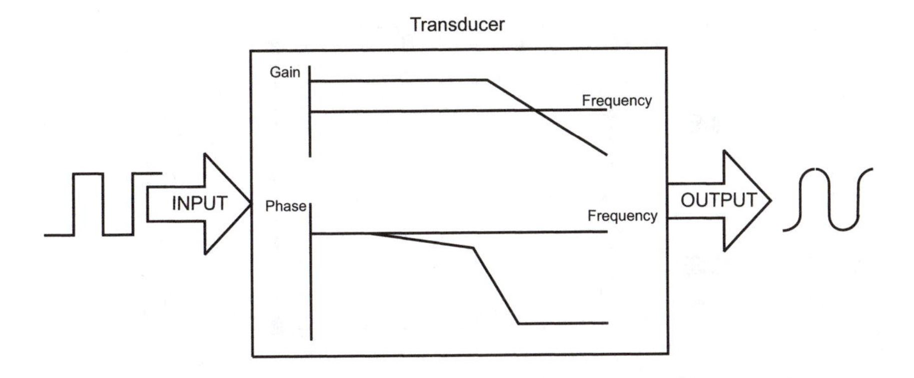

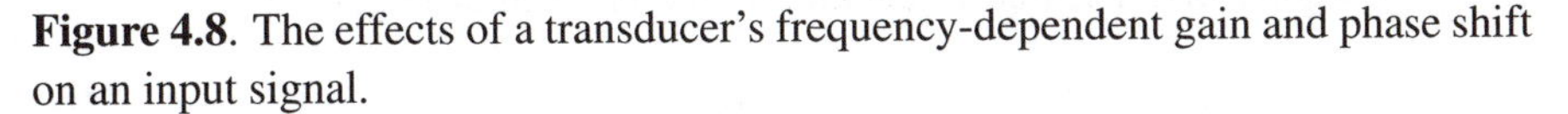

Figure 4.8. The effects of a transducer's frequency-dependent gain and phase shift on an input signal.

time. As the code changes, the switches will not change state at the same instant; this is particularly acute when the code changes from, say 01111 to 10000, where the output for 11111 may transiently appear. It is possible to add a deglitching function to a D/A converter by increasing the transfer time of the converter.

4.1.5 Dynamic performance

Only the static characteristics of transducers have been considered up to this point. However, if the measured signal is rapidly changing, the dynamic performance of the measurement system has to be considered. A transducer with a linear characteristic will achieve a constant performance for all inputs; but this is not true in a practical system, since the input will have a non-linear distortion caused by the transducer's frequency-dependent gain and the phase shift, Figure 4.8. The formal analysis of these effects can be conducted, and represented, by a first-order, linear, differential equation. The dynamic performance needs to be considered in the selection of any transducer; even if the speed or position changes slowly, to ensure that any transient effects are considered. A limited bandwidth transducer will seriously limit the overall system bandwidth, and hence its ability to respond to transients (such as the application or removal of torques from the load).

4.2 Rotating velocity transducers

While the velocity can be determined from position measurement, a number of transducers are able to provide a dedicated output which is proportional to the velocity.

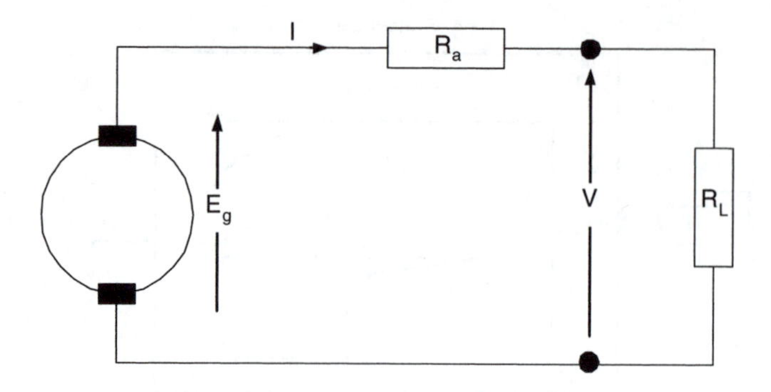

Figure 4.9. The equivalent circuit of a brushed tachogenerator.

4.2.1 Brushed d.c. tachogenerators

A brushed d.c. tachogenerator can be considered to be a precision d.c. generator, consisting of a permanent-magnet stator, with a wound armature. The output voltage, E_g, is related to the tachogenerator speed, N (rev min^{-1}), by the voltage constant, K_g (V rev^{-1} min)

$$E_g = K_g N \tag{4.10}$$

In a tachogenerator with a conventional iron-copper armature, a ripple voltage will be superimposed on the d.c. output because of the relatively low number of commutator segments; the frequency and the magnitude of this ripple voltage will be dependent on the number of poles, armature segments, and brushes. A ripple-voltage component with a peak-to-peak value of five to six per cent of the output voltage is typical for brushed tachogenerators. The ripple voltage can be reduced by the use of a moving-coil configuration which has a high number of coils per pole; this minimises the ripple voltage to around two to three per cent. The armature consists of a cylindrical, hollow rotor, composed of wires held together by fibreglass and a polymer resin and has a low moment of inertia which ensures that the system performance is not compromised, similar to that of the ironless-rotor d.c. machine, discussed in Chapter 5. In addition to the low inertia and the low ripple content of the output, the axial magnets ensure that the motor length is small. In practice, this could add as little as 1 mm to the length of the overall package. A further refinement is the provision of frameless designs: this allows the system designer to mount the tacho directly on the shaft to be measured, thus removing any coupling errors.

The performance of a brushed tachogenerator depends on it being used within its specified operating capabilities; the linearity of the output will suffer if the load resistance, R_L, is allowed to fall below the manufacturer's recommended value. From Figure 4.9

$$E_g = R_a I + R_L I \tag{4.11}$$

where R_a is the armature resistance; hence the terminal voltage, V, is given by

$$V = \frac{R_L K_g N}{R_a + R_L} \tag{4.12}$$

The load resistance should be as large as possible to ensure that the terminal voltage is maximised; however, the current which is drawn should be sufficiently high to ensure that the commutator surface does not become contaminated.

4.2.2 Brushless d.c. tachogenerators

With the increasing use of brushless d.c. motors in servo systems, motor speeds are no longer limited by brushes; this leads to shaft speeds approaching 100 000 rev min^{-1} in some high-performance machine tools. The maximum speed of a brushed tachogenerator is limited to the speed at which aerodynamic lifting of the brushes occurs, and by increased armature-core losses which result in the output linearity deteriorating. Brushless tachogenerators have been developed as a response to these problems. The principle of their operation is identical to that of brushless motors (as discussed in Chapter 6), with the switching between phases being controlled by stator-mounted Hall-effect sensors. If the tachogenerator is integral to the motor, the Hall-effect sensors can be used for both motor and tachogenerator control. The maximum operational speed is only limited by the physical construction of the rotor assembly. There are no moving parts other than the rotor; this leads to a high reliability device, suitable for remote applications.

4.2.3 Incremental systems

An incremental-velocity measurement system is shown in Figure 4.10. A slotted disc, located on the shaft whose speed is to be measured, is placed between a light source and a detector. The source is usually a light-emitting diode; these diode have a longer life, and they are more rugged than filament bulbs, but are restricted to a temperature range of -10 to +75°C. The output of the photodetector needs to be conditioned prior to the measurement to ensure that the waveform presented has the correct voltage levels and switching speeds for the measurement system. The frequency of the signal, and hence the speed of the shaft, can be measured by one of two methods. Firstly, the frequency can be measured, in the conventional fashion, by counting the number of pulses within a set time period. This is satisfactory as long as the speed does not approach zero, when the timing period becomes excessive. To overcome this, an enveloping approach (shown in Figure 4.10) can be used. Each half-cycle of the encoder output is gated with a high-frequency clock; the number of cycles which are enveloped is determined, and this value is used to calculate the shaft speed. It should be noted that even this method will prove difficult to use at very low speeds, because the number of cycles per half-cycle becomes excessive.

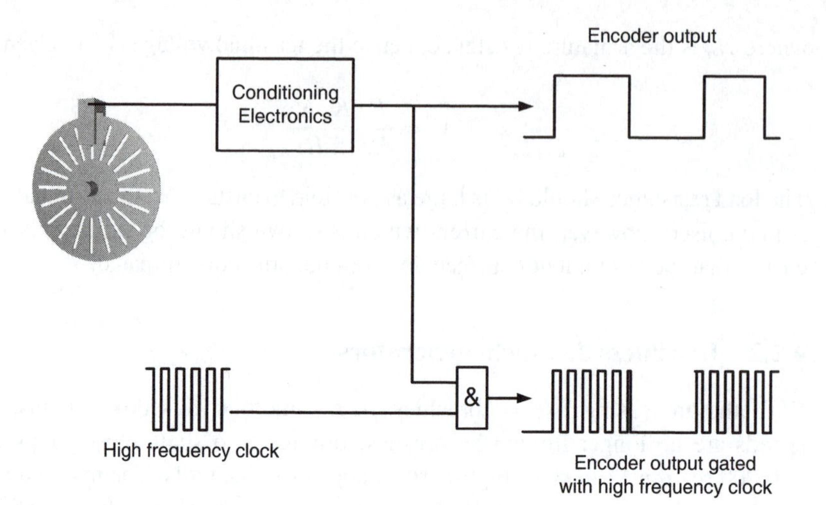

Figure 4.10. Speed-measurement using an incremental encoder. The output can be directly taken from the conditioning electronics, or to increase the resolution, the encoder waveform can be gated by a high frequency carrier.

The maximum operation speed of an incremental system is limited by the high-frequency characteristics of the electronics, and particularly by the opto-electronics. The resolution of the disc will determine the maximum speed at with the encoder can be operated, as shown in Figure 4.11.

4.2.4 Electromechanical pulse encoders

Using the counting techniques discussed above, it is possible to replace an optical encoder with an electromechanical system. A steel or a soft-iron toothed wheel is fitted to the shaft, and a magnetic, inductive, or capacitative proximity sensor is used to detect the presence of the teeth. While such a system is not normally capable of producing highly accurate speed measurements it can provide a rugged system which can be used in high-reliability applications such as over-speed/underspeed detectors for motors or generators.

4.3 Position transducers

Position transducers are available in three main types: incremental, semi-absolute, and absolute. A typical incremental encoder is an encoder that produces a set number of pulses per revolution, which are counted to produce the positional information. If the power is lost, or the data is corrupted, rezeroing is required to obtain the true information. An incremental encoder can be improved by the addition of a once-per-revolution marker; this will correct against noise in the system, but complete rezeroing will still be required after a power loss, because the count-

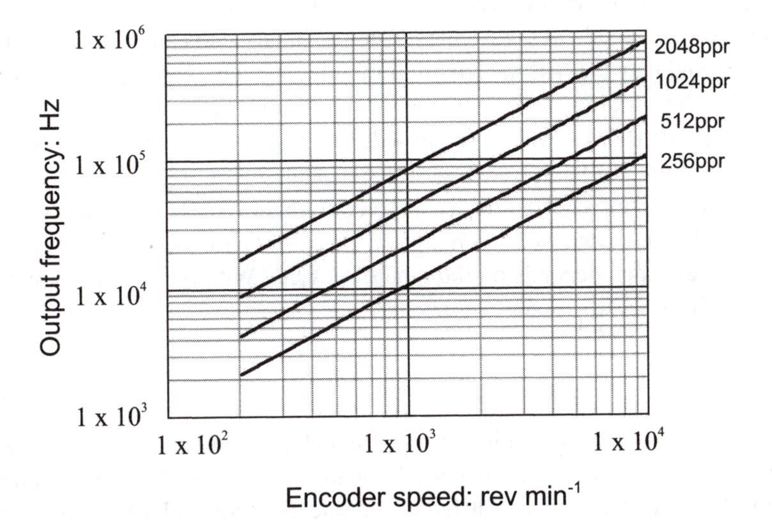

Figure 4.11. Encoder output frequencies as a function of speed for a range of incremental encoders, (ppr – pulses per revolution).

ing of the number of revolutions is also lost. An absolute transducer will maintain the zero and thus it will provide true information despite a loss of power for any length of time.

4.3.1 Brushed potentiometers

The principle of a potentiometer can be used in either a linear or a rotary absolute-position transducer in which the output voltage is a function of displacement. An excellent performance can be obtained if the drawbacks of the non-uniform track resistance and of the brush contact are considered to be acceptable. The accuracy of such a device will be dependent on regulation of the excitation voltage, which can be maximised by the use of a bridge circuit. A typical servo grade device will have a resolution of 0.05% of the full scale, with an accuracy of $\pm 0.1\%$. The maximum operating speed of a rotational version is typically limited to 500 rev min^{-1} by the brushes.

4.3.2 Linear variable differential transformers – LVDT

One of the most common methods of directly measuring a linear displacement to a high degree of accuracy uses a linear variable differential transformer (LVDT); the principal features of LVDTs are shown in Figure 4.12(a). The operation is based on a transformer in which the coupling between the primary and secondary coils (see Figure 4.12(b)), is determined by the position of a movable ferromagnetic core. The core is assembled using precision linear bearings to give low friction and wear. The most widely used design has a secondary winding which is split into two, on either side of the primary. The secondary coils are wound in opposite directions and they are half the length of the moving core. In order to achieve high accuracies the windings have to be identical both in length and in inductance otherwise an unwanted quadrature signal will be produced, leading to non-linearities in the measurement; values of 0.5% for the accuracy are typical for LVDTs, increasing to 0.1% on selected devices. To operate an LVDT, the primary winding is energised with a sinusoidal excitation voltage, in the frequency range 2–10 kHz; the exact frequency depends on the type of device. With the secondary windings connected in series, the output voltage is

$$V_{out} = V_1 + V_2 \tag{4.13}$$

When the core is in midposition, V_1 will equal V_2, and the output will be zero. As the core is displaced, the magnitude of the output rises linearly as shown in Figure4.12(c), with a 0° phase difference in one direction and a 180° phase difference in the opposite direction. Hence the magnitude of the output signal is proportional to the displacement of the central core, and the phase indicates the direction of travel. By the use of a suitable demodulator, a bipolar analogue voltage which is directly proportional to the displacement can be produced. Commercially available transducers can be obtained with displacements as small as 1 mm up to 600 mm in a variety of linearities and sensitivities. Because there is no physical contact between the core and the coils, the main mechanical components of the LVDT will not degrade with use. If precision bearings are used in the design, an almost infinite resolution, with zero hysteresis, is possible. The small core size and mass, and the lack of friction, mean that LVDTs have a high-response capability for dynamic measurements (for example, measurement of vibrations). Due to their rugged construction, it is possible to obtain LVDTs that are capable of operatinge in extreme environments, for example, ambient pressures up to 10^7 Pa and temperatures up to 700°C are not uncommon.

4.3.3 Resolvers

Resolvers are based on similar principles to LVDTs, but the primary winding moves relative to the two secondary windings rather than having a moving solid core, as shown in Figure 4.13(a). As the relative positions of the primary and secondary windings change, the output varies as the sine of the angle. By having two windings ninety electrical degrees apart and considering only the ratio of the outputs (Figure 4.13(b)), the variations due to the input voltage and the frequency changes become unimportant. The signals from the resolver are therefore relatively insensitive to an electrically noisy environment, and they can be transmitted over considerable distances with little loss in accuracy. In order to dispense with the need for sliprings, a separate rotary transformer is used to provide power to the rotating primary windings. The stator consists of the two output windings spaced

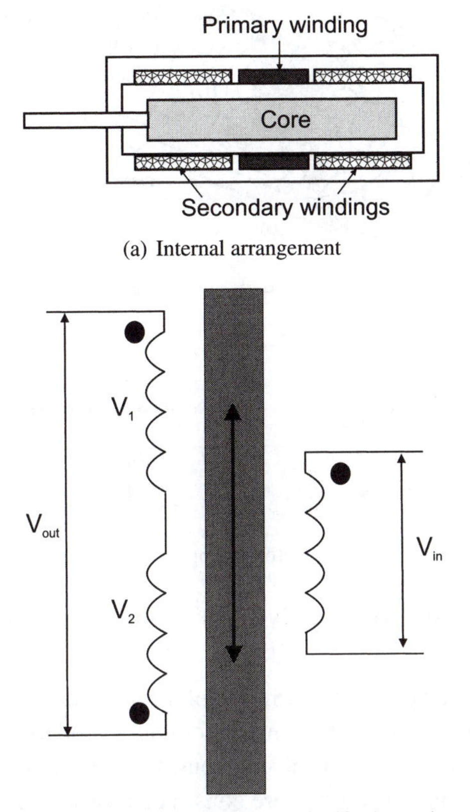

(a) Internal arrangement

(b) Electrical circuit, the dots signify the positive ending of the winding.

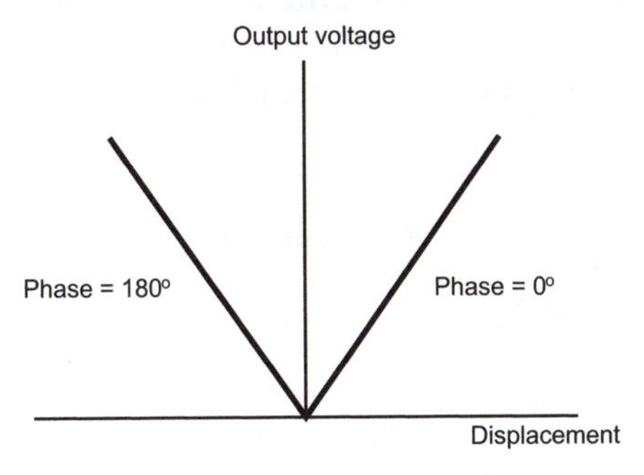

(c) Operational characteristics

Figure 4.12. The operation of the LVDT

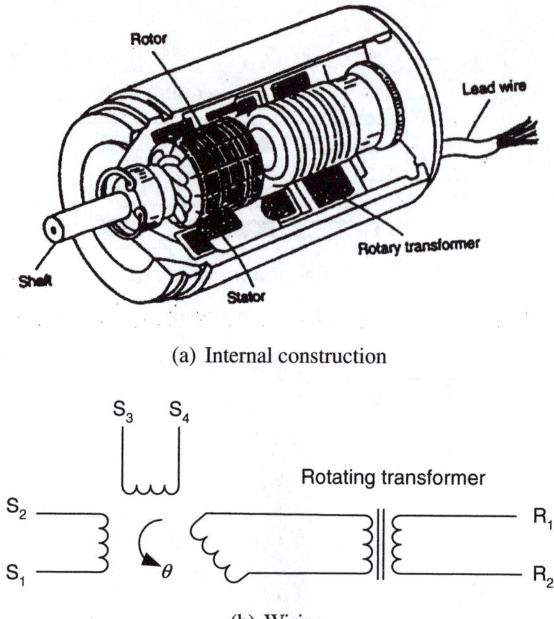

(a) Internal construction

(b) Wiring

Figure 4.13. Resolver construction and wiring

90 electrical degrees apart and the primary of a rotary transformer. The rotor also carries the secondary of the rotary transformer that is used to excite the rotor of the resolver. In the construction of resolvers, considerable care is taken to ensure that the cores, windings, and the air gap are constructed to an accuracy which ensures that non-linearity does not occur. In practice, errors can be caused by a number of factors including: a difference in the primary/secondary transformation ratio, an electrical phase shift, or a zero shift error between the two secondary windings and unequal loading of the windings by the external decoder. If the input to the resolver is

$$V = A \sin \omega t \tag{4.14}$$

the two outputs signals will be

$$V_{out1} = Ak_1 \sin \theta \sin(\omega t + \alpha) \tag{4.15a}$$

$$V_{out2} = Ak_2 \sin \theta \sin(\omega t + \alpha) \tag{4.15b}$$

where A is the amplitude of the excitation voltage, and k_1 and k_2 are the transformation ratios between the primary and the two secondary windings (which ideally should be equal), ω equals $2\pi f$ where f is the carrier frequency, and α is the rotor/stator phase shift (including any zeroing error).

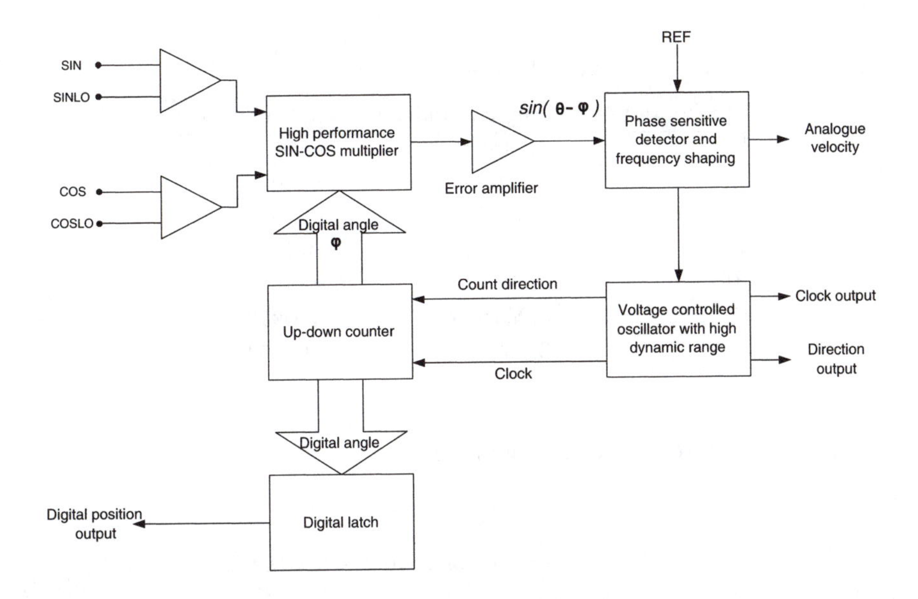

Figure 4.14. Internal function block diagram of a resolver-to-converter. The two stationary windings of the resolver are connected to SIN and SINLO, and COS and COSLO respectively. The resolver is powered by an external oscillator, which also provides the REF signal.

The output from the resolver can be used either directly as an analogue signal or after conversion to a digital signal. The advent of resolve-to-digital converters (RDC) has allowed digital data to be easily produced from resolvers. A modern RDC uses a ratiometric method; therefore, the system is not affected by changes to the absolute values of the signal to and from the resolver. This is of considerable importance if the transmission distance between the resolver and the RDC is large. It is current practice to provide a complete RDC as integrated circuits or as hybrid packages. This ensures that the best possible performance is obtained, with the packages' components optimised for temperature drift and other external sources of inaccuracy. A number of manufacturers provide devices that determine the resolver's velocity and position in a number of different formats: as a bipolar analogue signal or as a digital clock proportional to the speed, together with a logical direction signal. It is not uncommon for a device to haves 12-bit resolution up to 375 rev s^{-1}.

In this type of tracking converter, the two inputs (assuming a perfect resolver where $\alpha = 0$ and $k = k_1 = k_2$) are multiplied by the value held in a counter; if the output of the counter is assumed to be equivalent to an angle φ, then

$$V_1' = Ak \sin\theta \cos\varphi \sin\omega t \tag{4.16a}$$

$$V_2' = Ak \cos\theta \sin\varphi \sin\omega t \tag{4.16b}$$

Table 4.1. Resolution over 360°

Number of bits	Angle in radians	Angle in degrees
1	3.1415	180.00
2	1.5707	90.00
4	0.3927	22.5
8	0.02545	1.4063
10	0.00614	0.3516
12	0.001534	0.08789
16	0.000096	0.00549

and the difference from the error amplifier is

$$\begin{aligned} V_1' - V_2' =& Ak \sin \omega t (\cos \varphi \sin \theta - \cos \theta \sin \varphi) \\ =& Ak \sin \omega t \sin(\theta - \varphi) \end{aligned} \tag{4.17}$$

A phase-sensitive detector, a voltage-controlled oscillator, and a counter form a closed-loop control system that attempts to minimise $\sin(\theta - \varphi)$. At the zero point, θ will equal φ, and the output of the counter will equal the angle of the resolver. In the selection of a tracking RDC, two major parameters need to be considered: the resolution (see Table 4.1) and the accuracy, both static and dynamic. The dynamic accuracy depends on how fast the voltage controlled oscillator (VCO) input tracks the error signal, which is dependent on the excitation frequency of the resolver that is used as part of the phase sensitive detector. One of the most significant forms of error is the lag in the tracking converter as the system accelerates; these errors may need to be considered in very-high-performance systems.

While a single resolver is only absolute over one revolution, applications often require absolute measurements over a number of revolutions. One possible solution is to couple two resolvers by a gear system (see Figure 4.15) so that the second resolver will rotate once for n turns of the input shaft. While this solution is perfectly acceptable, accuracies can be compromised by the backlash and tooth wear in the gearing. If anti–backlash gears are used, these effects will be very small; but they could be significant if the full 16-bits accuracy is required. In an anti-backlash gear, two independent gears are mounted on the same hub with a spring between the two providing a constant full-tooth engagement with the mating spur gear, thereby eliminating backlash in the mesh.

While the mechanical approach is satisfactory, it is more convenient to use a multipole resolver, where up to 32 cycles of stator voltage can be produced within 360 mechanical degrees. To provide absolute angular information, a second, coarse (one speed), winding is provided. By cascading a number of resolver-to-digital converters together, very-high-resolution systems can be constructed.

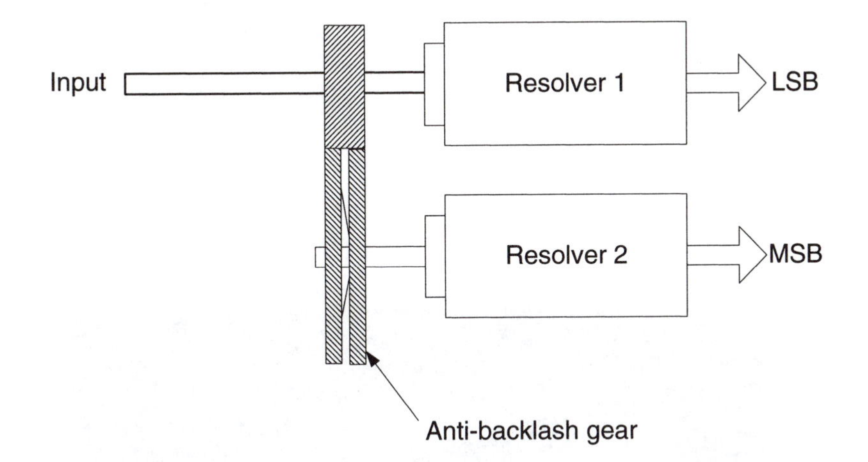

Figure 4.15. The use of anti-backlash gearing to increase the range of resolvers.

4.3.4 Rotary and linear Inductosyn

Inductosyn is the trademark of a position transducer manufactured by Inductosyn International of Valhalla, New York; the most the widely used version is based on an inductive principle. Linear Inductosyns can be fabricated in lengths of up to 40 m, or they can be manufactured in a rotary form up to 0.5 m in diameter (see Figure 4.16(a)). As Inductosyns are inductively coupled, non-contact transducers, they are very tolerant to changes in the local dielectric constant; therefore, their operation will not be affected by dust, oil, or pressure changes in a hostile environment. Inductosyns have applications in machine-tool, subsea, and aerospace areas, where very high resolution and accuracy are required.

Inductosyns can be considered to be planar resolvers; the rotor and stator elements consist of a high precision hairpin element printed as a track over the complete length of the device (see Figure 4.16(b)). The length of one complete cycle of the pattern is the pitch P. An alternating current in the primary will induce a signal in the secondary. The amplitude is dependent on the relative positions of the primary and secondary windings, giving

$$V_{out1} = kV \cos\left[\frac{2\pi x}{P}\right] \tag{4.18}$$

where V is the excitation voltage ($V = V_{pk} \sin \omega t$), k is the transformation ratio, and x is the displacement. If a second output winding is displaced by $\pi/2$ electrical degrees from the first winding, its output voltage will be:

$$V_{out1} = kV \sin\left[\frac{2\pi x}{P}\right] \tag{4.19}$$

As these output voltages have the same form as those of a resolver, an identical converter can be used to determine the displacement, x. In practice, the number

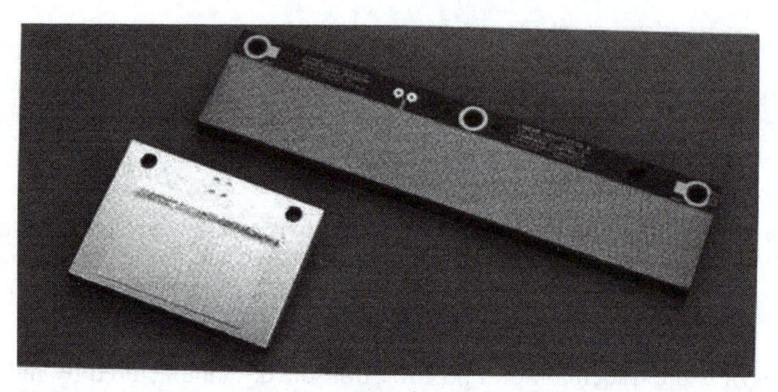

(a) A commercial linear system, photograph reproduce by permission of Inductosyn International, Farrand Controls, Valhalla, NY.

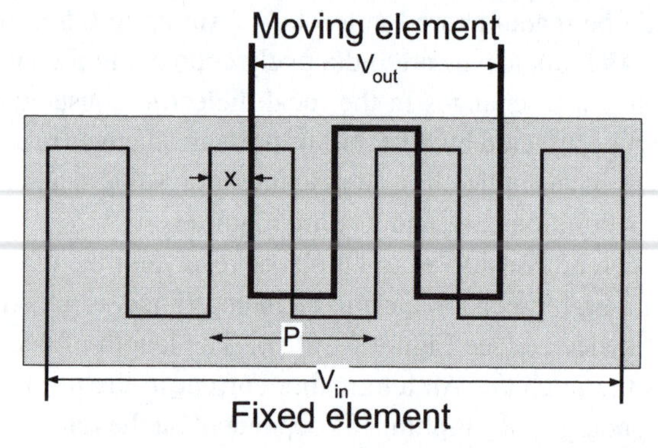

(b) The relationship between the fixed and moving elements in an Inductosyn

Figure 4.16. The linear Inductosyn.

of complete pitches are counted to determine the total distance move. The pitch of a metric linear Inductosyn is is such that a resolution of 5×10^{-6} m to be achieved. Rotary Inductosyns are supplied with pitch counts in the range 32–2048 per revolution, with achievable accuracies to $\pm 0.5''$.

4.3.5 Optical position sensors

Optically based encoders are widely used for position measurements in robots and machine tools. They can take one of three forms: absolute, semi-absolute, and incremental. Each of these types of encoder consists of three elements: an optical receiver, a light source, and a code wheel. The receiver is normally a phototransistor or diode which responds to the light intensity which is received. As discussed earlier, the light source can either be a solid-state light-emitting diode or a filament bulb. The difference between the types of encoders is characterised by the information contained on the code wheel and by how it is interpreted by an external control system.

Absolute optical encoders incorporate a code wheel that is encoded in binary, either in pure binary or in grey code, with one bit per track. The latter is preferred because only one bit changes between any two states. This prevents errors, since there is no way of guaranteeing that all the bits will change simultaneously at the boundary between two states, due to inherent manufacturing problems with the code wheel. For example, if pure binary is used (see Table 4.2), it would be possible to generate an output of 15 during the transition from 7 to 8. Code wheels are normally produced on glass substrates by photographic methods. This is costly for high resolutions; as will be readily appreciated, as the resolution of an absolute encoder increases, so does the size and complexity of the code wheel (see Figure 4.17(a)).

Semi-absolute and incremental optical encoders are identical in most respects, and they can thus be considered together. The construction of an incremental encoder is based on a code wheel which has a single track of equal-sized, opaque and translucent slots; and, as the wheel is rotated, an alternating signal is produced with a frequency which is proportional to the speed of rotation (see Figure4.17(b)). Semi-absolute encoders are incremental encoders with an additional output giving one pulse per revolution. As the output of these detectors is typically a distorted sine wave, the output needs to be suitably conditioned to produce a clean square wave for other electronic systems. This circuitry can be mounted in the encoder or it can form part of the external system. As the resolution of the encoder increases, the use of physical slots in the code wheel will become unreliable; hence, use is made of gratings. As the code wheel is moved, the whole field observed by the optical receiver goes dark as the lines move in and out of phase.

As previously discussed, an encoder with a single track will allow the magnitude of the speed to be measured; the direction of rotation can be determined by the

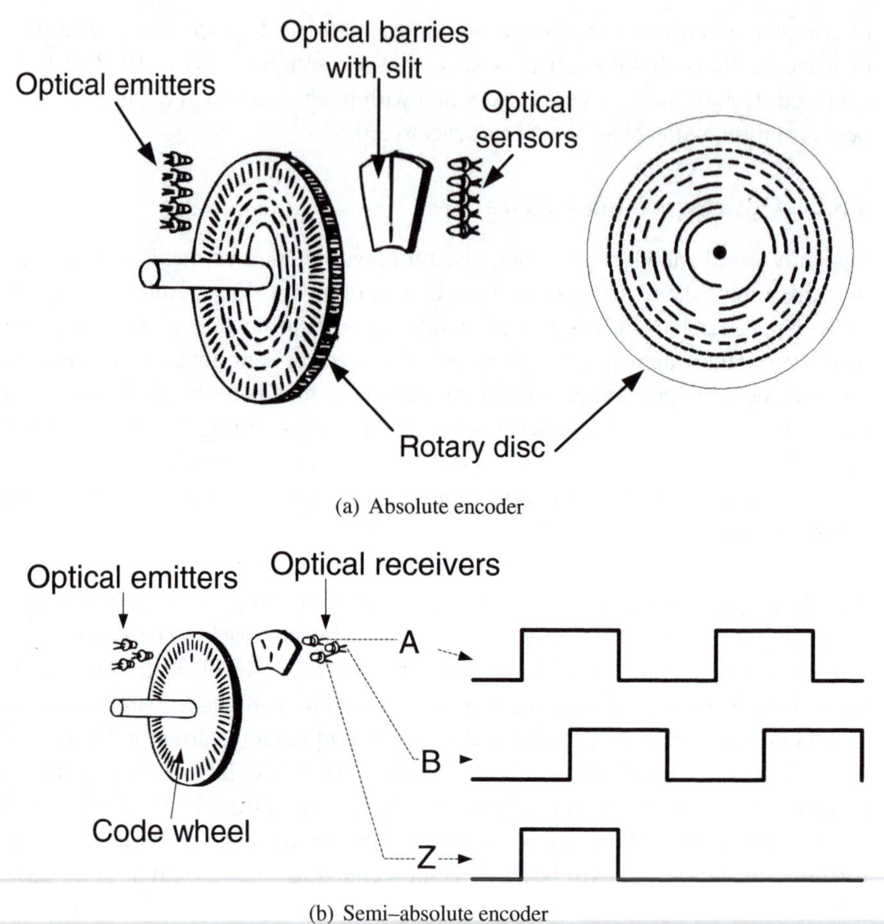

(a) Absolute encoder

(b) Semi–absolute encoder

Figure 4.17. Rotary optical encoders.

addition of a second track or an additional sensor to produce a quadrature signal. The two signals A and B shown in Figure 4.17(b) are displaced by 90 electrical degrees. As a result, if the encoder moves forward, channel A will lead channel B, and vice versa when the motion is reversed. A number of techniques can be used to detect the direction of motion; one possible technique is shown in Figure 4.18. Figure 4.19 shows the waveforms used to discriminate direction.

The encoder signal is used to generate a pulse from the monostable, which can be inhibited by the other channel; the resulting pulse is used to latch a flip-flop, whose output indicates the direction of motion. The speed and position are measured by pulse-counting techniques, the resolution being determined by the size of the counter and the encoder. An encoder is specified by the number of lines per rotation; however, since channels A and B are shifted by 90 electrical degrees it is possible to divide each encoder cycle in four, hence the resolution

Table 4.2. Pure binary and grey codes as used in an absolute rotary position encoder.

State	Pure binary	Grey code
0	0000	0000
1	0001	0001
2	0010	0011
3	0011	0010
4	0100	0110
5	0101	0111
6	0110	0101
7	0111	0100
8	1000	1100
9	1001	1101
10	1010	1111
11	1011	1110
12	1100	1010
13	1101	1011
14	1110	1001
15	1111	1000

of a 360 pulses per revolution (p.p.r.) encoder can be increased to 1440 counts per revolution by the addition of an electronic system. Since this increases the effective system resolution at a cost which is significantly lower than for encoders with four times the resolution, this can almost be considered to be a standard feature of position systems. Commercial systems are also available that will increase the encoder resolution by 8 and 12 times.

Linear optical encoders operate in an identical fashion to rotary incremental encoders, where a optical sensing head moves over the stationary grating, which is either a glass scale or a reflective steel strip. When the scale with a grating moves relative to another grating with an identical pattern – the index grating – the lines and gaps alternately align. The light-dark modulation produced is detected with optical sensors, a typical system is shown in Figure 4.20. The arrangement of the optical sensors and the signal processing required is a function of the design and the encoder resolution. It is possible to obtain reflective linear optical encoders in lengths of up to 50m, the performance depends on the care of the installation, particularly the alignment between the encoder track and the moving sensor head.

It is possible to purchase absolute linear encoders which have up to seven tracks, the information from which is combined to provide the absolute position. Due to the complexity of the process, these encoders are limited to lengths of 3 m or less, with a resolution of up to 0.1 μm.

As with rotary encoders a reference mark is provided on a second track, parallel to the incremental track, which are scanned, and used to locate the datum position,

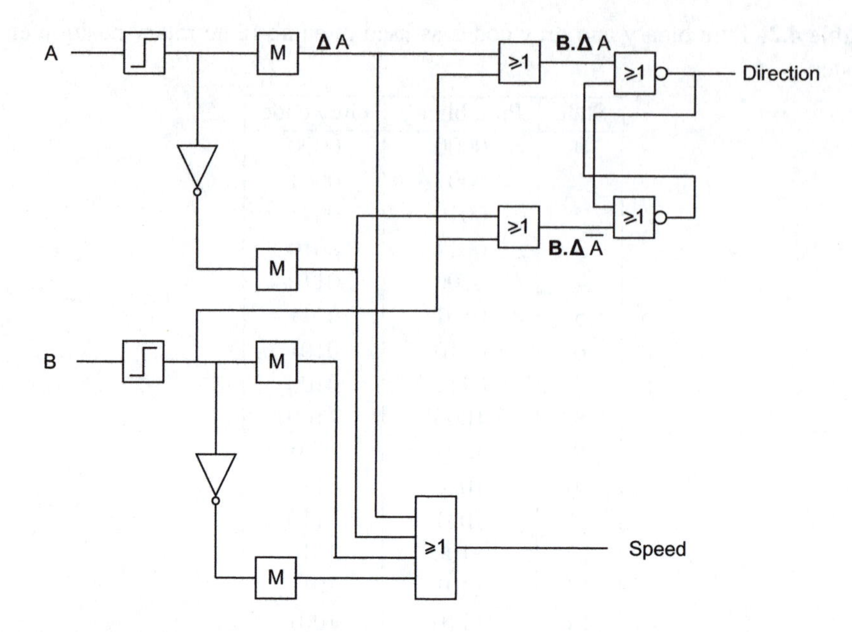

Figure 4.18. Position decoding for an incremental encoder, the blocks marked M are single-shot monostables, operating on the rising edge, the waveforms are shown in Figure 4.19.

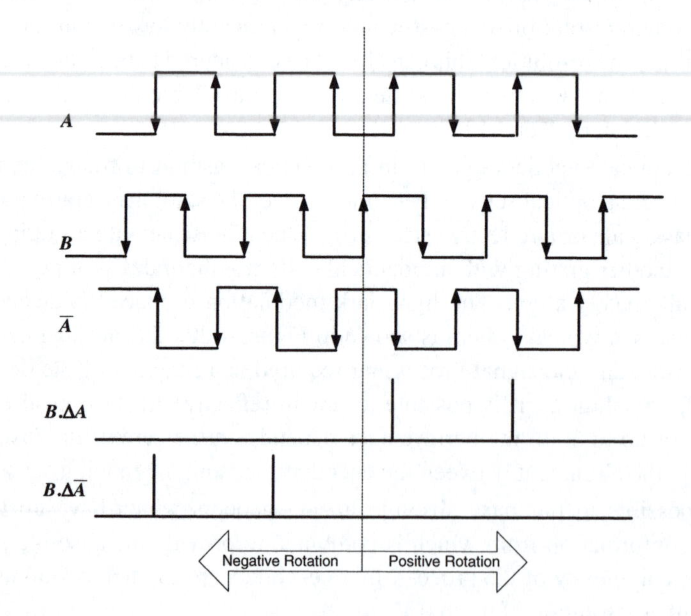

Figure 4.19. The discrimination of direction using an incremental encoder.

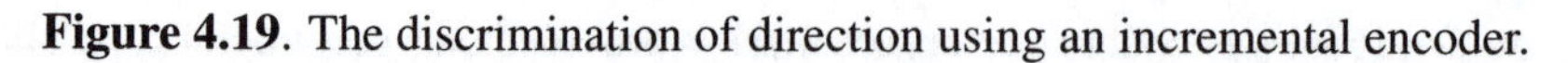

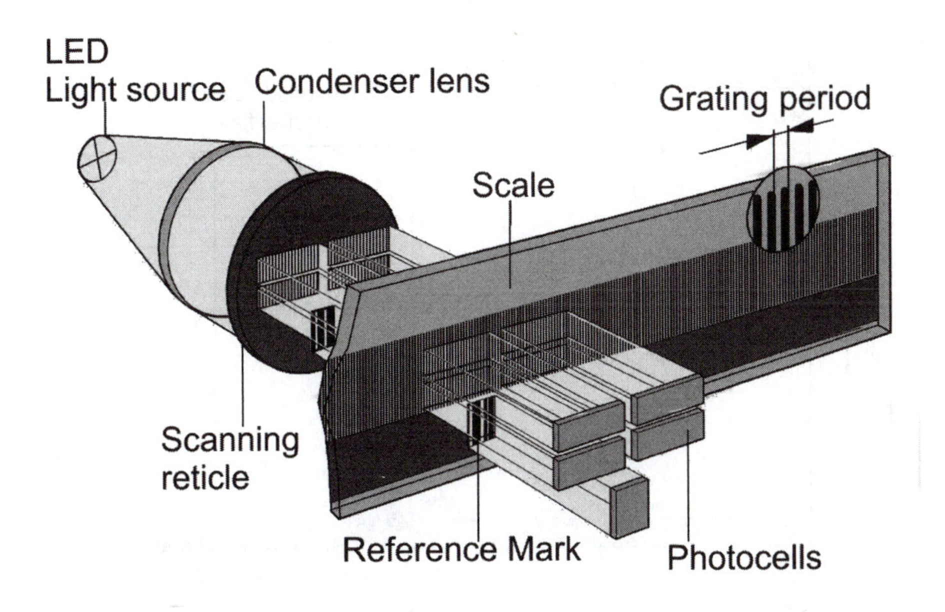

Figure 4.20. A linear optical encoder showing the grating arrangement. Image reproduced by permission of Dr. Johannes Heidenhain GmbH, Traunreut, Germany.

as discussed in section 4.4.3.

4.4 Application of position and velocity transducers

The correct installation of an encoder or transducer is critical to its satisfactory operation. During installation, particular consideration must be given to the mechanical aspects and to the connection to the system's measurement electronics.

4.4.1 Mechanical installation

The previous sections have described the operation of a range of velocity and position transducers. In practice, units are supplied either complete or as a set of components in a frameless design. Frameless transducers are supplied to allow direct integration into the mechanical structure of a system, therefore reducing, or eliminating, errors caused by windup in couplings or shafts and eliminating backlash in gears. A range of common sizes has been developed for resolvers and optical encoders; the more significant sizes are listed in Table 4.3. These standard sizes permit easy interchangeability between manufacturers' products. It should be noted that the shafts can either be solid or hollow, giving designers a number of integrations options for the design of a mechanical systems.

In coupling the motor or the load to a rotary transducer, care must be taken to ensure that the respective shafts are correctly aligned in all axes; if they are not

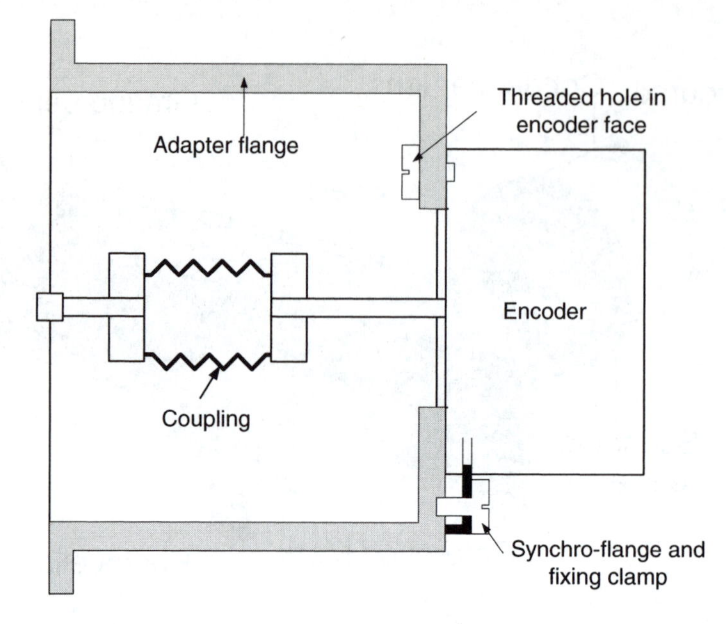

Figure 4.21. Connecting an encoder to a shaft using a bellow coupling and adapter flange. Threaded holes can be provided on the face of the encoder, or if the encoder is fitted with a synchro-flange, fixing clamps can be used.

Table 4.3. Standard encoder and resolver sizes, dimensions in mm

Type	Size	Diameter	Length	Shaft diameter
Frameless	15	36.83	25.4	6.35
Frameless	21	52.37	31.37	12.7
Housed	8	19.13	31.5	
Housed	11	27.05	40.39	

correctly aligned, a considerable load will be placed on the transducer bearings, leading to premature failure. Two methods of fixing are shown in Figure 4.21, during assembly is is normal practice to tighten the flange mounting screw last, to ensure that the input shaft and encoder shaft are in alignment. Since transducers can be supplied either with a conventional solid shaft or with a hollow shaft, the coupling which is used depends on the type of encoder and on the application. The use of bellows couplings will allow very small alignment errors to be eliminated, while still retaining a solid coupling between the motor and the transducer. If a hollow shaft encoder is used, the shaft can be coupled directly to the transducer, while its transducer itself is fitted to the system using a compliant mount. If a frameless transducer is used, its installation will be very specific to the unit and the application, and the manufacturer will normally supply details of the installation design where necessary.

In installing both linear and rotary transducers a number of additional requirements need to be considered:

- The operating temperature range specified by the manufacturer defines the limits within which the values given in the specifications for the encoders are maintained. Operating out side the limits will results in a degradation of performance.

- All encoders are subject to various types of acceleration during operation and mounting, again these will be detailed in the manufacturers' specifications. Due to presence of a glass code wheel or linear scale, encoders are considered fragile, particularly during the assembly of a system.

- Both linear and rotary encoders have internal friction, particularly if the design includes a seal, this is normally specified as a torque or force in the specifications.

- All types of encoders should not be subjected to vibration during operation: this can be more significant for a long linear encoder. To function properly, the more solid the mounting surface the better. It is recommended that linear encoders should not be mounted on hollow parts.

- As discussed in section 2.5.5 is a major factor in selecting drives and their associated components. For example linear encoders are normally protected IP 53 (see Table 2.3) provided that they are mounted with the linear seal is facing away from possible sources of contamination. If the encoder will be exposed to heavy concentrations of coolant and lubricant mist, the scale housing can be fed with compressed air to raise the IP rating to IP 64 effectively preventing the ingress of contamination.

4.4.2 Electrical interconnection

The wiring and connectors between the transducer and the processing electronics are critical to the operation of a system. If they are not satisfactory, in the case of a digital encoder, any electrical noise which is introduced will probably result in additional pulses being counted and hence in an increasing positional error. In analogue systems, electrical noise resulting from poor connections will result in a poor signal-to-noise ratio and hence in a degraded performance, These problems can be reduced by the use of twisted screened cables and high-quality connectors throughout the system. As shown Figure 4.11, at high speeds the encoder output frequency can exceed 100 kHz, and therefore the wiring and associated electronics must be designed to accommodate signals of this frequency; in particular stray capacitance must be minimised.

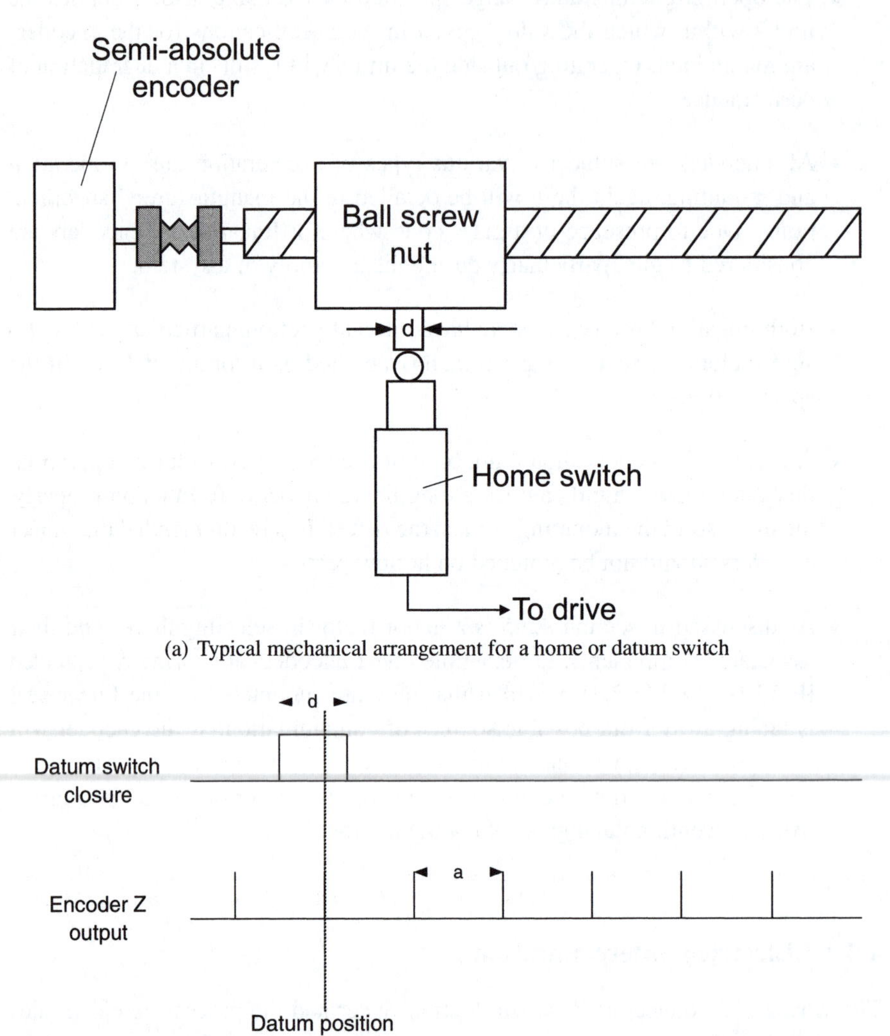

(a) Typical mechanical arrangement for a home or datum switch

(b) The relationship between the switch closure (d) and the distance between the output on the encoder's z-track, for reliability $a > d$.

Figure 4.22. The use of a home switch to define the datum position of a linear axis.

4.4.3 Determination of datum position

When a rotary incremental encoder is used in linear applications one of the design requirements is to accurately and repeatably determine the datum position. This is the point to where all measurements on a particular axis are referenced. As shown in Figure 4.17(b), it is common practice for an incremental to have track that provides one pulse per revolution. This is used to provide the exact datum point. To ensure that the datum point is selected within the correct encoder revolution, it is normal practice to provide a *home switch*, as shown in Figure 4.22. The distance moved once the switch is mode must be less than that which results in one encoder revolution, or the possibility exists that the datum position will have an error of plus/minus one encoder revolution.

4.5 Summary

This chapter has reviewed the range of velocity and position encoders that are currently available. In order to make a satisfactory selection, designers have to address a number of key question, including:

- What resolution and accuracy is required by the application, particularly as increasing both parameters directly affects the overall system cost?
- What are the environmental constraints at the location where the measurement system will be installed?
- How will the derived information be integrated into the system? The answer to this question will depend on the controller selected, and this question may need to be addressed at the completion the selection procedure.

The encoder or transducers which are selected will have a major effect on overall system performance; for if the wrong measurement is made, the system will never be able to produce the required result.

Chapter 5

Brushed direct-current motors

Direct-current (d.c.) brushed motors, either with a separately excited field or with a permanent-magnet rotor, have been used within variable speed drives for a considerable period of time. This class of motors has inherently straightforward operating characteristics, flexible performance, and high efficiency; these factors together with the long development history have resulted in brushed d.c. motors being used as a standard within many industrial applications. With recent developments in magnetic technology and manufacturing techniques, a wide range of d.c. brushed motors are available for use by servo system designers. Even with the latest developments in brushless d.c. and vector controlled a.c. motors, brushed d.c. motors have a number of advantages that will ensure their use by system designers for a considerable time to come.

This chapter reviews both the range of motors which are presently available and the options for their control. Brushed d.c. permanent-magnet motors can be obtained commercially in the following forms:

- Ironless-rotor motors.
- Iron-rotor motors.
- Torque motors.
- Printed-circuit motors.

Each of these motors has a number of advantages and disadvantages which need to be considered when selecting a motor for a particular application. Brushed d.c. motors, within certain constraints, can be controlled either with a linear or a switching amplifier. As discussed in Chapter 2, the motor and amplifier need to be considered as a combined system if the maximum performance is to be obtained.

5.1 Review of motor theory

The basic relationships for d.c., permanent-magnet, brushed motors whose equivalent circuit is shown in Figure 5.1(a) are given by

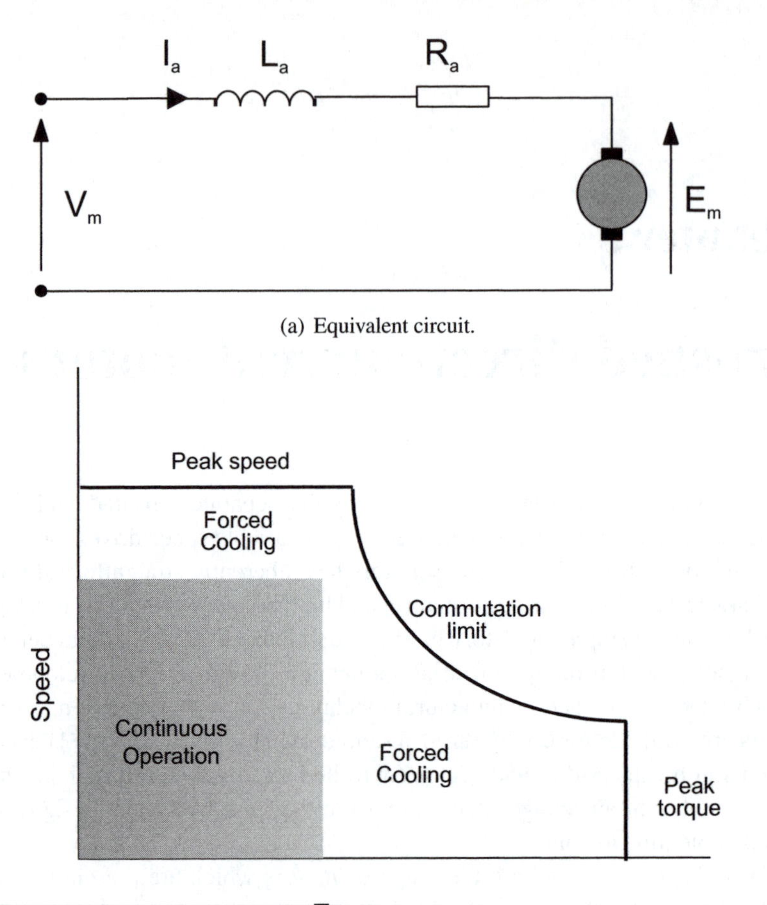

(a) Equivalent circuit.

(b) The speed-torque characteristics, showing the limiting values due to peak armature voltage, armature current and the commutation limit.

Figure 5.1. Brushed d.c. motor.

$$V_m = \omega_m K_e + I_a R_a + L_a \frac{dI_a}{dt} = E_m + I_a R_a + L_a \frac{dI_a}{dt} \tag{5.1a}$$

$$T = I_a K_t \tag{5.1b}$$

where I_a is the armature current, ω_m is the speed of rotation (rad s^{-1}), K_e is the motor's speed constant (Vrad^{-1}s), and K_t is the motor's torque constant (Nm A^{-1}).

The torque-speed envelope of a typical permanent-magnet motor is shown in Figure 5.1(b); it exhibits good low-speed torque-ripple characteristics through to a standstill, which makes it ideal for servo applications. However, the high speed characteristics are limited: as the rotational speed increases, the voltage between the commutator segments also increases; and if this is combined with a high armature current (that is, a high torque), a voltage breakdown between adjacent commutator segments will result in a motor flash-over. The result will be considerable damage to the motor and its drive. Therefore, the motor's absolute operational area is bounded in practice by the peak values of the armature current and the voltage (that is, the speed), and by the commutation limit. In addition, within these constraints, the thermal limits of the motor will dictate the area where continuous operation is possible; this area can be increased by the addition of forced ventilation.

5.2 Direct-current motors

5.2.1 Ironless-rotor motors

The construction of an ironless-rotor d.c. motor is shown in Figure 5.2. There are three elements: the rotor, the magnet assembly, and the brush assembly. The rotor is constructed as a self-supporting basket, with the conductors laid in a skewed fashion to minimise torque ripple and to maximise the mechanical strength. The conductors are bonded to each other and to an end disc or commutator plate (which supports the coil and the commutator segments) by an epoxy resin. This form of construction produces a rotor that is compact and of low weight and inertia. The motor is assembled around a central permanent magnet, which supports the main motor bearings and the outer housing. The outer housing protects the motor and it also acts as an integral part of the magnetic circuit. The commutators are located on a plate attached to the rear of the rotor, while the brush assembly is supported from the main housing. The brushes are manufactured from precious-metal springs resulting in a low-contact resistance throughout the motor's life and they ensure that the motor will start when a very low voltage is applied. Because of these design features, the ironless-rotor, d.c. brushed machines are limited to powers of less than 100 W; however, high output speeds are possible; and, depending on the motor type, speeds in excess of 10 000 rev $\min^{-1}$ are available.

The selection of an ironless rotor motor for an application is, in principle, no different than for any other type of motor; however, one important additional constraint is imposed by the self-supporting nature of the rotor. If the power rating is exceeded, the excessive rotor temperature will result in the degrading of the bonding medium, and the winding will separate at high speed. This can be prevented by careful consideration of both the thermal characteristics of the motor and its application requirements. The power, P_d, generated in the rotor is given by

$$P_d = I_a^2 R_a \tag{5.2}$$

where I_a is the root mean square (r.m.s.) armature current and R_a the armature resistance. From this value, the temperature rise of the rotor windings above the ambient temperature, T_r, can be calculated from

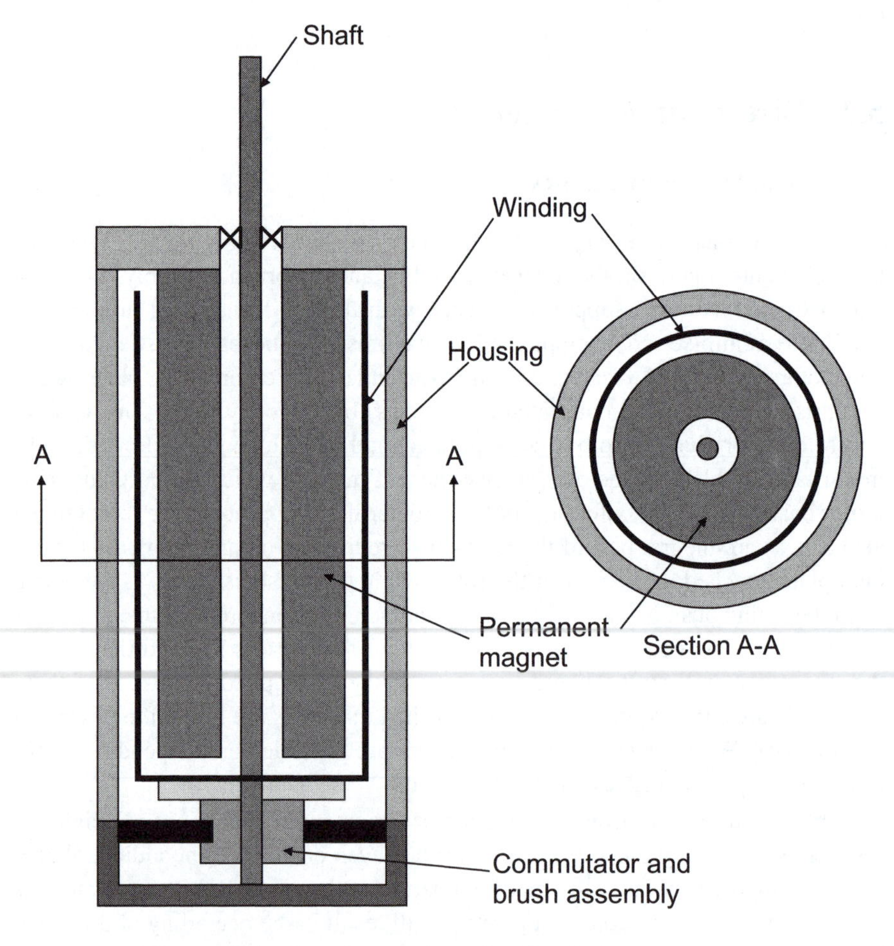

Figure 5.2. The construction of an ironless-rotor motor.

$$T_r = P_d(Rt_{r-h} + Rt_{h-a}) \tag{5.3}$$

The thermal resistance from the rotor to the housing, Rt_{r-h}, and from the housing to the ambient, Rt_{h-a} can be obtained from the manufacturer's data sheets. As long as the rotor's temperature is less than its specified maximum, no reliability problems will result. For a system designer, ironless rotor d.c. machines have a number of distinct advantages including:

- Linear speed-torque, voltage-speed, and load-current characteristics over the operational range of the motor.
- Due to the uniform magnetic field and the relatively large number of commutator segments there is no magnetic detent or preferred rotor position. In addition, this form of construction results in minimal torque ripple over the motor's speed range.
- The use of precious-metal brushes results in low brush friction, and hence a low starting torque. The high quality of the contact between the brushes and the commutator reduces the electromagnetic interference, EMI, and radio-frequency interference, RFI, to a minimum.
- The low mass of the rotor results in a low-inertia motor, permitting high accelerations to be achieved. Due to the low inductance of the rotor winding, this type of motor should be restricted to linear drives or very-high-frequency switched drives to reduce any ripple current to a minimum.

5.2.2 Iron-rotor motors

Permanent-magnet iron-rotor motors have evolved directly from wound-rotor designs and the design has been refined for servo applications. Due to the location of the magnets and the large air gap which is required, these motors tend to be relatively long with a small rotor diameter; this ensures that the motor's inertia is minimised. The manufacturers of these motors provide features that are designed to ensure ease of application; these features include the provision of integral tacho-generators, encoders, brakes, and fans, together with thermal trip indicators within the rotor windings. Due to the widespread application of these motors, a range of standard sizes and fixings have evolved; this considerably eases the procurement of the motors from a range of manufacturers.

5.2.3 Torque motors

As discussed in Section 2.1, the accuracy of a positioning system depends on the motor and gearbox being able to supply a constant torque from standstill to full speed with minimum backlash. However, certain applications requiring high-precision motion at very slow speeds (for example, telescope drives) conventional

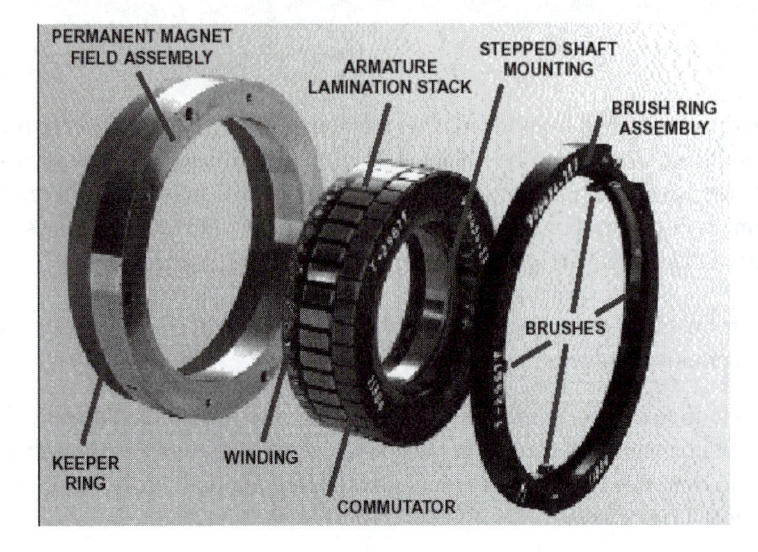

Figure 5.3. A exploded view of a brushed torque motor. Photograph courtesy of Danaher Motion, Kollmorgen.

motor-gearbox designs are unable to provide satisfactory results. In order to obtain the performance which is required, a torque motor can to be used, Figure 5.3. The operation of a torque motor is no different to that of an iron rotor machine; however, there are two significant constructional differences. Firstly, the number of commutator segments and brush pairs is significantly greater than is found in a conventional motor. The large motor diameters permit the use of a large number of commutator segments, with two or more sets of brushes. This design allows a machine to have a torque ripple which is considerably lower in magnitude and higher in frequency than a conventional brushed motor, and, depending on the motor size, this can be as low as 500 cycles per motor revolution at two per cent of the average output torque. Secondly, since the torque motors need to be directly integrated into the mechanical drive chain to maximise the stiffness, they are supplied as frameless machines (with the rotor, stator, and brush gear being supplied as separate items) and they are directly built into the mechanical system. This form of construction, while giving excellent performance, does require particular care in the design and fabrication of the system. The selection of a torque motor is no different from the selection of any other type of motor, and a detailed consideration of the torques and the speed is required. Since the motor is supplied as a frameless system, considerable care is required during the mechanical design and installation. In particular, the air gap must be kept at a constant size by minimising any eccentricity, and, during the installation, the stator's magnets must not be damaged or cause damage. Motor diameters in excess of 1 m are possible, and the present level of technology allows torque motors to provide speeds as low as one revolution in 40 days (1.17×10^{-5} rev min^{-1}) if a suitable drive system is used.

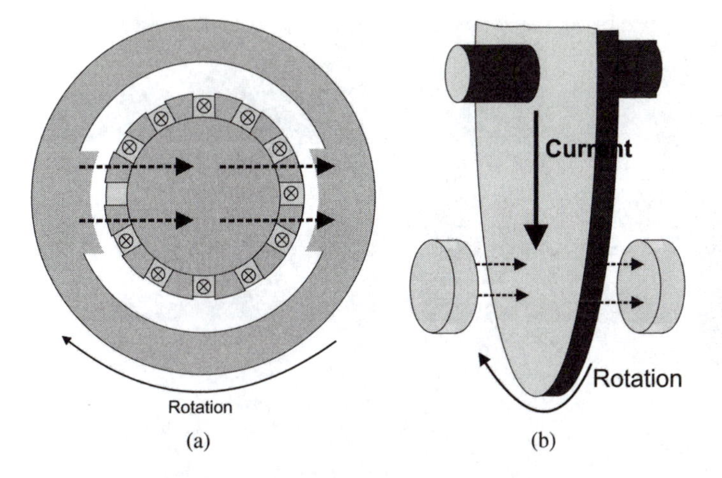

Figure 5.4. The principles of (a) radial and (b) axial field d.c. brushed motors.

5.2.4 Printed-circuit motors

The magnetic-flux path has been radial in the motors considered so far. This results in machines that are typically long and thin, with the actual size depending on their output power. However, the magnetic field is axial within a printed-circuit motor, leading to a very compact motor design, Figure 5.4. The magnets are mounted on either side of the rotor, and the magnetic path is completed by the outer casing of the motor. The commutators are located towards the centre of the rotor, with the brushes located on the rear of the motor case, Figure 5.5.

The main constraint on the length of the motor is the size of the magnets. The motor design could use either low-power ferrite magnets (to give a short motor) or Alnico magnets (to give a longer, more powerful, motor). The use of neodymium-iron-based magnetic materials has allowed high-power motors to be designed with minimum lengths. In addition, these materials are now stable up to 150°C and this, combined with their high coercivity rate, has made them highly suitable replacements for Alnico. However, there is a significant price penalty when these materials are used; but this is only one element in the total system cost and the enhanced performance of this class of motor must also be considered. The technical advantages of these materials over a conventional iron-rotor motor are summarised in Table 5.1. It can be concluded that the printed-circuit construction provides some significant advantages for system designers, including:

- A low-inertia armature, due to the low mass and thickness; this results in a motor with an exceptional torque-to-inertia ratio, and a typical motor can accelerate to 3000 rev min $^{-1}$, within 10 ms and 60° of rotation.

- With the high number of brushes and commutator segments, there is minimal torque ripple throughout the speed range.

Figure 5.5. Exploded view of a radial field pancake motor, the low-inertia rotor can be clearly seen. Photograph courtesy of Danaher Motion, Kollmorgen.

Table 5.1. Comparison between iron rotor and printed-circuit rotor d.c. machines.

Property	Conventional iron-rotor motor	Printed-circuit
Magnetic material	ferrite	neodymium
Rated torque	1.1 Nm	1.1 Nm
Rated speed	2700 rev min $^{-1}$	3000 rev min $^{-1}$
Peak torque	6 Nm	11.62 Nm
Inertia	1.1×10^{-3} kg m^2	1.3×10^{-4} kg m^2
Length	206 mm	27 mm
Diameter	102 mm	140 mm
Weight	5.1 kg	2.8 kg

- The very low inductance of the motor ensures a long brush life due to the absence of arcing at commutation this also allows high-speed, high-torque operation.

5.3 Drives for d.c. brushed motors

The principle and the implementation of brushed, d.c., motor controllers is amongst the simplest of all the motors considered in this book, with the motor speed being a direct function of the voltage that is applied between the two motor terminals. The commutation of the rotor current is undertaken by the mechanical arrangement of the commutator and brushes. In servo applications the motor's terminal voltage is normally controlled by a linear or switching amplifier. For completeness, static four-quadrant thyristor drives will be briefly considered; these drives are not considered to be servo drives, but they are widely used as spindle, tool, or auxiliary drives in machine-tool or robotic systems.

5.3.1 Four-quadrant thyristor converters

While not normally used in servo applications, four-quadrant thyristor converters are widely used in constant speed drives that use d.c. brushed motors. A single-phase converter can be used up to 15 kW; above this power, maintenance of the quality of the output, and the resultant supply harmonics, necessitates the use of a three-phase system. To permit four-quadrant operation, two identical converters connected in reverse parallel, Figure 5.6. Both converters are connected to the armature, but only one operates at a given time, acting as either a rectifier or an inverter. The other converter takes over whenever power to the armature current has to be reversed. Consequently, there is no need to reverse the armature or field. The time to switch from one converter to the other is typically 10 ms.

High performance industrial drives require precise speed and torque control down to and through zero speed. This implies that the converter voltage may at times be close to zero. At this operating point, the converter current is discontinuous, hence the motor's torque and speed tend to be erratic, and precise control is difficult to achieve. To resolve this problem, the two converters are designed to function simultaneously. When one functions as a rectifier, the other functions as an inverter, and vice versa. The armature current is the difference between the output currents from both converters. With this arrangement, the currents in both converters flow for 120°, even at zero armature current. As the two converters are continuously in operation, there is no delay in switching from one to the other. The armature current can be reversed almost instantaneously; consequently, this represents the most sophisticated control system available. In practice each converter must be provided with a large series inductor to limit the a.c. circulating currents, and the converters must be fed from separate sources, such as the isolated secondary windings of a 3-phase transformer. While these drives are highly efficient and reliable, their dynamic response is poor; with a 50 Hz supply and a three-phase converter, a current pulse occurs every 3.3 ms. This effectively limits the response of the drive to a change in demand or load by restricting the rate of the rise of the current.

5.3.2 Linear amplifiers

Linear amplifiers are widely used to control the speed of small, d.c. brushed motors. The basic principle is shown in Figure 5.7, where the difference between the required motor terminal voltage and the supply voltage is dissipated across a power device operating in a linear mode. Since the power which is dissipated is given by:

$$P_d = I_d(V_s - V_m) \tag{5.4}$$

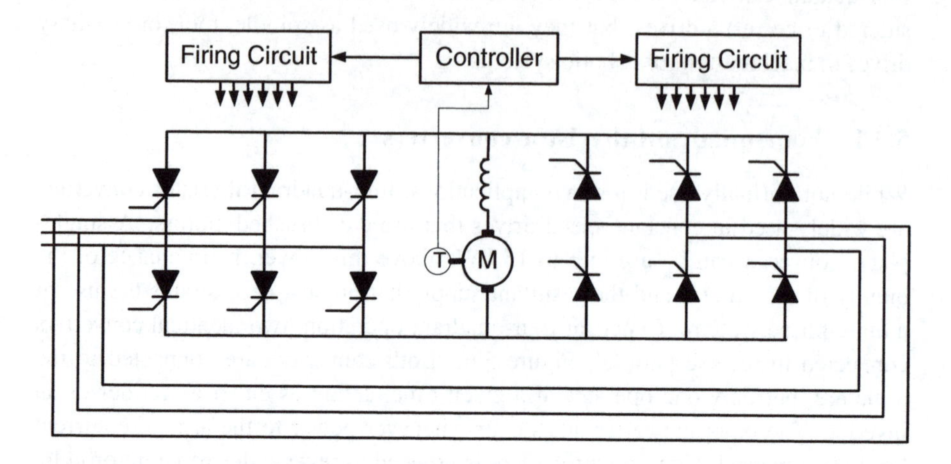

Figure 5.6. A four quadrant thyristor drive for d.c. brushed motors. This system as shown is not capable of handling circulating currents. To prevent circulating current the supplies must be isolated from each other through the use of a transformer.

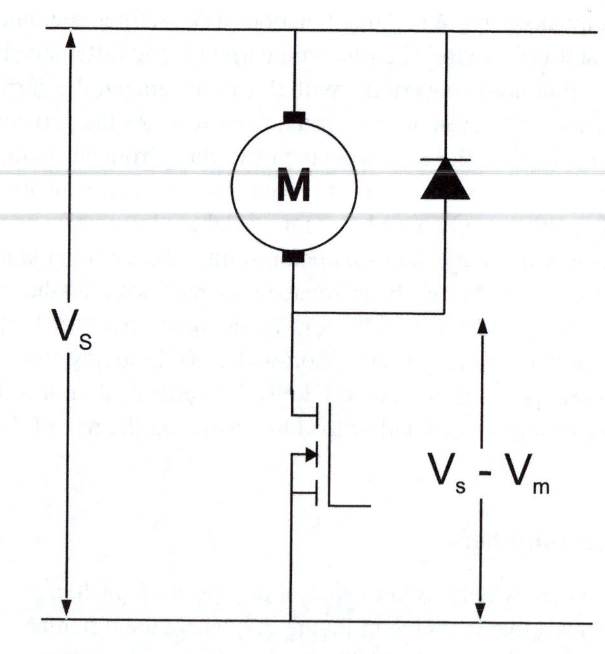

Figure 5.7. The principle of a linear, d.c. motor controller. The device operates in the linear mode, as opposed to a switching mode.

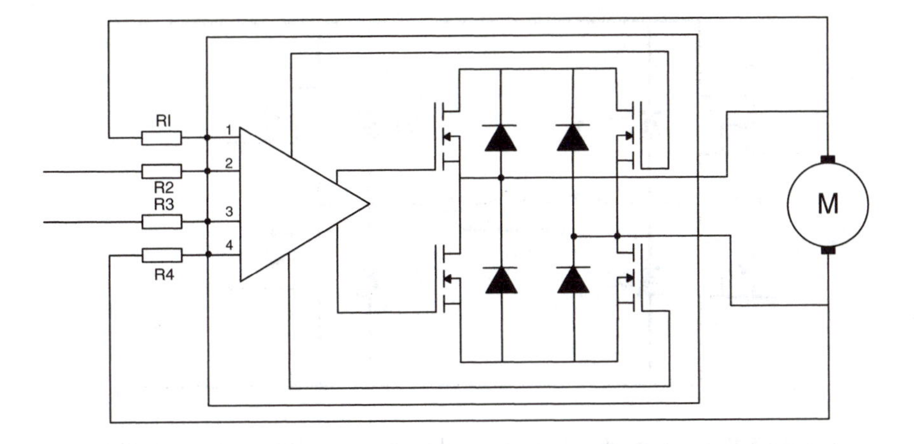

Figure 5.8. A linear amplifier connected as a voltage amplifier. The gain is set by R1 to R4, voltage feedback is via pins 1 and 4.

the overall system efficiency will vary between zero and one hundred per cent, depending on the speed and the torque of the motor. In order to achieve four quadrant operation, an H-bridge arrangement is used (see Figure 5.8) in which a linear amplifier can effectively be considered as an operational amplifier with a high-power output stage. The application of a linear amplifier is relatively straightforward, as it can be configured as either a voltage or a current amplifier, with adjustable gains. In the selection of the amplifier, considerable care must be taken to ensure that the maximum power rating of the package is not exceeded. The worst possible scenario combines a low speed with a high torque, particularly when the motor is at stall under load. Application of equation (5.4) will allow the power-dissipation requirements to be estimated and will allow comparison with the manufacturer's rating curves. Consideration should also be given to when the motor is decelerating – or plugging – in which case the motor's voltage is added to the output of the amplifier, the current being limited only by the armature's resistance or by the amplifier's current limit. The energy dissipated in the system can be determined by the application of equation (2.2). Apart from the energy that is dissipated in the motor's armature, all the energy is dissipated in the drive; if the motor is subjected to excessive speed reversals the power rating of the amplifier must be considered in detail. Therefore, in the selection of a linear amplifier, the thermal-dissipation problems are of considerable concern to the system designer. Commercial linear amplifiers are available in power ratings up to 1.5 kW and with output voltages of 60 V; this necessitates forced air cooling and derating of the power rating at high ambient temperatures. In most cases, a thermal trip circuit is provided to disable the amplifier if the temperature approaches the rated value.

The use of a linear amplifier gives the system designer considerable benefits over other forms of drives, which normally are based on a switching principle. In particular a linear drive may have very high bandwidths, typically greater than

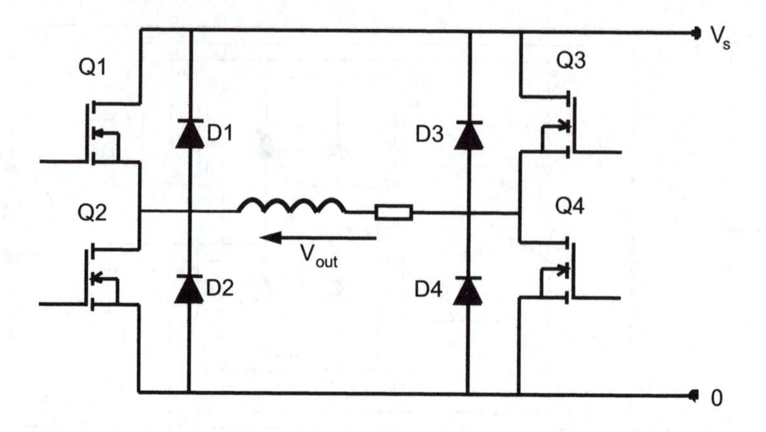

Figure 5.9. Four quadrant power bridge used in a PWM servo amplifier.

500Hz; this allows exceptional performances to be obtained with motors of low inertia and/or inductance, especially in ironless rotor and printed-circuit motors. Additional benefits include a low deadband that eliminates crossover distortion and low radiated acoustic and electromagnetic noise, due to the absence of switching devices.

5.3.3 Pulse width modulated servo drives

As noted earlier, when d.c., permanent-magnet, brushed motors are used in robotic or machine-tool applications, the overall performance of the drive system will significantly determine the accuracy and response of each motion axis. The thyristor or linear drives discussed so far are not suitable for the majority of applications: thyristor-based drives have a very low dynamic response and linear amplifiers have excessive power dissipation.

To control any load, including a d.c. motor, in all four quadrants a bidirectional current flow is required; this is achieved by using a basic four device, H-bridge, Figure 5.9. In order to achieve the maximum efficiency from this type of amplifier, the power devices operate in the switching mode rather than the highly dissipative linear mode. The four switching devices can be bipolar transistors, power MOSFETS (metal-oxide semiconductor field-effect transistors) or IGBTs (insulated-gate bipolar transistors) depending on the application's voltage and current ratings. In order to control the motor terminal voltage, the devices can be switching in a number of different ways; the most widely used method is to switch the devices at a constant frequency and to vary the on and off times of the devices. This is termed pulse-width modulation, PWM.

A number of different switching regimes can be used to control the amplifier's output voltage, three being discussed below. In each case, the switching pattern required to obtain a bipolar output voltage is considered in response to the amplifier's input command, V_c. The instantaneous amplifier terminal voltage, V_{out}, is

Table 5.2. Bipolar switching

State 1, Positive output voltage	State 2, Negative output voltage
Q_1, Q_4 on Q_2, Q_3 off $V_{out} = V_s$	Q_2, Q_3 on Q_1, Q_4 off $V_{out} = -V_s$

Table 5.3. Unipolar switching

Output voltage	Mode 1	Mode 2
Positive $V_c > 0$	Q_1, Q_4 on Q_2, Q_3 off $V_{out} = +V_s$	Q_2, Q_4 on Q_1, Q_3 off $V_{out} = 0$
Negative $V_c < 0$	Q_2, Q_3 on Q_2, Q_4 Off $V_{out} = -V_s$	Q_1, Q_3 on Q_2, Q_4 off $V_{out} = 0$

considered to be equal to the supply voltage, the voltage drop across the individual devices an be neglected. The device's switching delays are also neglected in this analysis, as are any time delays introduced by the control system.

Bipolar switching The output voltage, V_{out}, can be equal to either $+V_s$, or $-V_s$. The average value of the output voltage is controlled by the relative times spent in either state 1 or state 2, see Table 5.2.

Unipolar switching If the output voltage is required to be positive, Q_4 is turned on continuously, with Q_1 and Q_2 being used to control the magnitude of the load voltage, by PWM. When Q_2 is on, both the motor terminals are effectively connected to the negative supply rail, see Table 5.3.

Limited unipolar switching The bipolar and unipolar switching modes have the disadvantage that one pair of devices has to be switched off prior to a second pair being switched on. Because power semiconductors take a finite time to switch states, there is a danger of a short circuit across the power supply. This can be reduced by the introduction of a time delay between switching one pair of devices off and switching the second pair on. This delay is conventionally termed deadband. However, it is possible to provide a switching pattern that does not require the provision of a deadband; this is known as a limited-bipolar switching pattern. As in the other modes, the switching pattern depends on the polarity of the required output voltage; hence, if a positive voltage is required, Q_1 and Q_4 will be on and Q_2 and Q_3 will be off. In the limited unipolar mode, only one device, Q_1 is switched,

Table 5.4. Limited unipolar operation. The command voltage. V_c, is the voltage fed from any controller to power bridge to determine the output voltage.

Output voltage	Mode 1	Mode 2
Positive $V_c > 0$	Q_1, Q_4 on Q_2, Q_3 off $V_{out} = V_s$	Q_4 on Q_1, Q_2, Q_3 off $V_{out} = V_s$ if $I_a < 0$ $V_{out} = 0$ if $I_a > 0$ $0 < V_{out} < V_s$ if $I_a = 0$
Negative $V_c < 0$	Q_2 Q_3 on Q_1, Q_4 Off $V_{out} = -V_s$	Q_2 on Q_1, Q_2, Q_4 off $V_{out} = -V_s$ if $I_a > 0$ $V_{out} = 0$ if $I_a < 0$ $-V_s < V_{out} < 0$ if $I_a = 0$

hence the amplifier's terminal voltage depends on the instantaneous motor current. If the motor current, I_a, is positive, the current will flow via D_2 and Q_4 giving an amplifier's terminal voltage of zero. If the motor current is negative, the current flow is via D_1 and Q_4, therefore the motor terminal voltage equals the supply voltage. If no current is flowing in the armature the output voltage can be considered to be indeterminate. The switching pattern is shown in Table 5.4.

Comparison between bipolar and unipolar switching

The choice between the bipolar, unipolar, or limited unipolar switching regimes needs to be considered during the design of a switching amplifier. In particular, a unipolar amplifier requires that only one device is switched at any one time; this gives increased reliability. However, both unipolar switching regimes have two serious disadvantages:

- The control electronics are more complex because the switching pattern is different for a positive or negative demand.
- If a zero speed is required from a drive system, rapid changes from a positive to a negative motor terminal voltage will be required. However, unipolar amplifiers require a change to a switching pattern which will cause significant time delays and hence a poor dynamic response.

For these reasons bipolar switching is almost universally employed in PWM amplifiers, although the other configurations have sometimes been developed in drives for specific applications.

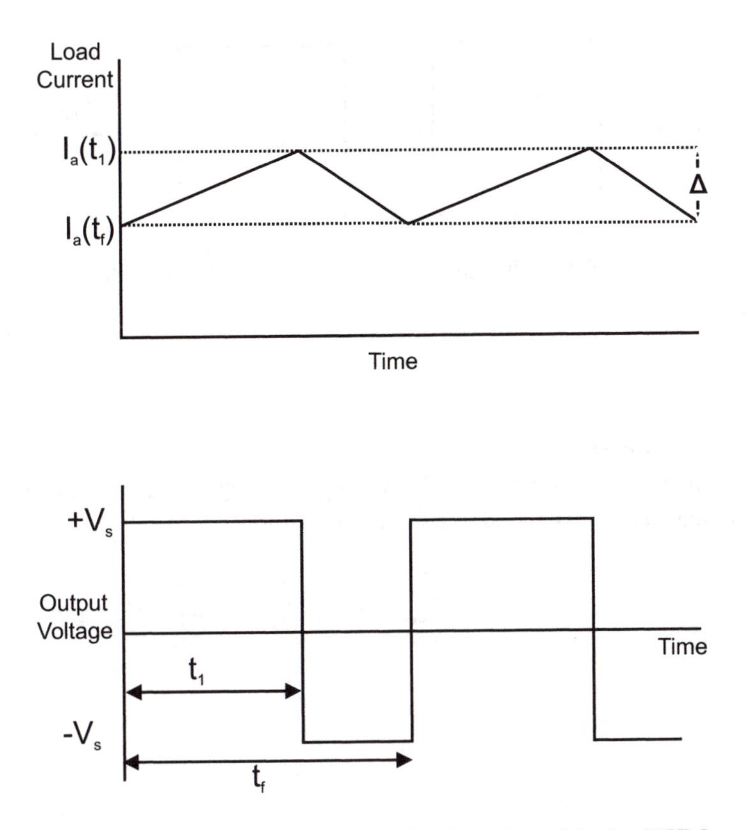

Figure 5.10. The load current and terminal voltage for a bipolar PWM amplifier.

5.3.4 Analysis of the bipolar PWM amplifier

In order to fully appreciate the interaction of a bipolar switching amplifier and its load, the output current has to be analysed. The effects of the load's inductance and of the amplifier's switching frequency on the load current also needs to be considered. As detailed in Table 5.2, the output voltage of a bipolar amplifier switches between $+V_s$ and $-V_s$. I practice the mark–space ratio that determines the voltage between the limits of $+V_s$ and $-V_s$ is determined by the amplifiers command's voltage, V_c.

The voltage and the current waveforms in the power bridge are shown in Figure 5.10. If the servo amplifier's input voltage is considered to be V_c with a peak value of V_{cpk} and is assumed to have a frequency which is lower than the switching frequency, then the load factor, ρ, is given by

$$\rho = \frac{V_c}{|V_{cpk}|} \tag{5.5}$$

Since the command voltage is bipolar, the load factor is limited to $-1 < \rho < +1$. Therefore, it follows that the length of the *on* phase, t_1, is given by

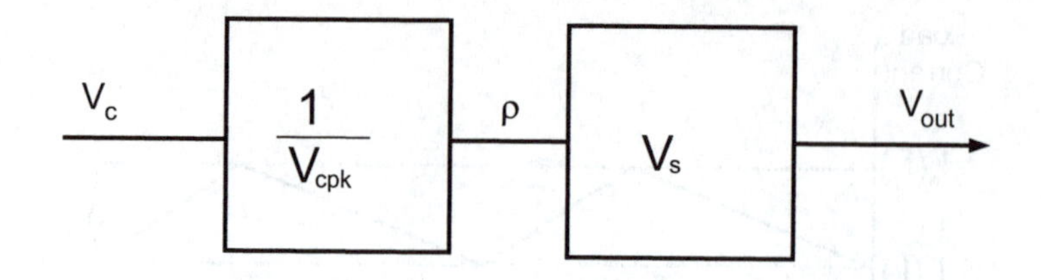

Figure 5.11. The equivalent circuit of a PWM amplifier.

$$t_1 = \frac{t_f(1+\rho)}{2} \tag{5.6}$$

where t_f is the switching period of the amplifier.

The output voltage of a bipolar PWM amplifier, neglecting any time delays in the switching, can be expressed as a Fourier transform

$$V_{out} = a_o + \sum_{n=1}^{\infty} a_n \cos\left[\frac{2\pi nt}{f_s} + \phi_n\right] \tag{5.7}$$

where

$$a_o = \rho V_s = \frac{V_s V_c}{|V_{cpk}|} \tag{5.8}$$

and

$$a_n = \frac{4V_s}{n\pi} \sin\left[\frac{n\pi(1+\rho)}{2}\right] \tag{5.9}$$

The load voltage therefore consists of a slowly varying component, a_o, which is dependent on the load factor together with high frequency components defined by a_n. The amplifier's switching frequency is selected so that the lowest harmonic frequency which is present is greater than the loads's bandwidth inductance of the load, and only the slowly varying components have to be considered. It is therefore possible to represent a PWM amplifier by the equivalent block diagram shown in Figure 5.11, where the equivalent gain of the amplifier is

$$V_{out} = \frac{V_s V_c}{V_{cpk}} \tag{5.10}$$

where V_{out}, is the amplifier's output voltage and V_s is the supply voltage.

If we now consider a motor-drive application, Figure 5.12, and if the switching period is considerably smaller than the motor's time constant and if the motor speed is constant over one switching cycle, then the motor's voltage equation is given by

$$V_m = R_a I_a + K_e \omega_m = R_a I_a + E_m \tag{5.11}$$

and the average armature current is given by

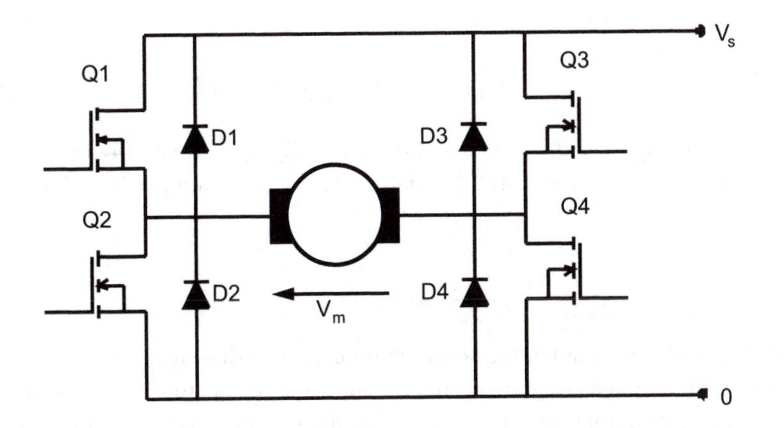

Figure 5.12. A four quadrant power bridge used to control a motor's voltage and current.

$$I_a = \frac{1}{t_f} \int_o^{t_f} I(t)dt \tag{5.12}$$

These equations can be further simplified by linearisation, and by neglecting the armature resistance and the brush-voltage drops, to give, for a bipolar amplifier,

$$L_a \frac{di_a}{dt} = V_s - E_m \quad \text{for} \quad 0 \leq t \leq t_1 \tag{5.13a}$$

$$L_a \frac{di_a}{dt} = -V_s - E_m \quad \text{for} \quad t_1 \leq t \leq t_f \tag{5.13b}$$

The solution to these equations results in the triangular waveform which is shown in Figure 5.10 and is expressed as

$$i_a(t) = I_a(0) + \left(\frac{V_s - E_m}{L_a} \right) t \quad \text{for} \quad 0 \leq t \leq t_1 \tag{5.14a}$$

$$i_a(t) = I_a(t_1) - \left(\frac{V_s - E_m}{L_a} \right) (t - t_1) \quad \text{for} \quad t_1 \leq t \leq t_f \tag{5.14b}$$

During steady-state operations, the current is periodic and $I(0) = I(t_f)$, hence

$$\left(\frac{V_s - E_m}{L_a} \right) t_1 - \left(\frac{V_s + E_m}{L_a} \right) (t_f - t_1) = 0 \tag{5.15}$$

If equation (5.15) is combined with equation (5.6) and with

$$\rho = \frac{E_m}{V_s} \tag{5.16}$$

then the total current variation, Δ, can be determined to be

$$\Delta = I_a(t_1) - I_a(0) = \frac{V_s t_f}{2L_a}(1 - \rho^2) \quad (5.17)$$

It should be recognised that this equation is equally applicable to both motor and non-motor loads. The peak-current variation will occur when $\rho = 0$, and it is given by

$$\Delta_{max} = \frac{V_s t_f}{2L_a} \quad (5.18)$$

where L_a is either the load or the armature inductance. Since the motor's torque is a function of the average armature current, while armature heating is a function of the r.m.s. value of the armature current, it is important to note that with a bipolar servo amplifier, even at zero mean current, there is current flowing through the motor, leading to armature heating. This has to be minimised to allow the best possible frame size of motor to be selected. The quality of the waveform is measured by the form factor, which is given by

$$\text{Form factor} = \frac{I_{rms}}{I_{average}} \quad (5.19)$$

For a switching amplifier with an output waveform as shown in Figure 5.10:

$$\text{Form factor} = \frac{\sqrt{I^2 + \frac{\Delta^2}{12}}}{I} \quad (5.20)$$

and on substituting for Δ, equation (5.18)

$$: \text{Form factor} = \sqrt{1 + \left(\frac{V_s}{6.9 L_a f_s I}\right)^2} \quad (5.21)$$

The armature or load current's form factor is a function of the motor's armature inductance, L_a, the amplifier switching frequency, f_s, and the supply voltage, V_s. Since the form factor is also a function of the current I, it is convenient to specify a minimum load inductance for a specific drive at its maximum average current. A form factor of 1.01 is considered typical for a PWM drive.

Example 5.1

Consider a drive that requires a PWM amplifier to drive a printed circuit motor with an armature inductance of 40 μH. If the drive switches at 20 kHz, calculate the addition inductance required, to maintain a form factor, F, of 1.01, at an average current of 10 A. The drive operates from a 50 V d.c. supply.

Rearranging equation 5.21 allows the minimum armature inductance to be calculated.

$$L_{min} = \frac{V_s}{6.9 f_s I_a \sqrt{F^2 - 1}} = 0.256\text{mH}$$

Given that the motor inductance is 40 μH, an external inductance of 0.216 mH is required to maintain the armature current at the required form factor.

If the additional inductance was not added, using equation (5.21), a form factor of 1.38 results; if this is substituted into equation (5.20), the peak-to-peak current is 16 A, compared to 2.5 A with the additional inductance.

The analysis of the current waveform given above has shown that the quality of the waveform is significantly determined by the inductance of the motor; a low inductance will give a high-current ripple leading to excessive motor heating which requires a larger frame size and perhaps forced cooling. This is of particular concern with the application of very-low-inductance printed circuit and ironless-rotor motors. The form factor can be improved by the use of very-high-frequency switching (greater than 40 kHz), or, in certain cases, additional inductance has to be added to the armature circuit this will improve the form factor but it may degrade the dynamic performance of the system.

5.3.5 PWM amplifiers

A block diagram of an analogue PWM amplifier is shown in Figure 5.13. The design consists of three main elements: an analogue servo loop, digital control logic, and the main power bridge.

The major feature of the operation of a PWM servo amplifier is the generation of the switching waveform. This can be achieved in a number of ways, of which two predominate:

- Current-controlled hysteresis
- Sub-harmonic modulation.

In a current-controlled-hysteresis system, the power devices are controlled by the load current (see Figure 5.14). When the load current exceeds a predetermined value, the power devices are switched, allowing the current to reduce. The frequency of the PWM waveform is uncontrolled. However, this approach is simple to implement and it is used on some small drives. In addition, it has found acceptance in a wide range of integrated circuit drives, for both d.c. brushed and brushless motors.

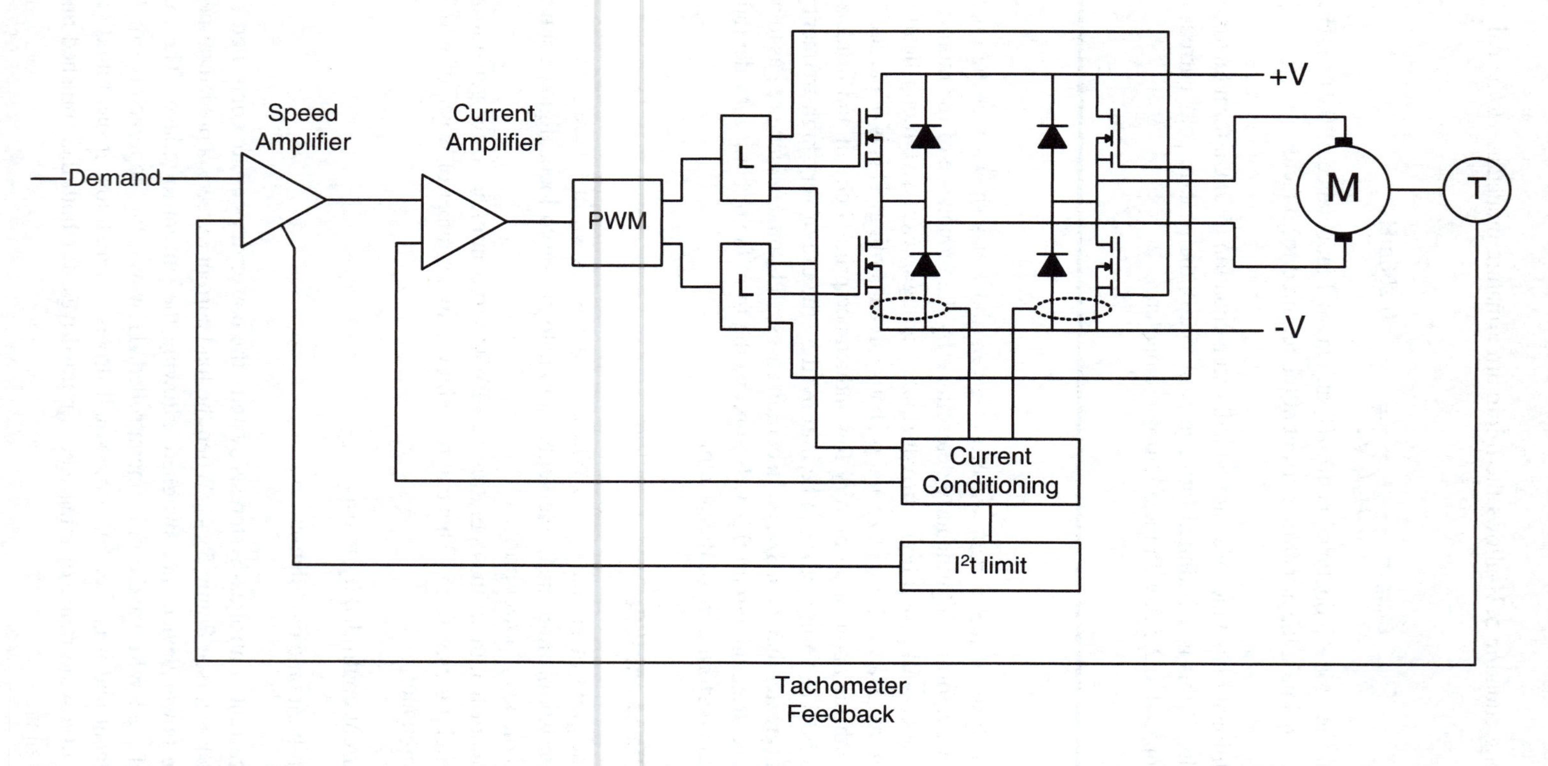

Figure 5.13. Block diagram of a commercial analogue PWM amplifier. The blocks, L, contain the logic and drives for the individual power devices.

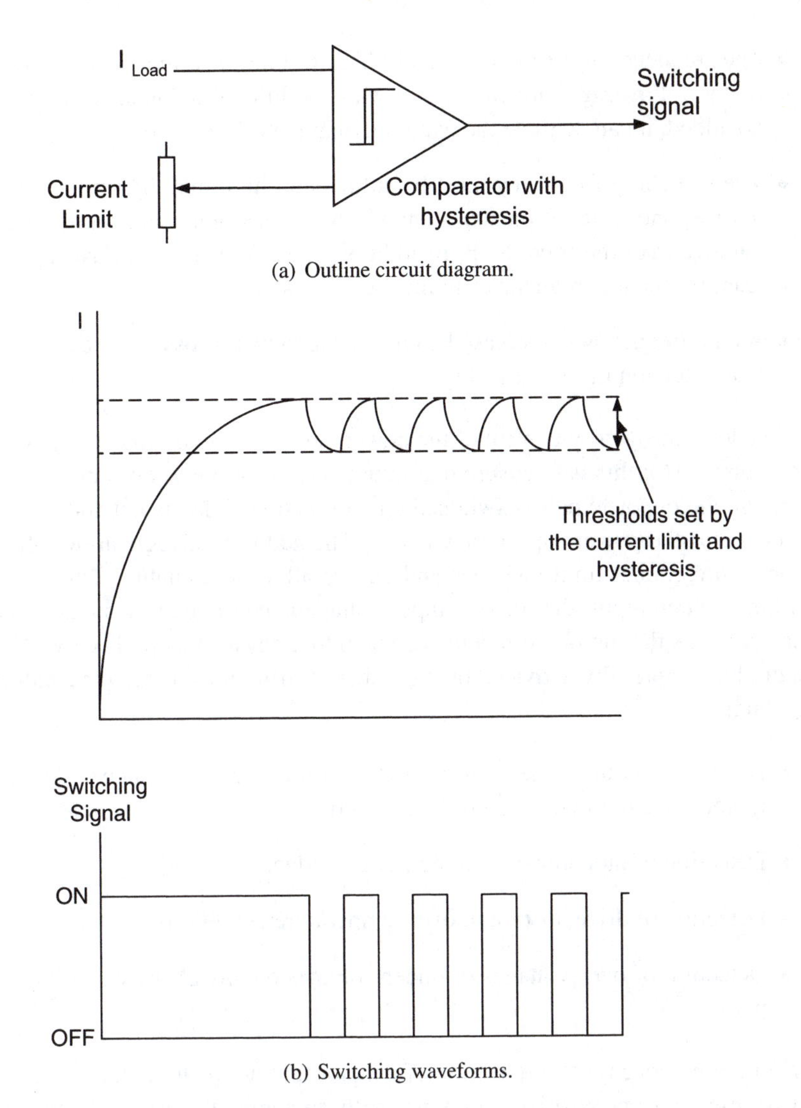

(a) Outline circuit diagram.

(b) Switching waveforms.

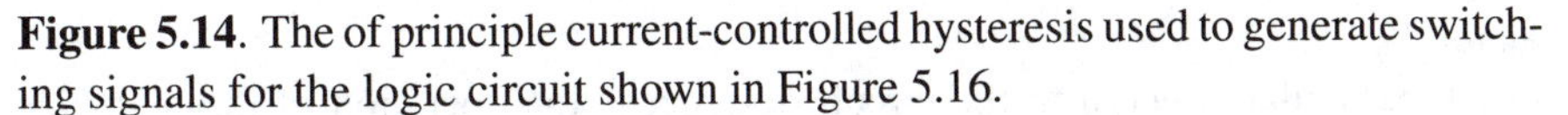

Figure 5.14. The of principle current-controlled hysteresis used to generate switching signals for the logic circuit shown in Figure 5.16.

The more conventional approach uses a modified form of subharmonic modulation which is similar to that used by the telecommunications industry (see Figure 5.15). The frequency of the triangular-waveform carrier sets the switching frequency of the drive. As the input changes, the mark/space ratio of the output changes. The switching frequency can be in excess of 40 kHz. As a general rule, the higher is the power output of the drive, the lower is the switching frequency. The selection of the amplifier switching frequency is critical to the operation of the drive, and it needs to be considered both by its designers and, at a later date, by its users. The main considerations are:

- The frequency must be sufficiently high to minimise the current ripple for motors of average inductance and while additional inductance can be (and is) added, it can degrade the overall dynamic performance.

- The switching frequency must be sufficiently high that the servo loop does not respond to it. Typically, it should be at least ten times higher than the bandwidth of the servo loop. In addition, it has to be greater than any significant resonance frequencies in the mechanical system.

- As the frequency is increased so the semiconductor losses increase, leading to a reduction in the reliability.

To allow satisfactory control of the power devices, a small time delay is added (see Figure 5.16); this is to ensure that time is given for one set of devices to turn off before the opposite pair is switched on. If this time delay is not added, a short circuit across the power supply may result. The addition of logic allows the provision of directional limit switches and an overall system enable. On activation of a limit-switch input, the motor's input voltage is restricted to a single polarity, which prevents the motor from being driven into a physical stop. The overall system disable is normally activated by a number of different sources (or conditions) including:

- During power up of the external controller it is normal practice to disable the motor-drive to prevent any transient motion.

- Detection of fault currents in the power bridge.

- Detection of drive, or motor, over-temperature conditions.

- Detection of over-voltages or under-voltages on any of drive's voltage supplies.

It is normal practice that on detection of a major fault to latch the fault, which can only be reset by some positive action, normally switching the drive off and then on again.

A particular concern to the circuit designer is the operation of the high-side device; the isolation and the drive method which are selected all depend on the size and the proposed application of the drive. In addition to these devices, the power bridge will include a current-sensing device for the current servo loop and device protection. As with any power electronic system, the mechanical construction must be carefully considered, particularly to ensure adequate heat dissipation, and a minimal lead length to minimise switching transients.

A design for an analogue control loop used in a servo drive is shown in Figure 5.17 and incorporates a conventional speed and current amplifier. Facilities are provided for multiple inputs and for the adjustment of gains and stability. The output of the amplifiers can be clamped to prevent transients during powering up

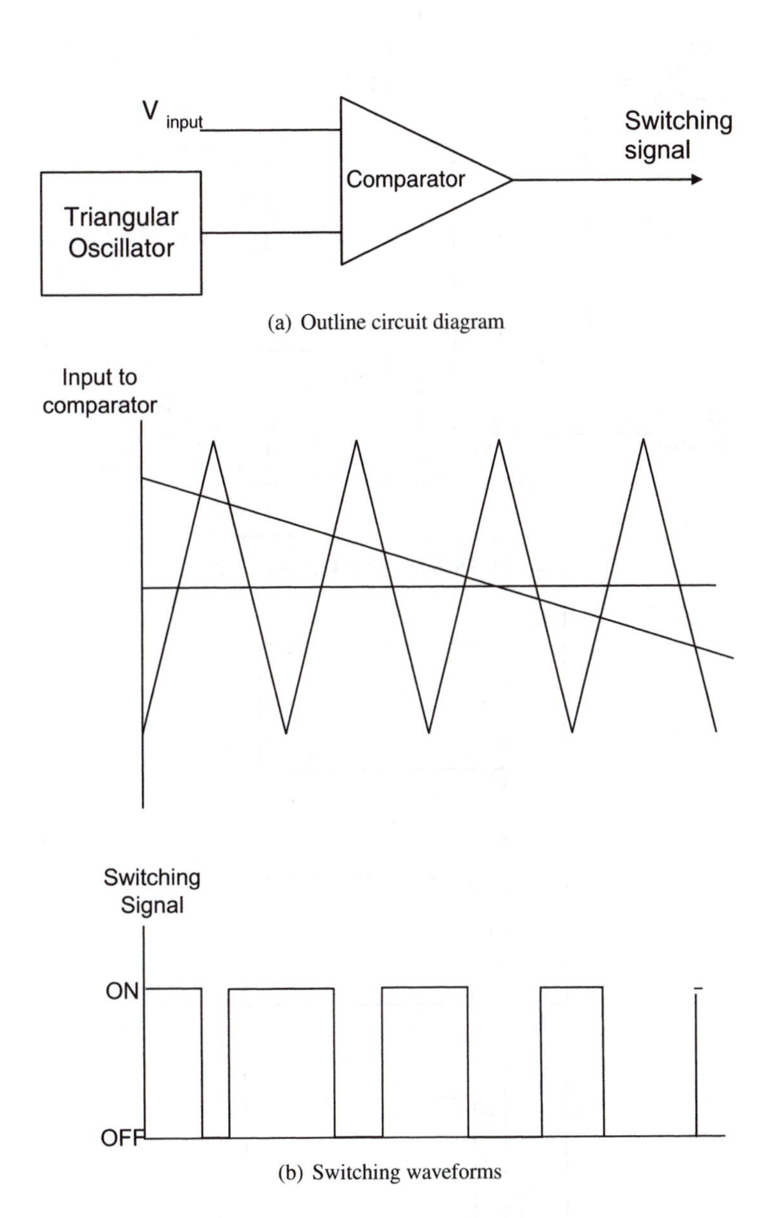

Figure 5.15. The principles of a sub-harmonic PWM system used to generate switching signals for the logic circuit shown in Figure 5.16.

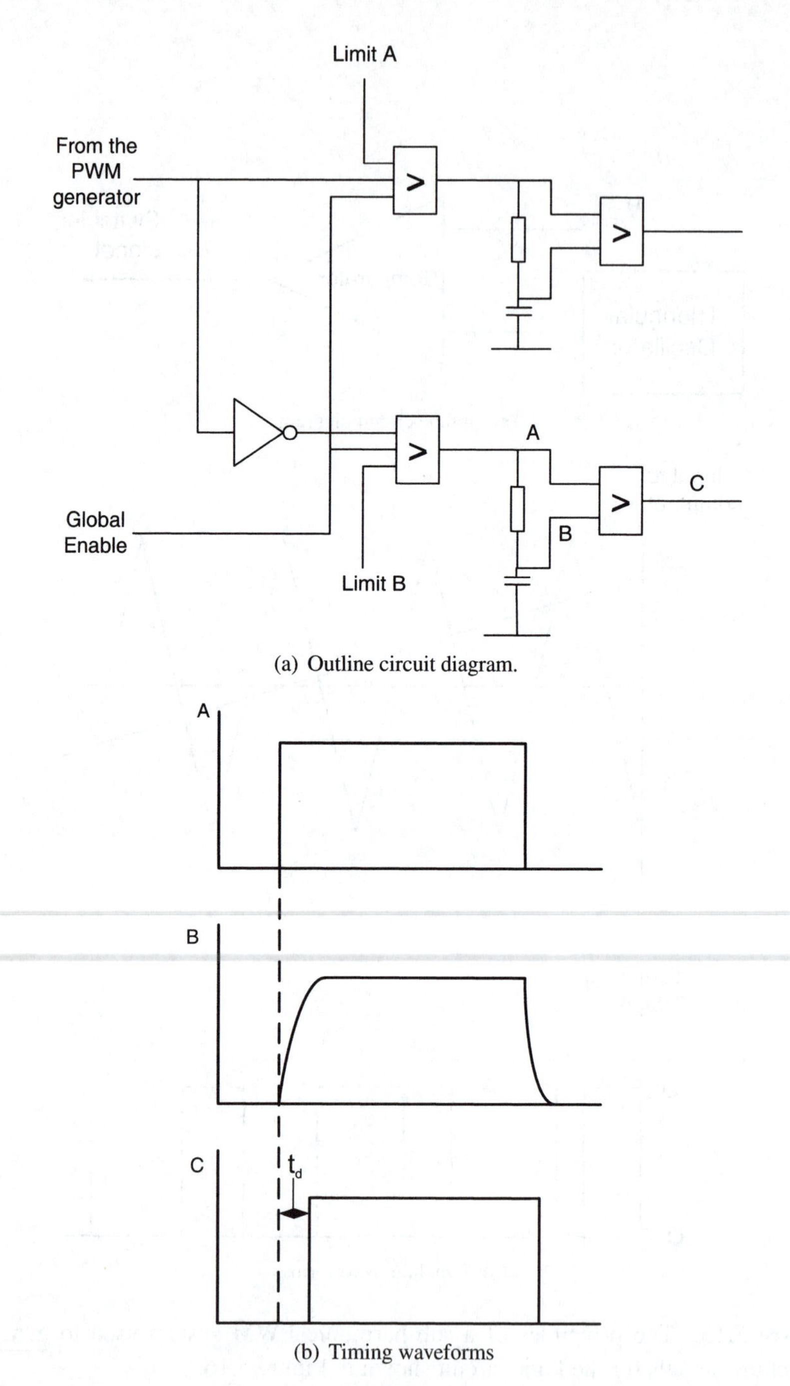

(a) Outline circuit diagram.

(b) Timing waveforms

Figure 5.16. The logic stage of a typical PWM analogue amplifier, showing the provision of the global and direction limits, and the generation of the time delay, t_d, to prevent bridge shoot through. The value of t_d is determined by the types of devices used in the power bridge, but is typically in the order of $1 - 5$ μs.

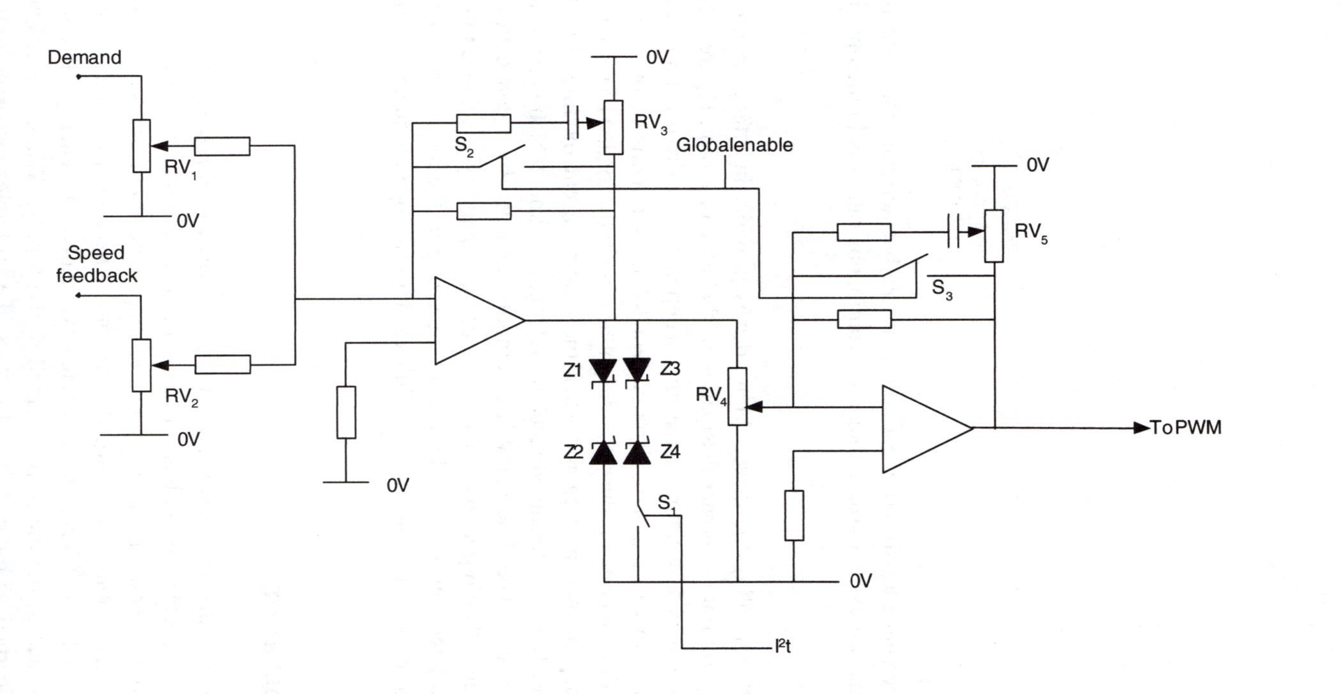

Figure 5.17. Details of an analogue stage for a PWM amplifier suitable for d.c. motor control. The CMOS analogue switches are used to control the I^2t limit (S_1) and the amplifier disables(S_2) and (S_3). The peak current is set by Z_1 and Z_2, with the I^2t foldback limit set by Z_3 and Z_4.

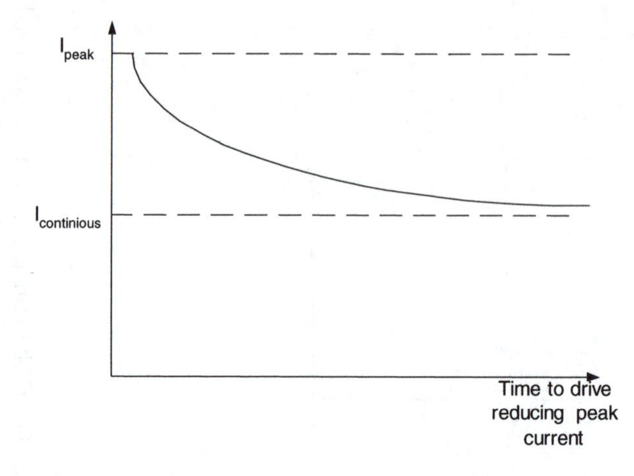

Figure 5.18. The characteristics of the I^2t current fold back circuit. The time that the drive remains supplying current above the continuous rating is limited by the curve shown.

and following the enabling the drive. Many drives provide, in effect, an r.m.s. current foldback to limit the power dissipation of the drive or motor, Figure 5.18. The operation is achieved by clamping the current demand to the drive's continuous rating, typically with zener diodes. The current foldback circuits incorporate a square-law function that reflects the rotor current heating in the load. The reduction in the current dependence is triggered when the motor current modified by the square load exceeds a predetermined value for a length of time, which is typically two to three seconds. The time delay is commonly provided by a conventional analogue integrator. The reduction in the peak-current capability is achieved by a second, lower-voltage, set of zener diodes. In the sizing of the motor-drive, care must be taken to ensure that the current foldback does not operate under normal operation.

5.4 Regeneration

Up to this point, consideration has been limited to drives which provide power to the motor either to accelerate the load or to hold it against a torque at a constant speed. The rapid decelerations that are found in servo-type applications can result in regeneration; that is, energy is returned to the motor from the load and is either absorbed by the motor or it has to be dissipated externally. The energy can either be dissipated as heat in the motor or returned to the supply. While this chapter is primarily concerned with d.c. brushed motors, the principles discussed in this section are also applicable to other forms of motor-drive systems. For the case of a motor being decelerated to a standstill, the motor's terminal voltage, speed, and current are shown as a function of time in Figure 5.19. Prior to deceleration from a speed of ω_{int}, the terminal voltage is equal to $K_e\omega_{int} + I_aR_a$ immediately following the application of the negative braking current, the terminal voltage drops to $K_e\omega_{int} - I_aR_a$. The terminal voltage will decrease as the speed decreases until it equals zero when $\omega_m K_e = I_R R_a$ at $t = t_o$. The motor will continue decelerating until reaching a standstill at time, $t = t_z$. During the period of regeneration, the energy supplied to the drive by the motor is given by

$$E_r = \int_0^{t_z} v_m(t) i_a(t) dt \tag{5.22}$$

where

$$v_m(t) = K_e\omega_m(t) + I_R R_a \tag{5.23}$$

Between t_0 and the time at which the motor reaches a standstill, energy is being supplied from the drive to the motor to overcome the motor's internal resistance. To ensure peak deceleration, the drive holds the motor current at I_R until the motor reaches zero speed. The deceleration is considered to be complete when the speed has dropped to zero, and hence equation (5.23) can be rewritten as

$$v_m(t_z) = I_R R_a \tag{5.24}$$

The motor's terminal voltage during the deceleration can then be written as

$$v_m(t) = \frac{-\omega_{int} K_e}{t_z} + \omega_{int} K_e + I_R R_a. \tag{5.25}$$

The time t_0 can be determined by equating the terminal voltage to zero, giving

$$t_0 = t_z \left(1 + \frac{I_a R_a}{\omega_{int} K_e}\right) \tag{5.26}$$

since I_R has a negative value for $t_0 < t_z$.

On solving for E_r, it is possible to determine that the total regenerative energy that is returned to the drive when a rotating system is decelerated from ω_{int}, to standstill is

$$\begin{aligned} E_r &= \int_0^{t_0} \left[\left(\frac{-\omega_{int} K_e}{t_z} t + \omega_{int} K_e + I_R R_a\right) I_R\right] dt \\ &= \frac{-t_0(\omega_{int} K_e I_R + I_R^2 R_a)}{2} \end{aligned} \tag{5.27}$$

The above equations expresses the regeneration energy in terms of the time for the motor's terminal voltage to reach zero. The total time for the system to reach zero speed can be determined from the deceleration torque, T_d, and the system inertia, I_{tot}, using,

$$\alpha = -\frac{\omega_{int}}{t_z} \tag{5.28}$$

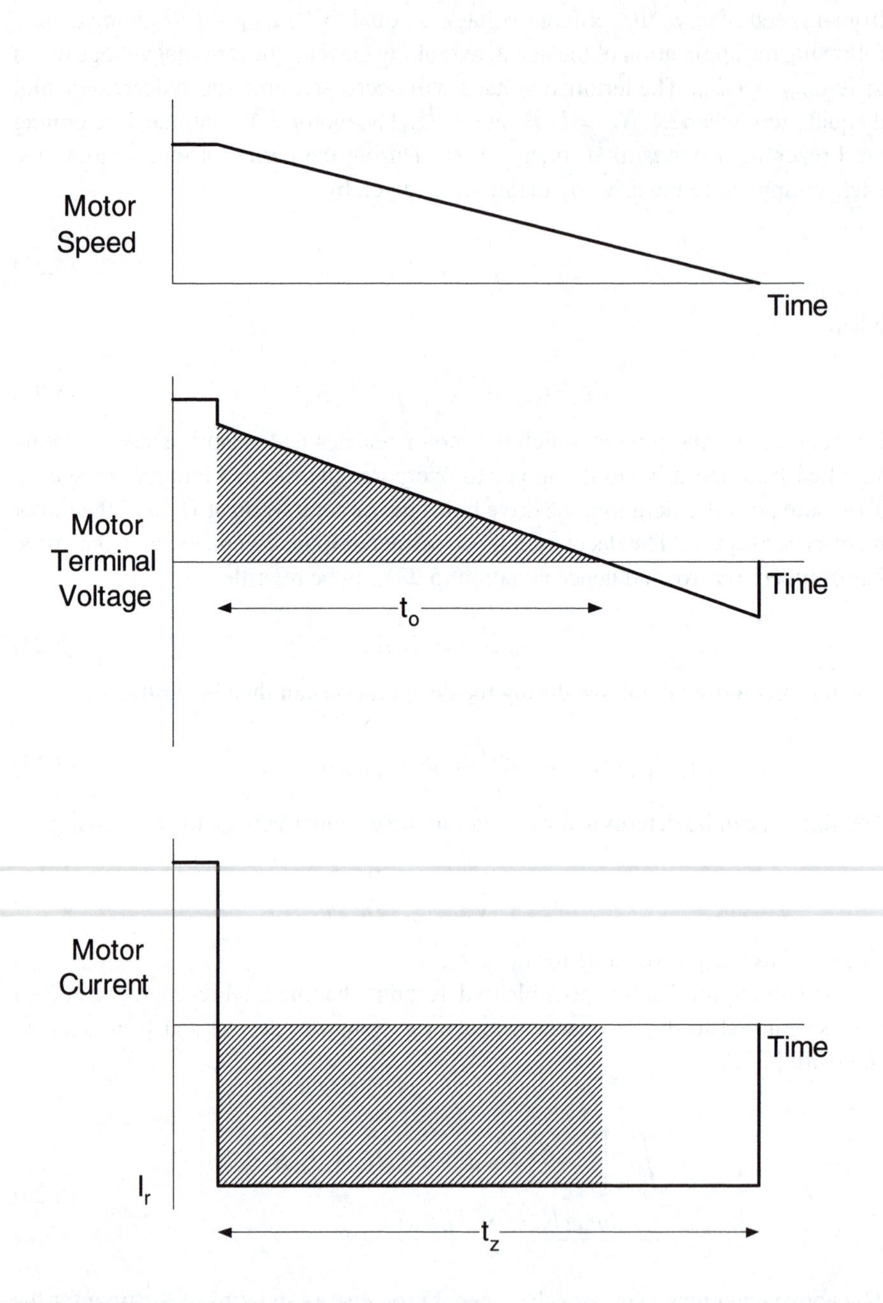

Figure 5.19. A motor's speed, terminal voltage, and armature current during high-speed regeneration. The shaded areas indicates when energy is being returned to the supply from the motor.

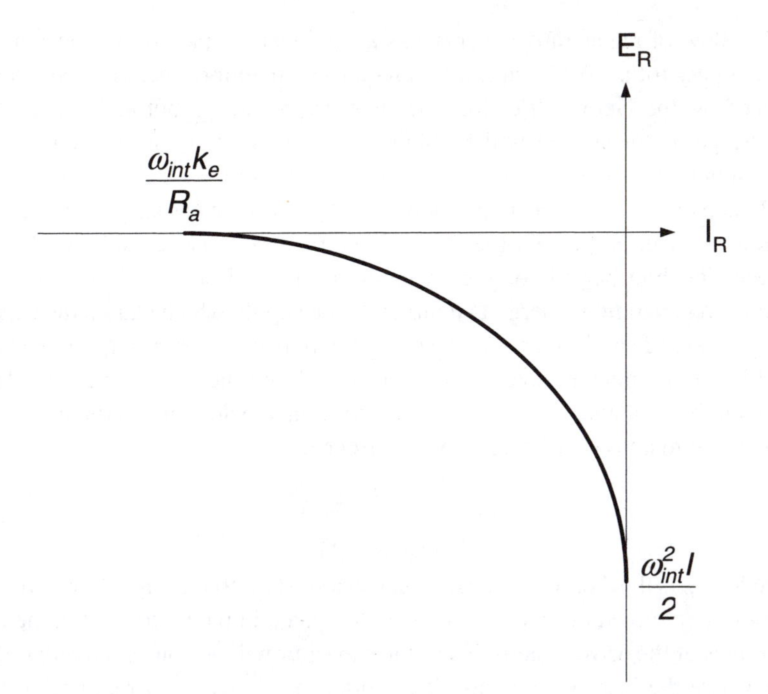

Figure 5.20. The regenerative energy as a function of the armature current.

$$K_t I_R = T_d = I_{tot}\alpha = -\frac{I_{tot}\omega_{int}}{t_z} \tag{5.29}$$

Hence it is possible to determine that

$$E_r = -\frac{I_{tot}\omega_{int}}{K_t}\left(\frac{\omega_{int}K_e}{2} + \frac{I_R^2 R_a^2}{2\omega_{int}K_e} + I_R R_a\right) \tag{5.30}$$

The first two terms in the parentheses are negative and represent the energy dissipation outside the motor, while the third term is positive and represents the energy dissipated in the armature resistance. The regenerative energy is plotted against the regenerative current, Figure 5.20, and the following points should be noted:

- Zero regenerative energy will be returned to the supply when the regenerative current equals $\omega_{int}K_e/R_a$.
- The maximum energy is returned when I_R is zero, when the returned regenerative energy is returned is equal to the total rotational energy of the system.
- The regenerative energy returned to the drive will increase as the regenerative current decreases.

The flow of regenerative energy has a significant impact on the design of the power supply for a PWM amplifier. In motoring operations there is a bidirectional current flow; the average direction is from the supply to the motor. At certain times the average motor current will be higher than the average current drawn by the power supply; this is accommodated by the energy-storage capability of the power-supply capacitors. The capacitor, acting both as a filter and an energy reservoir, needs to be sized to prevent the d.c. voltage rising to unacceptable levels during the free-wheeling periods of the switching cycles. However, during periods of sustained regeneration, energy is returned to the supply which causes the capacitor voltage to significantly increase. Using the anticipated energy requirements, it is possible to determine the size of the capacitor which is necessary to ensure that the maximum bus voltage, V_{max}, is restricted to a value below the continuous voltage rating of the main power-bridge devices. Hence

$$C \geqslant \left| \frac{-2E_r}{V_{max}^2 - V_{norm}^2} \right| \tag{5.31}$$

where V_{norm} is the d.c. link voltage under steady-state conditions. As noted earlier, some of the returned energy, E_R, will be dissipated in the motor and in the power electronics of the drive system; hence the capacitor will be conservatively sized. In cases where the load being controlled is high speed inertia, or requires sustained regenerative braking, the value of this capacitor will become excessive; therefore, a bus-voltage regulator will have to be considered.

The conventional solution to the use of large power-supply capacitors is the use of a shunt voltage regulator (shown in Figure 5.21). The regulator consists of a resistor and a semiconductor switch fitted directly across the d.c. bus; switching of the device is controlled by a comparator. The comparator's threshold voltage must be selected to ensure that the regulator is not switched on when the supply voltage is at its maximum permitted value during normal operation. In the determination of this value, the effects of the supply's voltage regulation have to be considered. The value and the ratings of the dissipative element and the semiconductor switch need to be selected to ensure satisfactory operation. Since the power supply voltage must not exceed V_{max}, during deceleration, the resistor must be capable of sinking the peak regenerative current at that voltage. The maximum size of the resistor is given by

$$R \leqslant \frac{V_{max}}{I_R} \tag{5.32}$$

and the average power that has to be dissipated can be determined from a knowledge of the load and the duty cycle

$$P_{resistor} = -\frac{E_r}{\text{Total cycle time}} \tag{5.33}$$

where the total cycle time is the sum of the acceleration, constant speed, deceleration, and standstill times. The resistors used in bus-voltage regulators are rated

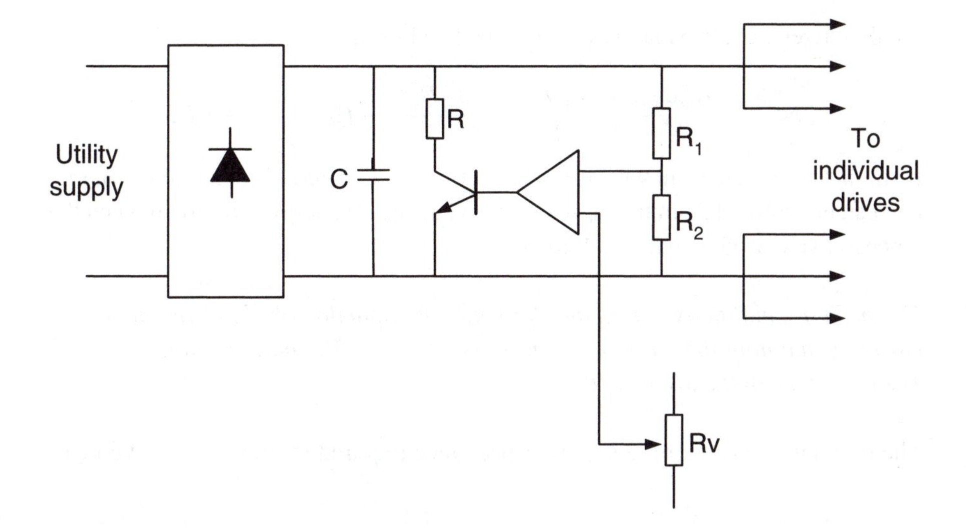

Figure 5.21. A single phase power supply for a drive system, including a shunt voltage regulator. The switching voltage is determined by the varible resistor, Rv. The shunt resistor is identified as R.

for a peak power which is considerably in excess of their continuous rating; this makes the use of wire-wound resistors widespread. The sizing of the semiconductor switch is based on peak current and load requirements, and its cooling requirements are determined by the expected load duty cycle.

Example 5.2

Consider a system with inertia, $I_{tot} = 2.5 \times 10^{-4}\ kg\ m^2$ *running at* $\omega_{int} = 3000\ rev\ min^{-1}$. *The system is driven by a motor with the following parameters;* $R_a = 2\ \Omega$, $K_e = 0.6\ V\ rad^{-1}s$ *and* $K_t = 0.6\ NmA^{-1}$. *If the regenerative current is held at −30 A, determine the time the system takes to reach standstill, and the energy returned to the supply.*

Using equation (5.29) it is possible to calculate the time to reach standstill

$$t_z = \frac{I - tot\omega_{int}}{K_t R_a} = 4.36 \times 10^{-3}\ \mathrm{s}$$

this in turn allows the determination of the time at which the motor terminal voltage equals zero, from equation (5.26)

$$t_0 = t_z \left(1 + \frac{I_R R_a}{\omega_{int} K_e}\right) = 2.97 \times 10^{-3}\ \mathrm{s}$$

finally the regenerative energy can be calculated using

$$E_r = -\frac{I_{tot}\omega_{int}}{K_t}\left(\frac{\omega_{int}K_e}{2} + \frac{I_R^2 R_a^2}{2\omega_{int}K_e} + I_R R_a\right) = -5.7 \text{ J}$$

If the limiting values shown in Figure 5.20 are calculated, the maximum energy that can be returned from the drive is -12.34 J, and no energy is returned when the regenerative current equals -94.25 A.

Given these conditions determine the value of capacitor or shunt resistance required to maintain the bus voltage at less that 110 V. The normal rating is 100 V. The cycle time of the drive is 10 s.

The minimum values of the supply capacitance required to prevent over–voltages is

$$C \geqslant \left|\frac{-2E_r}{V_{max}^2 - V_{norm}^2}\right| \geqslant 5460\ \mu\text{F}$$

If a shunt voltage regulator, Figure (5.21) is to be used, the maximum value of the resistor required will be

$$R = \frac{V_{max}}{I_R} = \frac{110}{30} = 3.7\ \Omega$$

The average power dissipation of the resistor will be

$$P_{average} = \frac{5.7}{10} = 0.57 \text{ W}$$

this compares with the instantaneous power rating of 3.3 kW.

5.5 Summary

Direct-current brushed motors are widely used in the servo-drive industry due to their simplicity of operation and reliability. This chapter has reviewed the basic electrical and mechanical relationships of the four main types of d.c. brushed motor, together with their control. The pulse width modulated servo amplifier have been considered in detail. In addition the problems of regenerative energy have been addressed, together with methods used to ensure that damage is not caused to the semiconductors with the power amplifier.

Chapter 6

Brushless motors and controllers

The application of brushless motors, of all types, is becoming of considerable importance to the designers of machine tools and robotic systems. This is largely due to the benefits that these types of motor bring to a system, particularly in improvements in the reliability. If the term is used in its broadest sense, this classification will include permanent-magnet motors, stepper motors, and alternating current (a.c.) induction motors. This chapter is concerned with permanent-magnet brushless motors and their associated controllers; a.c. asynchronous induction motors and stepper motors will be considered in Chapters 7 and 8, respectively.

Within the market place, there appears to be a degree of confusion in the naming of motors, with given constructions. To prevent confusion, the following motors and their associated controllers are considered in this chapter.

- Permanent-magnet synchronous motors with a trapezoidal winding distribution, commonly known as d.c. brushless motors. The associated controller consists of a conventional three-phase bridge, whose switching pattern is determined by a low-resolution rotor-position sensor.

- Sinewave-wound permanent-magnet synchronous motors incorporate windings with an approximately sinusoidal distribution which are supplied with sinusoidal currents. These motors are normally controlled by a version of vector control, which has considerable similarities with the method used to control the asynchronous induction motors discussed in Chapter 7. In order to achieve the required control resolution, these motors are fitted with a resolver or a similar high-resolution position transducer.

- Linear brushless motors are making significant inroads to the drives market; based on conventional brushless technology, they made an ideal replacement for high performance applications that would have been previously based on leadscrews and ballscrews.

As will be discussed, the operation of these permanent-magnet brushless motors is totally dependent on their associated electronics; so as the reliability and

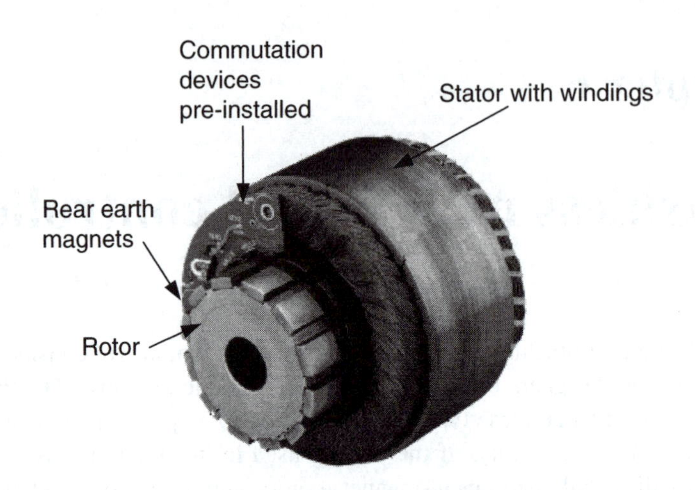

Figure 6.1. Key features of a frameless brushless dc motor.

availability of power electronic devices and specialist integrated circuits has improved over the last twenty years the number of applications has also increased. Due to these motors' high reliability and low-maintenance requirements, they are ideally suited to a wide range of applications including computer cooling fans and robotic drives.

As with brushed motors, the brushless motor can be obtained in a number of versions, not only in is electromagnetic characteristics as summarised above, but also in its mechanical construction. The motors can be supplied either as a conventional framed or a frameless motor. The general arrangements of frameless brushless motors is shown in Figure 6.1. A three-phase stator winding is constructed in a similar fashion to that of an a.c. induction motor; it is wound to give a trapezoidal air-gap flux in the case of a d.c. brushless motor, or with a sinusoidal distribution in the case of a sinewave-wound, permanent-magnet, synchronous motor. The rotor consists of a number of high-performance permanent magnets rigidly fixed to the rotor's core structure; the arrangement, shapes, and location of the magnets can be modified to give a range of motor characteristics. One of the problems of the construction of a permanent magnet rotor is the possibility of the failure of the magnet-rotor bond at high rotational speeds and accelerations. The preferred solutions to this problem include encasing the rotor in a thin stainless-steel jacket, or binding the outer surface with a glass-fibre or similar non-metallic yarn. In addition, a suitable adhesive should be used; and, to prevent problems, the compound which is selected should be thermally stable and it should have a linear-expansion coefficient which is close to that of the magnets and the rotor material.

The magnetic material selected for a motor is largely determined by the required output specifications; in high-performance motors neodymiumiron-boron, NdFeB, usually has the highest energy product of commercially available magnets, typically 20 kJm^{-3}, whereas ferrite has an energy product of 200 Jm^{-3} . To obtain a high flux density in the air gap, the magnet's flux density and the pole face area need to be considered in considerable detail during the electromagnetic design process for the motor. In practice, the limiting factors are the volume of the magnetic material required for ferrite magnets, or the high cost of the material for NdFeB; hence careful optimisation of the design is necessary. The net result is a small permanent-magnet motor, when compared with brushed d.c. or a.c. induction motors with a similar power output.

When compared with the brushed d.c. motors, the advantages of the brushless design are readily apparent:

- The construction of the motor, with the heat-generating stator windings on the outside of the motor frame, allows direct heat dissipation to the environment, without heat flowing through bearings and across the air gap.
- Any possibility of sparking is eliminated by the removal brushes; this allows the motors to be used in hazardous environments, and there is a considerable reduction in the radio-frequency interference (RFI) which is generated.
- Maintenance costs are reduced, both for brush replacement and for problems resulting from the dust which is generated by brush wear.
- The speed-torque restrictions caused by the commutation limit, as found in d.c. brushed motors, are eliminated.

However, these advantages do not come without a corresponding set of disadvantages. In a d.c brushed motor, the commutation of rotor currents is undertaken by the mechanical arrangement of the commutator and brush gear. In brushless motors this mechanical system is replaced by an electronic commutator comprising a three-phase power bridge, a rotor-position encoder with a suitable resolution, and commutation logic to switch the bridge's devices in the correct pattern to produce a motoring torque (see Figure 6.2).

As part of any selection procedure, it is necessary to compare brushless motors against their main competitors. Compared with brushed d.c. motors they are more expensive, but they are also smaller, easier to maintain, and more reliable, and there is the additional complexity of the three-phase drive. Compared with induction motors they are again smaller and more expensive, but the power electronic design is identical. For the majority of motor types, the speed and torque performance are almost identical when complemented by high-performance controllers. Hence permanent-magnet brushless motors are widely used for power outputs of up to 20 kW; above this power level, vector-controlled induction motors

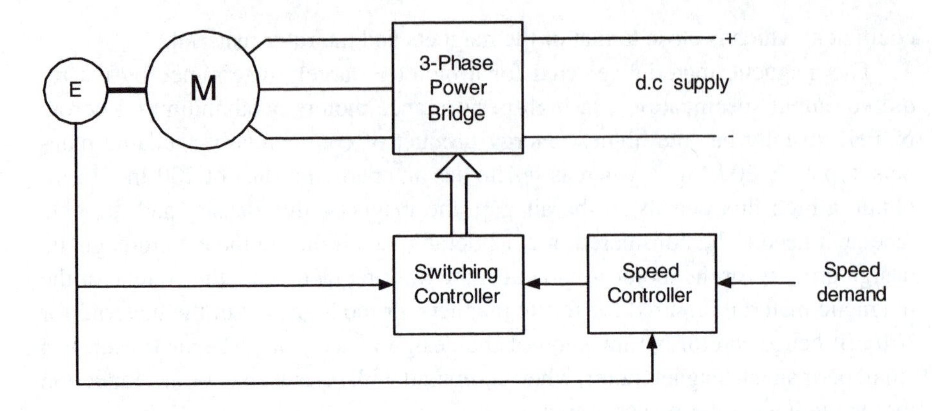

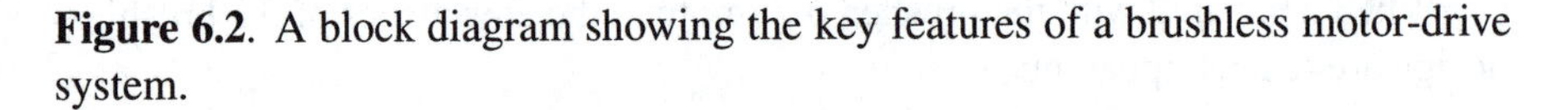

Figure 6.2. A block diagram showing the key features of a brushless motor-drive system.

are predominating for specialist applications. As with all design problems, the final selection of the system is left to the system designer, who is able to balance the relative advantages and disadvantages on an objective basis.

6.1 The d.c. brushless motor

The basic torque and voltage equations of d.c. brushless motors closely resemble those of d.c. brushed motors. This section presents a simple analysis for determining a motor's characteristics; this also allows an appreciation of its limitations. In this analysis, a simple two-pole motor is considered (see Figure 6.3); the key features of this design are the rotor's magnetic pole arc of 180°, and a three-phase stator winding with two slots per phase and N turns per slot, (Miller, 1989). The air-gap flux, neglecting any fringing, can be considered to be a square wave (see Figure 6.4(a)). As the rotor is rotated, a voltage is induced within the stator windings, and the flux linkage varies linearly as a function of the rotor position, with the maximum positive linkage for the winding occurring at $\theta = 0°$ and the maximum negative flux linkage occurring at $\theta = 180°$.

If a single phase is considered, the total flux linkage can be determined by integrating the contribution of individual turns, to give a maximum flux linkage of

$$\psi_{max} = NB_g \pi r l \tag{6.1}$$

where B_g is the air-gap flux density and l and r are the length and radius, respectively, of the stator. If the rotor's position is now considered, the flux linkage is given as a function of position by

$$\psi(\theta) = \left(1 - \frac{\theta}{\pi/2}\right) \psi_{max} \tag{6.2}$$

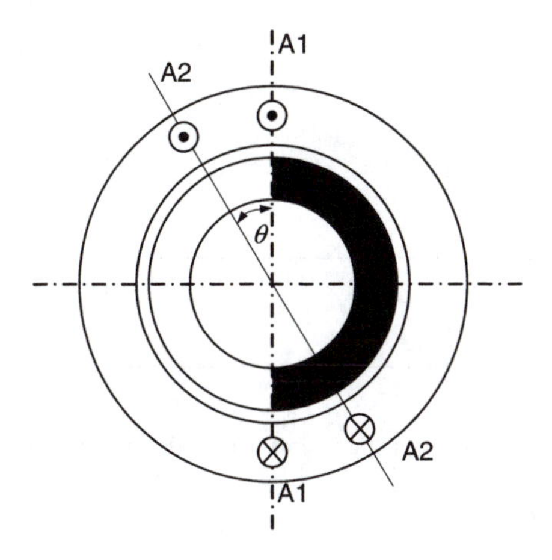

Figure 6.3. Cross section of idealised brushless motor, showing two coils for a single phase, $\theta = 30°$.

$\psi(\theta)$ is shown in Figure 6.4(b) for both coils. From the flux linkage, the instantaneous e.m.f. induced in the coil can be determined in the conventional manner as

$$\begin{aligned} e = & -\frac{d\psi}{dt} = -\frac{d\psi}{d\theta}\frac{d\theta}{dt} = -\omega_m \frac{d\psi}{d\theta} \\ = & N B_g \pi r l \omega_m \end{aligned} \tag{6.3}$$

where ω_m is the mechanical rotational speed in rad s^{-1}. To obtain the total back electromotive force (e.m.f.) for the individual windings, the contribution from both coils needs to be considered; this gives

$$e_p = N_p B_g \pi r l \omega_m \tag{6.4}$$

where N_p is the number of turns per phase; it is equal to $2N$ for this particular motor. The phase e.m.f. is shown as a function of the rotor position in Figure 6.4(c); it is the sum of the voltages for the two windings, which are displaced by 30°, Figure 6.4(d). In the case of the motor under consideration; the length of constant portion of the e.m.f. waveform is theoretically 150°, but due to the construction of the motor and magnetic fringing, it is in practice closer to 120°.

To control the power to a brushless d.c. motor, a three-phase bridge, Figure 6.5(a), is used. With the motor star connected, only two phases can carry a current at any one time, and hence only two devices need to conduct in any one switching period. The idealised phase currents are shown in Figure 6.5(c); they are 120° wide, with a peak magnitude of I. The switching pattern is arranged to give a current flow against the e.m.f.; a positive current is defined as a motoring current. The device switching sequence in Figure 6.5(d) is arranged to produce a balanced three-phase motor supply.

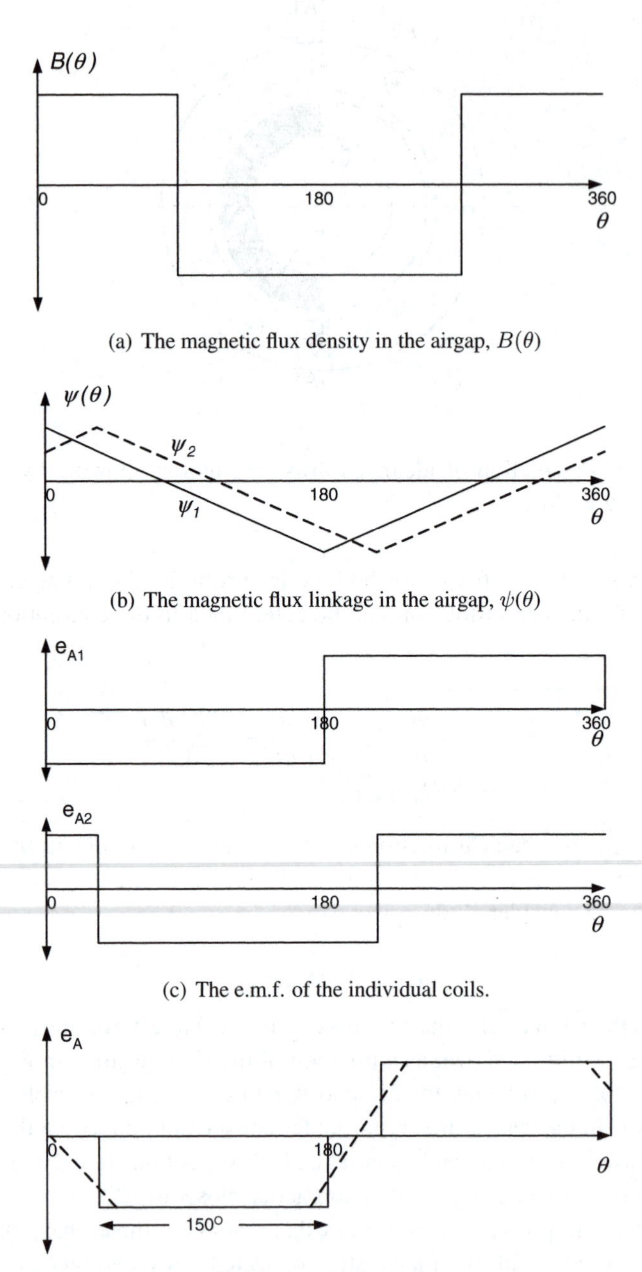

(a) The magnetic flux density in the airgap, $B(\theta)$

(b) The magnetic flux linkage in the airgap, $\psi(\theta)$

(c) The e.m.f. of the individual coils.

(d) The e.m.f. of a phase.

Figure 6.4. The waveforms for an idealised brushless motor, adapted from Miller (1989).

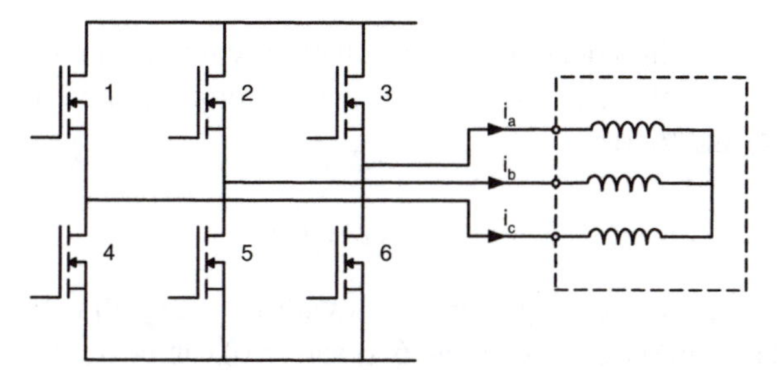

(a) The relationship between the power bridge and motor windings.

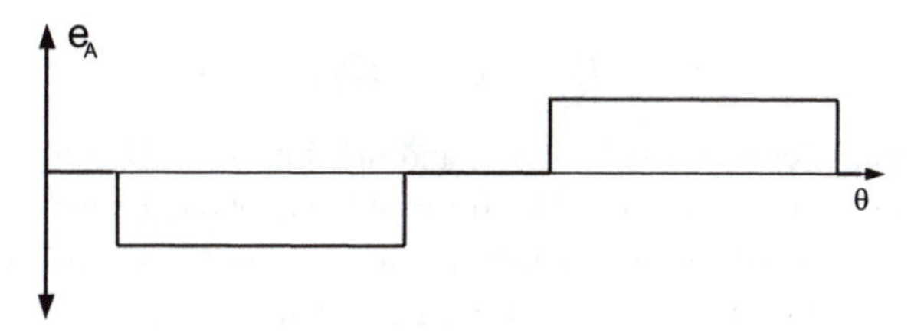

(b) The e.m.f. for phase A as a function of poition.

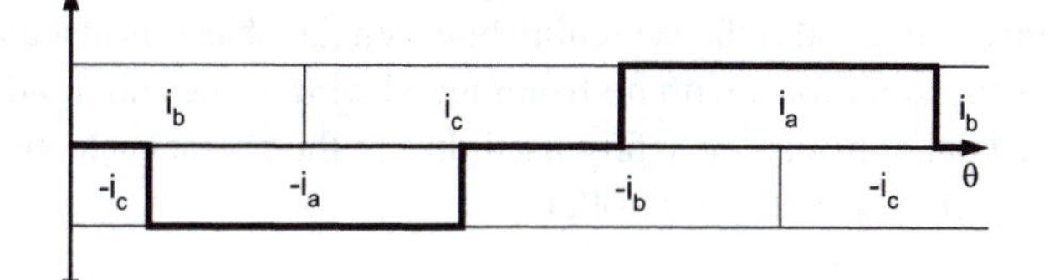

(c) Ideal current for the *A* phase.

2,6	2,4	3,4	3,5	1,5	1,6	2,6

(d) The switching pattern for the three phase bridge.

Figure 6.5. The switching requirements for an ideal brushless motor, adapted from Miller (1989).

The torque and the speed constants can be determined by considering the balance between the motor's mechanical output power and its electrical input power over each conduction period; that is

$$P = \omega_m T_e = 2e_p I \tag{6.5}$$

The factor of two in this equation is the result of the current simultaneously flowing through two motor phases. The electromagnetic torque can be determined from equations (6.4) and (6.5)

$$T_e = 4N_P B_g lrI \tag{6.6}$$

Rewriting the voltage and torque equations with $E = 2e_p$, to represent the e.m.f. of any two phases in series, equations (6.4) and (6.6) can be rewritten as

$$E = k\psi\omega_m \equiv K'_v\omega_m \tag{6.7a}$$

$$T_e = k\psi I \equiv K'_T I \tag{6.7b}$$

where the armature constant, $k = 4N_P$, and the flux, $\psi = B_g\pi rl$, are determined by the construction of the motor. The form of these two equations is very similar to the corresponding equations for brushed d.c. motors (see equation (5.1)); this explains, to a large extent, why these motors are called brushless d.c. motors within the drives industry, when they are more correctly described as permanent-magnet synchronous motors with a trapezoidal flux distribution. In practice, the equations above will only hold good if the switching between the phases is instantaneous, and if the flux density is uniform with no fringing; while this does not hold true for real motors, these equations can be safely used during the normal selection procedure for a motor and its associated controller.

6.1.1 Torque-speed characteristics

Using the relationships above, the torque-speed characteristics of an ideal d.c. brushless motor can be determined; it is assumed that the commutation and back e.m.f. voltage waveforms are perfect, as shown in Figure 6.5(b) and Figure6.5(c). If the star-connected motor configuration is considered, the instantaneous voltage equation can be written as

$$V_s = E + IR \tag{6.8}$$

where R is the sum of the individual phase resistances, V_s is the motor's terminal voltage (neglecting semiconductor and other voltage drops), and E is the sum of two phase e.m.f.'s. Using the voltage, torque, and speed equations discussed above, the motor's torque-speed characteristics can be determined. The torque-speed relationship is given by

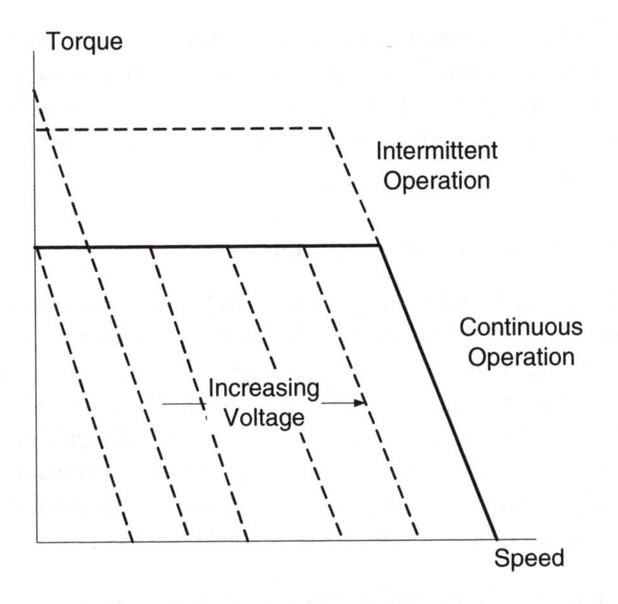

Figure 6.6. The torque speed characteristics of an ideal brushless d.c. motor.

$$\omega_m = \omega_0 \left(1 - \frac{T}{T_0}\right) \tag{6.9}$$

where the no-load speed is defined by,

$$\omega_0 = \frac{V_s}{k\psi} \tag{6.10}$$

The stall torque of the motor is given by,

$$T_0 = k\psi I_0 \tag{6.11}$$

where,

$$I_0 = \frac{V_s}{R} \tag{6.12}$$

The resultant characteristics, shown in Figure 6.6, are similar to those for a conventional d.c., brushed, shunt motor. The speed of the motor is determined by the terminal voltage, and under load it will be a function of the winding resistance. As the terminal voltage is modified, the family of curves shown in Figure 6.6 will result; in a practical application, terminal-voltage control is normally achieved by the use of pulse-width modulation. As with any other motor, the continuous and intermittent operational limits are determined by the maximum power dissipation of the motor, and the temperature limits imposed by the insulation of the winding. In contrast to d.c. brushed motors, which have limits imposed by the commutator's operation, the peak torques of brushless motors can be developed to the peak speed, subject to any power-dissipation restrictions. The characteristics will be degraded in real motors by the effects of winding inductance, armature reactance and non-uniform flux distribution. In order to undertake a full analysis of the characteristics of a d.c. brushless motor, a full electromagnetic analysis must be undertaken using a computer-aided design (CAD) package.

6.1.2 Brushless d.c. motor controllers

As discussed above, a brushless d.c. motor controller is based on a standard three-phase, six-device power bridge, while high-powered systems will use bridges constructed from discrete components; advances in power-electronics integration techniques have resulted in a range of smart power controllers for small-drive applications. In all cases the control of brushless d.c. motors depends on an ability to control the winding currents relative to the rotor's position, to obtain the switching pattern shown in Figure 6.5(d). The block diagram of a suitable controller is shown in Figure 6.7; the controller consists of the following elements:

- A low-resolution, rotor-position measurement system.
- Commutation logic to determine the main power device's switching pattern.
- Speed controller incorporating the pulse-width-modulator.
- A three-phase power bridge.

Rotor-position measurement

Since a motor's output performance largely depends on the accuracy of a power bridge's switching, relative to the phase voltages, a reliable and accurate rotor-position-measurement system is required. This can be achieved by the use of Hall-effect devices, by terminal voltage measurement for a limited range of applications, and by resolvers and encoders for specialist high-performance applications. In practice, Hall-effect devices are the most widely used.

Hall-effect devices are low-cost magnetic sensors, whose principle of operation is shown in Figure 6.8(a). When a magnetic field is applied to a piece of a current-carrying semiconductor, a voltage which is proportional to the magnetic field applied is generated, as long as the current is held constant. The directions of the applied magnetic field, the current, and the Hall voltage are mutually perpendicular. The digital devices used in brushless d.c. motors consist of Hall-effect devices enhanced by the addition of a Schmitt-trigger to give a known switching point (see Figure 6.8(a)); the resultant hysteresis ensures a positive switching characteristic (see Figure 6.8(b)). The resultant digital output is suitable for use with the commutation logic with a minimal interface. Hall-effect devices have a number of advantages, including the capability of operating at frequencies in excess of 100 kHz, high reliability, and low cost. In most applications, Hall-effect devices

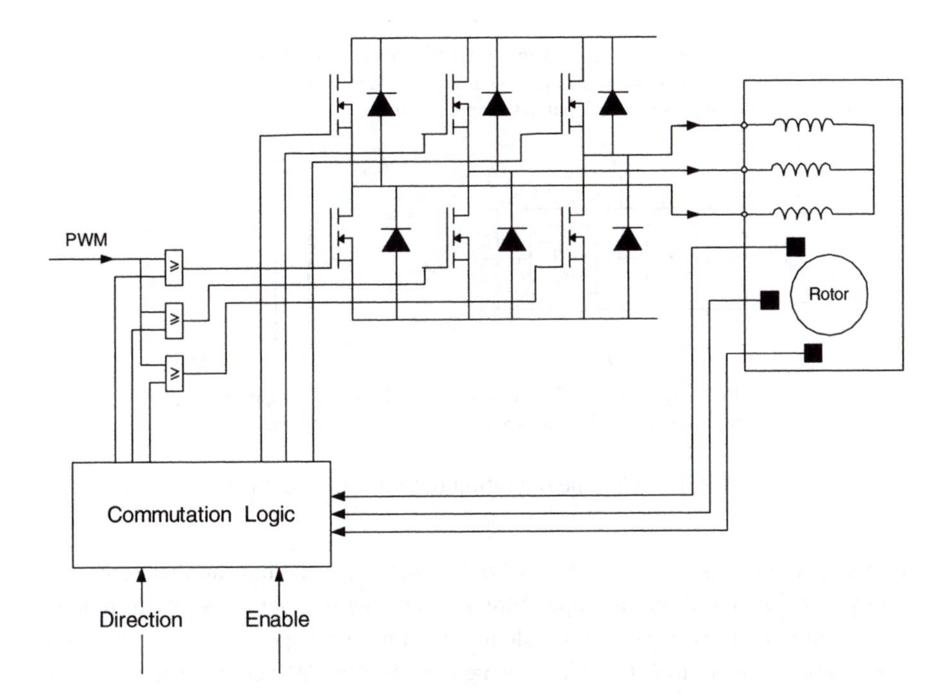

Figure 6.7. Block diagram of a controller for a brushless d.c. motor. The speed controller has been omitted for clarity, the circuit used can be based on that shown in Figure 5.17.

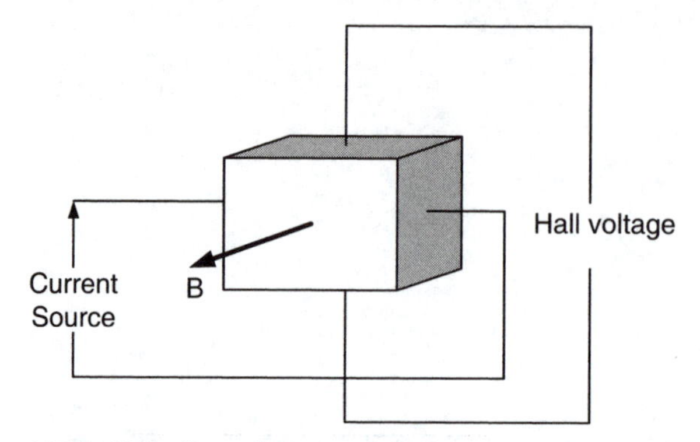

(a) The Hall-effect, when a magnetic flux is applied to a piece of semiconductor carrying a current, a voltage results. The Hall voltage is proportional to the applied flux.

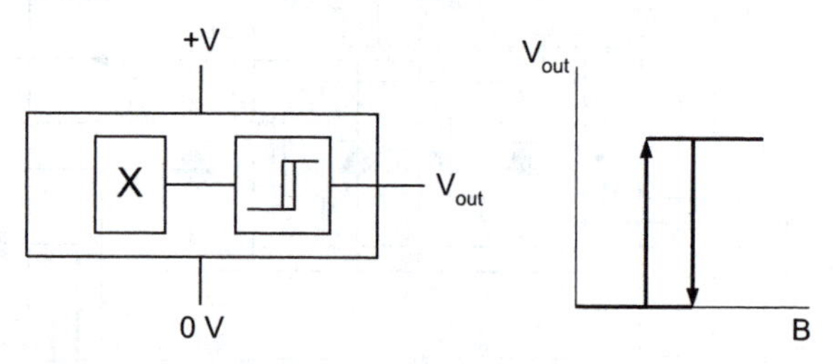

(b) A digital Hall-effect devise, a Schmitt trigger is applied to the output of the Hall device to give a clean waveform.

Figure 6.8. The operation of Hall-effect devices.

normally directly sense the rotors's field; however, in certain applications (normally associated with high-temperatures), a separate magnet assembly is attached to the rotor shaft. While it is possible to obtain military-specification devices that are capable of operating at 150°C, in general Hall-effect devices have temperature-sensitivity problem, due to their physical characteristics.

As noted earlier, Hall-effect devices are used in combination with the commutation logic to provide the required drive signals for the power bridge. Satisfactory operation requires the mechanical separation of the three sensors to be given by:

$$\frac{360^\circ}{\text{Number of phases} \times \text{Number of pole pairs}} \tag{6.13}$$

In the case of a two-pole, three-phase, d.c., brushless motor, a mechanical displacement of 120° between individual Hall-effect devices is required. In practice Hall-effect devices are normally mounted on a small printed circuit board that is factory fitted to the motor; while this is a robust construction, the relative position between the sensors and the magnets is fixed and it cannot be optimised at a later date.

While d.c. brushless commutation logic normally only requires the determina-

Table 6.1. Typical commutation sequence which should be used in conjunction with Figure 6.7, for device identification

Hall device			Active power device			
A	B	C	Forward		Reverse	
0	0	1	2	4	1	5
1	0	1	2	6	3	5
1	0	0	1	6	3	4
1	1	0	1	5	2	4
0	1	0	3	5	2	6
0	1	1	3	4	1	6

tion of three points within 360° electrical degrees, resolvers or encoders are used in high-performance d.c. brushless-motor applications. With the use of high resolution position measurement, it is possible to correct for any inaccuracies resulting from the manufacture of the motor, and therefore to maximise the output. The operation of resolvers and absolute optical position encoders was considered in Chapter 4. The use of this approach, while beneficial in certain applications, does negate the relative simplicity of d.c. brushless motors. If a very-high performance approach is required then a sinewave-wound machine should be considered.

Commutation logic

Commutation logic is used to determine the switching sequence of the power circuit. The pattern is developed from the design of the motor, from the required direction of motion, and from the positional information obtained from the rotor-position-measurement system. An example of a suitable truth table is given in Table 6.1. In this example, the Hall-effect devices are separated by 120 electrical degrees, for a two-pole motor.

The logic to decode the sensor output can either be implemented as discrete logic, or more commonly as a customised logic-gate array. In a commercial device, a number of additional features are normally provided, particularly the ability to operate with d.c. brushless motors of different construction, for example, three-phase motors with the Hall-effect devices separated by either 30, 60, or 240 electrical degrees, and four-phase motors with a separation of 90 electrical degrees. In addition, most commutation-logic devices provide a facility for totally disabling the power bridge, and are capable of directly driving power-bridge devices with a minimum of additional circuitry.

Speed controller

The speed control of a d.c. brushless motor is undertaken by the control of the motor's terminal voltage; this is normally achieved by PWM of the supply voltage.

If a d.c.-brushless-motor-commutator integrated circuit is used, the PWM switching waveform can be directly gated with the commutation switching pattern; but, in practice, only the lower devices need to be controlled. As with d.c.-brushed-motor servo amplifiers, PWM can be undertaken by either subharmonic or current-controlled hysteresis techniques, as discussed in Section 5.3.3. As discussed earlier, the characteristics of brushless d.c. motors are very similar to those of brushed motors; hence it is possible to control these motors over a wide speed and torque range using a conventional analogue control loop.

Power-bridge circuits

The power circuit for a d.c brushless drive consists of a conventional six-device, three-phase, power bridge, as shown in Figure 6.7, the devices used will depend on the rating of the drive, but for small applications MOSFETs (metal-oxide semiconductor field-effect transistors) predominate. As with the PWM bridge discussed in Section 5.3.5 the bridge is provided with a number of auxiliary circuits to ensure protection against over-voltages, under-voltages, fault currents, and excessive device temperatures. When the motor regenerates, the energy which is returned will cause the bus voltage to rise; this excess energy can be dissipated by the use of a conventional bus-voltage regulator, as discussed in Section 5.4.

6.2 Sinewave-wound brushless motors

Sinewave-wound permanent-magnet brushless motors have a number of significant differences when compared with the trapezoidally wound d.c. brushless motors which affect their detailed construction and analysis. These motors' main characteristics are (Miller, 1989):

- The air gap flux is sinusoidal, generated by a number of specially shaped rotor magnets.
- The windings have a sinusoidal distribution.
- The motor is supplied with three-phase sinusoidal current.

While the analysis is more complex than for d.c. brushless motors, it must be considered in some detail if the operation of these motor is to be fully understood. The cross section of an idealised sinewave-wound brushless motor is shown in Figure 6.9. If the number of turns per pole, N_P, is given by

$$N_p = \frac{N}{2} \tag{6.14}$$

and given an ideal winding distribution, the number of conductors within the angle $d\theta$ of a p-pole-pair motor, at a position θ, is given by

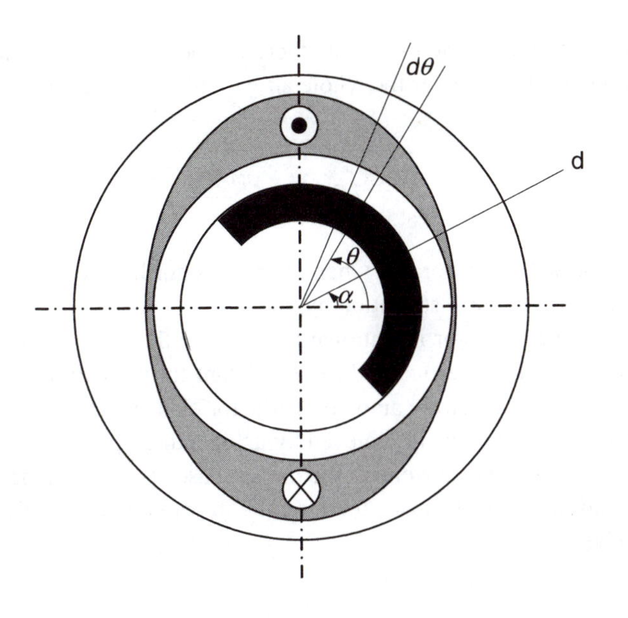

Figure 6.9. Cross section of an ideal sinewave-wound permanent magnet machine.

$$\frac{N \sin p\theta d\theta}{2} \tag{6.15}$$

The sinusoidal flux distribution within the air gap is provided by specially shaped rotor magnets. The rotor-flux distribution is centred on the north axis, which can be considered to be displaced by α radians from the axis of the stator winding; and is given by

$$B(\theta) = \hat{B} \cos(p\theta - \alpha) \tag{6.16}$$

Using the above relationships, it is possible to determine the torque and speed characteristics of a sinusoidally wound permanent-magnet motor by the application of conventional electromagnetic theory.

6.2.1 Torque characteristics

The force on a group of stator conductors of length L, within the angle dθ of the stator, is determined by the product of the flux and the stator current, i;

$$F = \hat{B}il\frac{N \sin p\theta \cos(p\theta - \alpha)d\theta}{2} \tag{6.17}$$

The resultant torque on a rotor of radius r, including the contribution of the opposite winding element, is

$$T = -2Fr \tag{6.18}$$

The total output torque of the motor can therefore be determined by integration of the elemental contribution over the whole air gap; giving for a *p*-pole pair motor

$$T = -\int_0^{\pi/2} 2Frd\theta = \frac{\pi r \hat{B} l i N \sin\alpha}{2} \tag{6.19}$$

The peak torque will be generated when the rotor's north axis lags the axes of the stator's ampere-conductor distribution by 90°.

The case where the rotor is stationary relative to the stator was considered in the analysis above. In order to produce a constant torque with the rotor rotating at a constant speed, the stator's ampere-conductor distribution must rotate in synchronism with the rotor. This is achieved by using a three-phase winding supplied with a three-phase balance current. If the r.m.s. phase current is *I*, then for a motor where the winding are 120° apart, the rotating ampere conductor distribution can be shown to be

$$\begin{aligned} &\hat{\imath}\cos\omega t\frac{N}{2}\sin p\theta \quad + \\ &\hat{\imath}\cos\left(\omega t - \frac{2\pi}{3}\right)\frac{N}{2}\sin\left(p\theta - \frac{2\pi}{3}\right) + \\ &\hat{\imath}\frac{N}{2}\sin\left(p\theta + \frac{2\pi}{3}\right)\cos\left(\omega t + \frac{2\pi}{3}\right) = \frac{3\sqrt{2}I}{4}N\sin(p\theta - \omega t) \end{aligned} \tag{6.20}$$

The rotating magnetic-flux distribution is given by

$$B(\theta) = \hat{B}\cos(p\theta - \omega t - \alpha) \tag{6.21}$$

If this is combined with equation (6.19), the output torque can be calculated to be

$$T = \frac{3}{2}\sqrt{2}I\frac{\pi r l \hat{B} N}{2}\sin\beta \tag{6.22}$$

The angle β, which equals $-\alpha$, is termed the torque angle, and it is held positive for motoring; any variation in β will require adjustment of the phase current to hold a constant torque. This equation shows that the peak torque, and hence a motor's efficiency, is optimal when $\beta = \pi/2$. To ensure that the ampere-conductor distribution remains in synchronism with the rotor's magnetic field, the stator's supply frequency, f, is made equal to the rotor's rotational frequency, ω_s, hence

$$\omega_s = 2\pi f \tag{6.23}$$

which is related to the motor's mechanical angular velocity, ω_m, by

$$\omega_m = \frac{\omega_s}{p} \tag{6.24}$$

In order for the torque angle to be kept constant (and hence for the stator's ampere distribution to be kept in synchronism with the rotor), use is made of a vector or field oriented approach to the control; this requires a resolver or a similar high-performance position-measurement system to be fitted to the motor.

6.2.2 Voltage characteristics

The e.m.f. of a sinusoidally wound brushless motor can be determined in a manner which is similar to that used for the determination of the torque relationships. For the same elemental group of conductors in a machine with *p*-pole pairs, the contribution to the back e.m.f. of an elemental portion of the winding is given by

$$\delta e = \frac{B(\theta) l \omega_m N \sin p\theta \delta\theta}{2} \tag{6.25}$$

Using the flux relationship given in (6.21) and integrating, the r.m.s. phase e.m.f. is given by

$$E_p = \frac{N r \hat{B} l \omega_m \pi}{2\sqrt{2}p} \tag{6.26}$$

and the line-to-line voltage by $\sqrt{3}E_p$.

6.2.3 Torque-speed characteristics

Ideal motors were considered in the analysis above; in practice, the construction of the stator windings, and particularly the effect of the stator's slots, has a significant effect on the motor's performance and characteristics. In addition, the location of the magnets, either mounted on the surface or within the body of the rotor, has to be considered in detail. It is normal to undertake a detailed modelling of this type of motor using electromagnetic CAD packages (Hendershot and Miller, 1994).

If the analytic method considered above is extended to include the effects of the construction of the windings, it can be shown that the r.m.s. phase e.m.f. is given by

$$\frac{E_p \omega_s \psi_m}{\sqrt{2}} \tag{6.27}$$

and the output torque is given by

$$T = \frac{3\pi p I \sqrt{2} \psi_m \sin\beta}{8} \tag{6.28}$$

The flux linkage provided by the rotor-mounted magnets is given by

$$\psi_m = k N_p B_m \tag{6.29}$$

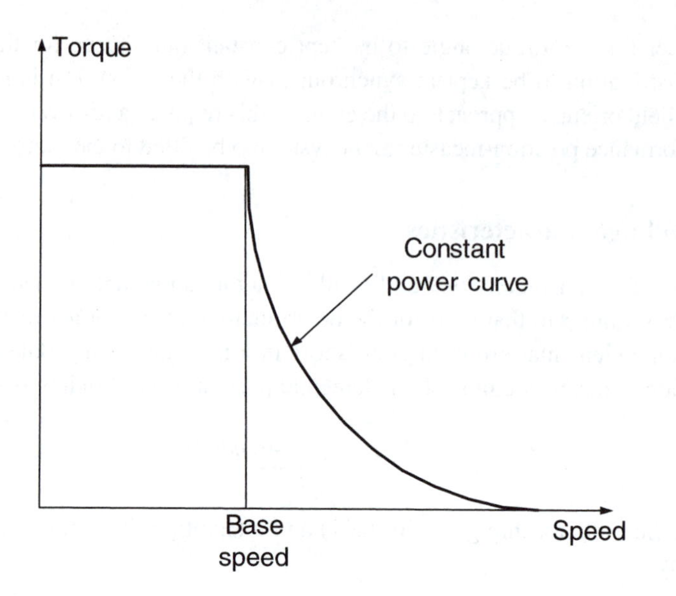

Figure 6.10. The torque speed characteristics of a sinewave wound machine. Above base speed β can be adjust to increase motor speed, however a constant power limit as defined by equation (6.31) will apply.

where k is a constant which is introduced to accommodate the physical construction of the stator windings, and B_m is the air-gap flux density. The torque equation can also be expressed in the form;

$$T = \frac{pE_p I \sin\beta}{\omega_s} \tag{6.30}$$

and hence,

$$\omega_m T = 3E_p I \sin\beta \tag{6.31}$$

This verifies that the product of the back e.m.f. and the phase current is equal to the input power at $\beta = \pi/2$; therefore, the ability to control this angle is considered to be critical to the satisfactory performance of the motor.

The overall torque-speed characteristics of the motor derived from this equation is shown in Figure 6.10. The peak torque can be maintained up to the base speed. Above this speed, by modifying β, the motor will effectively enter a field-weakening mode, allowing an increase in the speed at the expense of the peak torque. The motor's efficiency is reduced in this region because the motor is being supplied with the peak current.

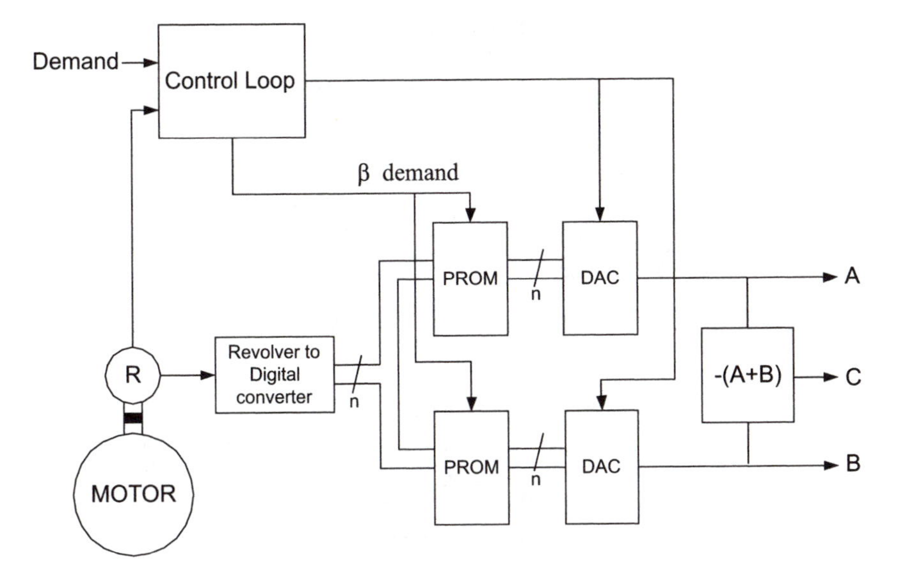

Figure 6.11. The block diagrams of a sinewave wound machine controller, using a resolver as the position transducer. The control loop provides two signal, an analogue input to the multiplying DACs and β as digital word to the PROMs to allow operation above base speed. The three outputs (A,B and C) are fed via a suitable power amplifier to the motor.

6.2.4 Control of sinewave-wound brushless motors

The block diagram for a simple hardware-based controller for a sinewave-wound motor is shown in Figure 6.11; it can be seen to be superficially similar to that for a d.c. brushless motor. The major difference is in the type of position encoder which is employed and in the interpretation of its data. To synchronise the winding currents with the rotor's position and to hold β constant at the required value, a number of different techniques can be used; Figure 6.11 shows one approach. The digital output of the motor's shaft encoder or resolver is used to address a programmable read-only memory (PROM) which holds a number of synthesised sinewaves. Due to the symmetry of a three-phase supply, only two sinewaves are stored; the third sinewave can be computed. As the position of the motor changes, the sinewaves are read out sequentially, ensuring that the motor's current remains in synchronism with the motor's position. By digital addition, it is possible to move the supply waveform away from the optimum value, effectively adjusting the value of β. The analogue current demand determined by the servo amplifier modulates the sinewave which is generated by using a multiplying digital-to-analogue converter. The winding current can be produced either by direct amplification of the analogue demand using a linear amplifier, or by PWM within a conventional three-phase power bridge. The use of a linear-amplification current waveform with minimal harmonics results in exceptional performance, but this requires more com-

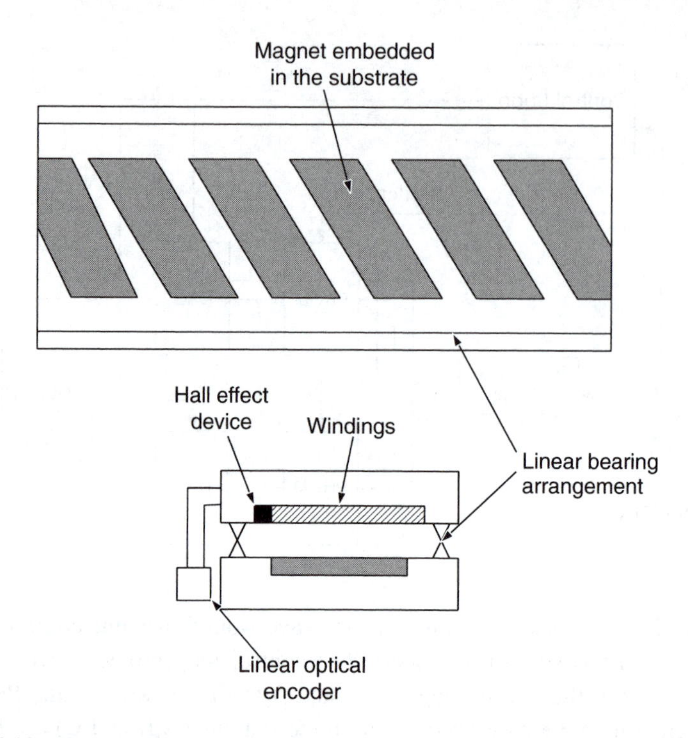

Figure 6.12. The construction of a linear brushless motor. The upper figure shows the plan of the substrate, the lower shows a cross section through the substrate and the carriage that contains the windings.

plex electronics and a highly dissipative linear amplifier. In practice, this approach is restricted to critical applications, for example, in the manufacture and testing of magnetic media.

An additional approach to the control of sinusoidally wound machines is to use vector control in an identical fashion to that used with a.c. induction motors, the implementation of which is discussed in Section 7.3.2.

6.3 Linear motors

A linear motor operates with a conventional d.c. brushless motor amplifier operating in the force, instead of torque mode as used in rotary machines. As with a conventional brushless motor the operation requires the use of a position feedback to control position, velocity and acceleration. In principle the operation of the motor is identical to that of its rotary equivalent. The magnets are mounted on the stationary track, with the coil and sensors fitted to a moving assembly, Figure 6.12. In the figure the encoder track for the linear encoder is glued to the side of the stationary substrate, (see Section 4.3.5). As with the rotary machine the windings are 120° electrical degrees apart. For a trapezoidal winding, Hall-effect devices are mounted adjacent to the coils to control the commutation pattern. If sinewave commutation is used, the linear encoder used for the position feedback is also to control the commutation of the motor. A phase finding strategy based on Hall-effect encoders is required on power-up, then the motor phases are incrementally advanced on each encoder pulse.

6.4 Summary

This chapter has discussed the rotary d.c. brushless motors and sinewave-wound motors. While the application requirements will normally dictate which is the best option, a number of comparisons can be drawn:

- If the torque per r.m.s. ampere is compared using equations (6.6) and (6.22), for identical peak motor fluxes, the torques of square-wave-wound motors exceed those of sinewave-wound motors by a factor of 1.47. This effectively determines the relative sizes between two motors with comparable characteristics.

- The control systems for square-wave-wound d.c. brushless motors are considerably simpler than those required for sinusoidally wound motors. Direct-current brushless motors only require low-cost position encoders, whereas sinewave-wound motors require high-precision systems.

While both types of motor give performances in excess of those of brushed motors, they do so at a cost premium, which needs to be considerable. Brushless motors are being more widely used as the cost of motors, and their essential electronics, continues to fall as the technology matures. With the present technology, the performance is exceptional; this has led to the introduction of electric drives in applications which have been the preserve of hydraulics. However, as with all decisions of drive-system selection, the additional complexity of these types of drives has to be balanced against their high reliability and performance. The final point needs to be considered in some detail when a linear motor is being used to replace a ball screw driven by conventional rotary machine.

Chapter 7

Induction motors

As noted in the introduction, this book is primarily concerned with motor-drives that are capable of being used in a wide range of low- to medium-power closed-loop servo applications. With the recent advances in microprocessor technology, it is now possible to develop commercially viable drives that allow alternating-current (a.c.) asynchronous induction motors to be controlled with the accuracy and the response times which are necessary for servo applications. The importance of this development should not be underemphasised. Induction motors are perhaps the most rugged and best-understood motors presently available. Alternating-current asynchronous motors are considered to be the universal machine of manufacturing industry. It has been estimated that they are used in seventy to eighty per cent of all industrial drive applications, although the majority are in fixed-speed applications such as pump or fan drives. The main advantages of induction motors are their simple and rugged structure, their simple maintenance, and their economy of operation. Compared with brushed motors, a.c. motors can be designed to give substantially higher output ratings with lower weights and lower inertias, and they do not have the problems which are associated with the maintenance of commutators and brush gears. The purpose of this chapter is to briefly review the operation of advanced induction-motor-drive systems which are capable of matching the performance other servo motor-drives.

While induction motors are widely used in fixed-speed applications, variable-speed applications are commonplace across industry. Therefore, as an introduction to induction motors, this chapter will first briefly consider speed control using both fixed-frequency/variable-voltage and variable-voltage/variable-frequency supplies; this approach is termed *scalar control*. In order to achieve the performance required by servo applications, induction motors have to be controlled using *vector controllers*.

The key features that differentiate between scalar and vector control are:

- Vector control is designed to operate with a standard a.c., squirrel-cage, asynchronous, induction motor of known characteristics. The only addition to the motor is a rotary position encoder.

- A vector controller and its associated induction motor form an integrated drive; the drive and the motor have to be matched to achieve satisfactory operation.

- A vector-controlled induction motor and drive is capable of control in all four quadrants through zero speed, without any discontinuity. In addition, the drive is capable of holding a load stationary against an external applied torque.

- The vector-controlled-induction-motor's supply currents are controlled, both in magnitude and phase in real time, in response to the demand and to external disturbances.

7.1 Induction motor characteristics

Traditionally, a.c. asynchronous induction motors operated under constant speed, open-loop conditions, where their steady-state characteristics are of primary importance, (Bose, 1987). In precision, closed-loop, variable-speed or position applications, the motor's dynamic performance has to be considered; this is considerably more complex for induction motors than for the motors which have been considered previously in this book. The dynamic characteristics of a.c. motors can be analysed by the use of the two-axis d-q model.

The cross section of an idealised, a.c., squirrel-cage induction motor is shown in Figure 7.1. As with sine-wave-wound permanent-magnet brushless motors, it can be shown that if the effects of winding-current harmonics caused by the non-ideal mechanical construction of the motor are ignored, and if the stator windings ($a_s\, b_s\, c_s$) are supplied with a balanced three-phase supply, then a distributed sinusoidal flux wave rotates within the air gap at a speed of N_e rev min^{-1}, which is given by

$$N_e = \frac{60 f_e}{p} \tag{7.1}$$

where f_e is the supply frequency and p is the number of pole pairs. The speed, N_e, is called the induction motor's synchronous speed. If the rotor is held stationary, the rotor conductors will be subjected to a rotating magnetic field, resulting in an induced rotor current with an identical frequency. The interaction of the air gap flux and the induced rotor current generates a force, and hence it generates the motor's output torque. If the rotor is rotated at a synchronous speed in the same direction as the air-gap flux, no induction will take place and hence no torque is produced. At any intermediate speed, N_r, the speed difference, $N_e - N_r$, can be expressed in terms of the motor's slip, s

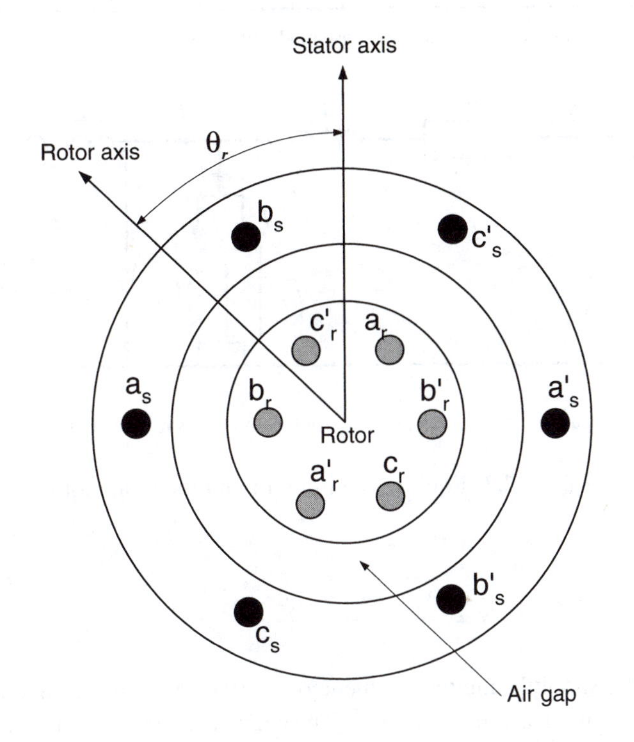

Figure 7.1. Cross section of an idealised three-phase, two-pole induction motor. The rotor and stator windings are represented as concentrated coils. The rotor's speed is ω_r, and the lag between the rotor and stator axes is θ_r.

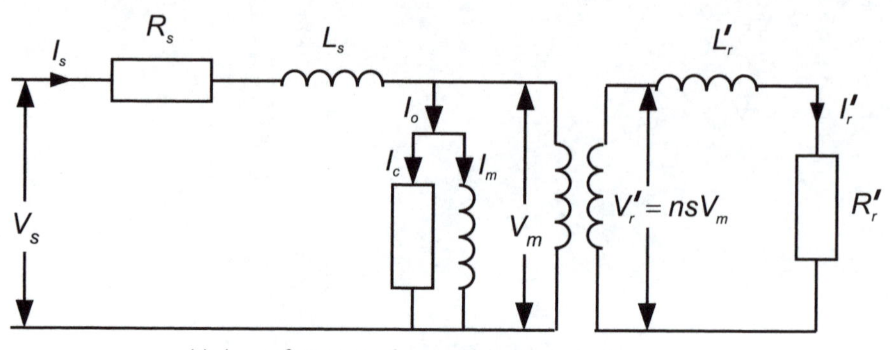

(a) A transformer per phase model of the induction motor.

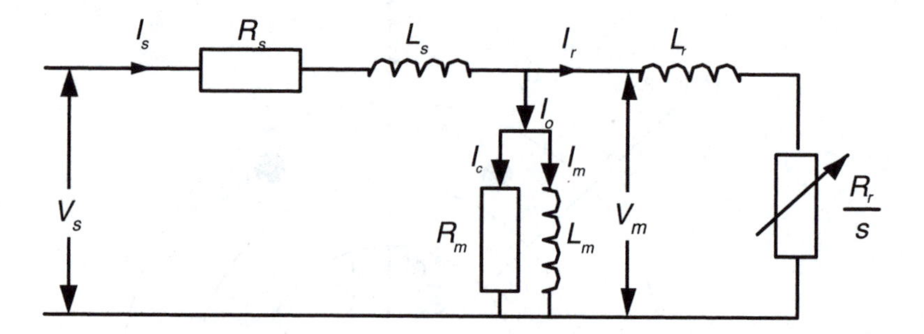

(b) Induction motor model with all rotor components referred to the stator.

Figure 7.2. Equivalent circuit of an induction motor.

$$s = \frac{N_e - N_r}{N_e} = \frac{\omega_e - \omega_r}{\omega_e} = \frac{\omega_s}{\omega_e} \tag{7.2}$$

where ω_e (the supply's angular frequency), ω_r (the rotor speed), and ω_s (the slip frequency) are all measured in rad s^{-1}. The equivalent circuit for induction motors is conventionally developed using a phase-equivalent circuit (see Figure 7.2(a)). The stator's terminal voltage, V_s differs from V_m by the voltage drop across the leakage resistance and the inductance. The stator current, I_r comprises an excitation component, I_m and the rotor's reflected current, I'_r. The rotor's induced voltage, V'_r (because of the effective turns ratio, n, between the rotor and stator, and the slip) is equal to snV_m. The relative motion between the rotor and the rotating field produces a rotor current, I'_r, at the slip frequency, which in turn is limited by the rotor's resistance and leakage impedance. It is conventional to refer the rotor circuit elements to the stator side of the model, which results in the equivalent circuit shown in Figure 7.2, where the rotor current is

$$I_r = nI'_r = \frac{n^2 sV_m}{R'_r + j\omega_s L'_r} = \frac{V_m}{R_r/s + j\omega_e L_r} \tag{7.3}$$

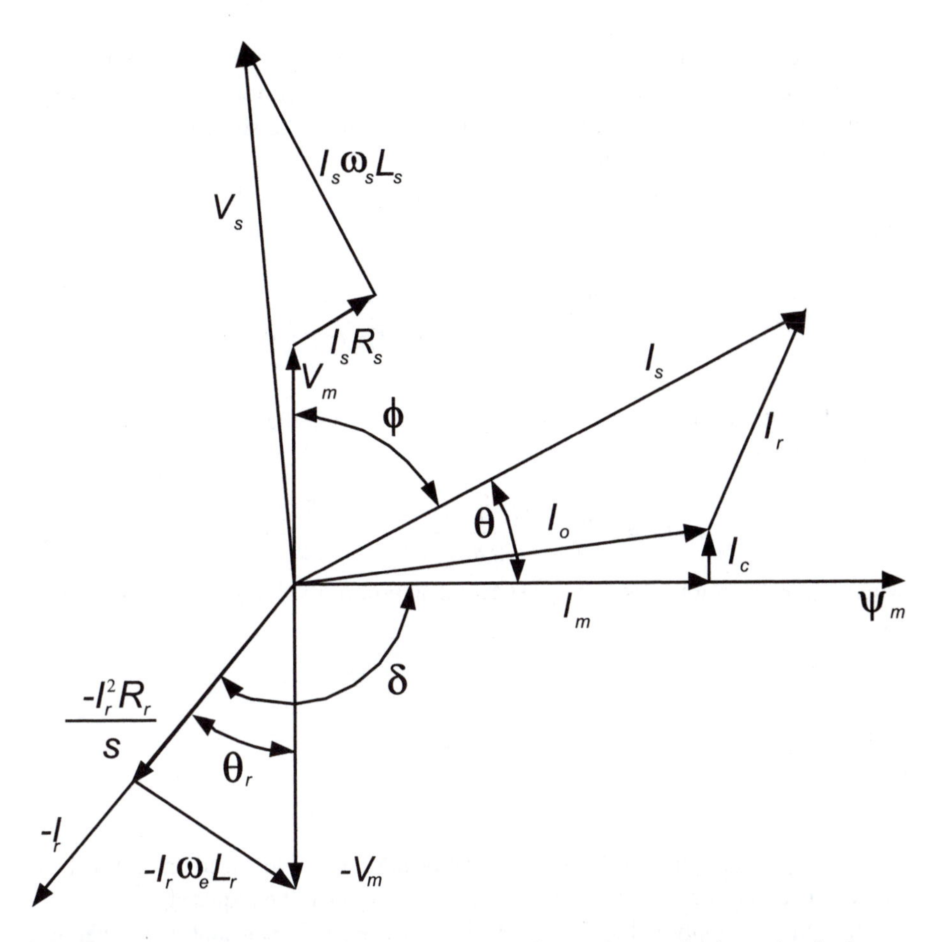

Figure 7.3. The phasor diagram for the induction motor equivalent circuit shown in Figure 7.2(b).

The air-gap flux which is rotating at the slip frequency, relative to the rotor, induces a voltage at the slip frequency in the rotor, which results in a rotor current; this current lags the voltage by the rotor power factor, θ_r. The phasor diagram for the motor whose equivalent circuit is shown in Figure 7.2(b) is given in Figure 7.3. The derivation of the electrical torque as a function of the rotor current and the flux is somewhat complex; this derivation is fully discussed in the literature (Bose, 1987). The torque can be expressed in the form

$$T_e = K_T|\psi_m||\hat{I}_r|\sin\delta \tag{7.4}$$

where K_T is the effective induction-motor torque constant, $|\psi_m|$ is the peak air-gap flux, $|\hat{I}_r|$ is the peak value rotor current, and $\delta = 90 + \theta_r$. The torque constant, K_T, is dependent on the number of poles and on the motor's winding configurations.

At a standstill, when the motor's slip is equal to unity, the equivalent circuit corresponds to a short-circuited transformer; while at synchronous speed, the slip, and hence the rotor current, is zero, and the motor supply current equals the stator's excitation current, I_0. At subsynchronous speeds, with the slip close to zero, the rotor current is principally influenced by the ratio R_r/s.

From this equivalent circuit of the induction motor, the following relationships apply

$$\text{Input power} = P_i = 3V_s I_s \cos\phi \tag{7.5a}$$

$$\text{Output power} = P_o = \frac{3I_r^2 R_r(1-s)}{s} \tag{7.5b}$$

Since the output power is the product of the speed and the torque, the generated torque can be expressed as

$$T_e = \frac{P_o}{\omega_m} = \frac{3I_r^2 R_r(1-s)}{s\omega_m} = \frac{3I_r^2 R_r p}{s\omega_e} \tag{7.6}$$

where ω_m is the rotor's mechanical speed. The power loss within the rotor is given by

$$P_{loss} = I_r^2 R_r \tag{7.7}$$

and the power across the air gap is given by

$$P_{gap} = P_o + P_{loss} \tag{7.8}$$

where P_{loss} is dissipated as heat. If the motor has a variable-speed drive, this heat loss can become considerable, and forced ventilation will be required.

If both the supply voltage and the frequency are held constant, the generated torque, T_e, can be determined as a function of the slip; giving the characteristic shown in Figure 7.4. Three areas can be identified: plugging $(1.0 \leqslant s \leqslant 2.0)$, motoring $(0 \leqslant s \leqslant 1.0)$, and regeneration $(s \leqslant 0)$. As the slip increases from zero, the torque increases in a quasilinear curve until the breakdown torque, T_b, is reached. In this portion of the motoring region, the stator's voltage drop is small while the air-gap flux remains approximately constant. Beyond the breakdown torque, the generated torque decreases with increasing slip. If the equivalent circuit is further simplified by neglecting the core losses, the slip at which the breakdown torque occurs, s_b, is given by

$$s_b = \pm\frac{R_r}{\sqrt{R_s^2 + \omega_e^2(L_s + L_r)^2}} \tag{7.9}$$

The values for the breakdown torque and the starting torque can both be determined by substitution of the corresponding value of slip into equation (7.6).

In the plugging region, the rotor rotates in the opposite direction to the air-gap flux; hence $s > 1$. This condition will arise if the stator's supply phase sequence is reversed while the motor is running, or if the motor experiences an overhauling

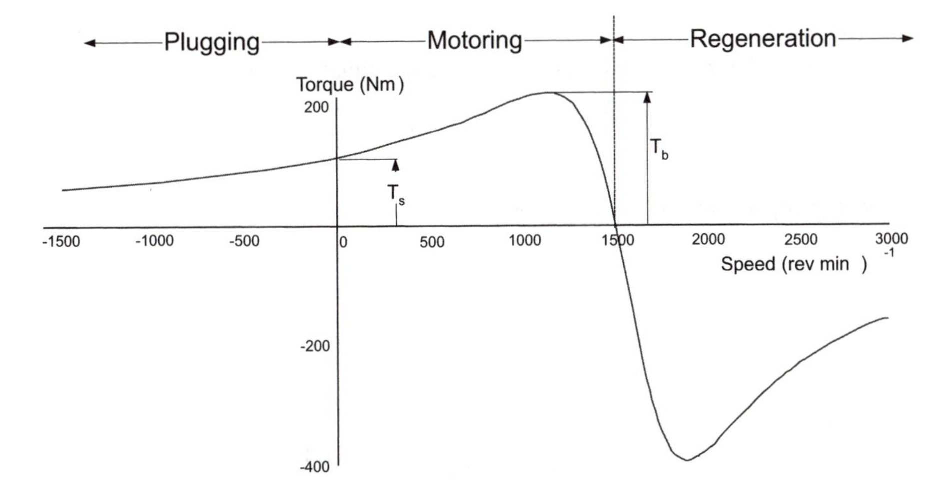

Figure 7.4. Torque-speed curve for a 2-pole induction motor operating with a constant-voltage, 50 Hz supply. T_s is the starting torque, and T_b is the breakdown torque.

load. The torque generated during plugging acts as a braking torque, with the resultant energy being dissipated within the motor. In practice, this region is only entered during transient speed changes-because excessive motor heating would result from continuous operation in the plugging region.

In the regenerative region, the rotor rotates at super-synchronous speeds in the same direction as the air-gap flux, hence $s < 0$. This implies a negative value to the rotor resistance term, R_r/s. As positive resistances are defined as resistances that effectively consume energy (for example, during motoring), negative values can be considered to generate energy. This energy flow will result in a negative or regenerative braking torque. Since the energy is returned to the supply, the motor can remain in the regenerative region for extended periods of time; this forms an important part of the control required for an induction motor in variable-speed applications.

Example 7.1

Determine the starting torque, and breakdown slip and torque for a 2-pole Y wound induction motor operating at 50 Hz. The motor's parameters with reference to Figure 7.2(b) are $R_s = 0.43\ \Omega$, $X_s = 0.51\ \Omega$, $R_m = 150\ \Omega$, $X_s = 31\ \Omega$, $R_r' = 0.38\ \Omega$ *and* $X_r' = 0.98\ \Omega$. *The supply voltage is 380 V line-to-line.*

Starting torque
At zero speed the rotor speed is zero, hence using equation (7.6)

$$T_e = \frac{3I_r^2 R_r p}{s\omega_e}$$

where the slip at standstill is given by

$$s = \frac{\omega_e - \omega_r}{\omega_e} = 1$$

$$\omega_e = 2\pi f = 158 \text{ rad s}^{-1}$$

$$I_r = \frac{219.4}{Z_s + Z_r + \frac{Z_s Z_m}{Z_m}} = 116 + 63i \text{ A}$$

hence

$$T_e = 118.4 \text{ Nm}$$

Breakdown slip and torque
The value of s_b can be determined by using equation (7.9)

$$s_b = \pm \frac{R_r}{\sqrt{R_s^2 + \omega_e^2 (L_s + L_r)^2}} = \pm \frac{R_r}{\sqrt{R_s^2 + (X_s + X_r)^2}} = \pm 0.245$$

A slip of ± 0.245 equates to a speed of 1132 rev min^{-1} and 1868 rev min^{-1} as shown in Figure 7.4.

From the slip values, the torque can therefore be calculated, giving 226 Nm at a slip of +0.245 and 393 Nm at a slip of -0.245.

7.2 Scalar control

A wide range of induction-motor-speed-control strategies exist, including voltage control, voltage and current-fed variable-frequency inverters, cycloconverters, and slip-energy recovery. However, within the application areas being considered, the use of voltage- and current-fed inverters predominates; additional speed-control systems are widely discussed in the literature (Bose, 1987; Sen, 1989).

The torque-speed curve of an induction motor can be modified by using a variable-voltage supply, where the motor's supply voltage is controlled either by a variable transformer or by a phase-controlled anti-parallel converter in each supply line, as shown in Figure 7.5(a) (Crowder and Smith, 1979). By examination of equation (7.9), it can be seen that the slip at which breakdown occurs is not dependent on the supply voltage. Only the magnitude of the torque is affected, and this results in the family of curves which are shown in Figure 7.5(b). When the load's torque-speed characteristic is also plotted on the same axes, the characteristics of speed control under voltage control can be seen. This form of control is only suitable for small motors with a high value of the breakdown slip; even so, the motor losses are large, and forced cooling will be required even at high speeds.

The more commonly used method of speed control is to supply the motor with a variable-frequency supply, using either a voltage- or a current-fed inverter. Since current-fed inverters are used for drives in excess of 150 kW, they will not be discussed further. A block diagram of a voltage-fed inverter drive is shown in Figure 7.6. The speed-loop error is used to control the frequency of a conventional three-phase inverter. As the supply frequency decreases, the motor's air gap will saturate; this results in excessive stator currents. To prevent this problem, the supply voltage is also controlled, with the ratio between the supply frequency and the voltage held constant.

In the inverter scheme shown in Figure 7.6, a function generator, operating from the frequency-demand signal, determines the inverter's supply voltage. The function generator's transfer characteristic can be modified to compensate for the effective increase in the stator resistance at low frequencies. Typical torque-speed curves for a motor-drive consisting of a variable-frequency inverter and an induction motor are shown in Figure 7.7. Since an inverter can supply frequencies in excess of those of the utility supply, it is possible to operate motors at speeds in excess of the motor's base speed (that is, the speed determined by the rated supply frequency); however, the mechanical and thermal effects of such operation should be fully considered early in the design process. If the inverter bridge is controlled using pulse-width modulation (PWM), the direct-current (d.c.) link voltage can be supplied by an uncontrolled rectifier bridge, allowing the motor's supply voltage and frequency to be determined by the switching pattern of the inverter bridge. However, it should be noted that, as with d.c. drives, the use of an uncontrolled rectifier requires the regenerative energy to be dissipated by a bus voltage regulator, rather than being returned to the supply. The method used to generate the PWM waveform is normally identical to the approach which is used in d.c. brushed and brushless drives, as discussed in Section 5.3.5.

Since the supply waveform to the motor is nonsinusoidal, consideration has to be given to harmonic losses in an inverter driven motor. In the generation of the PWM waveform, consideration must be given to minimising the harmonic content so that the motor losses are reduced. Except at low frequencies, it is normal practice to synchronise the carrier with the output waveform, and also to ensure that it is an integral ratio of the output waveform; this ensures that the harmonic content is

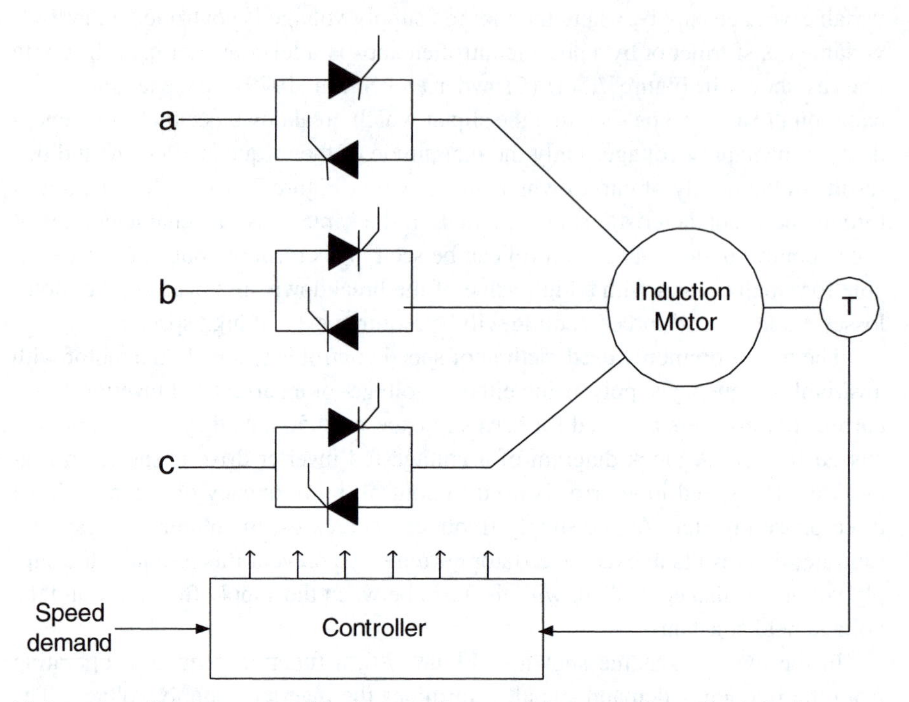

(a) An anti-parallel arrangement of thyristors used to control the stator voltage of an induction motor.

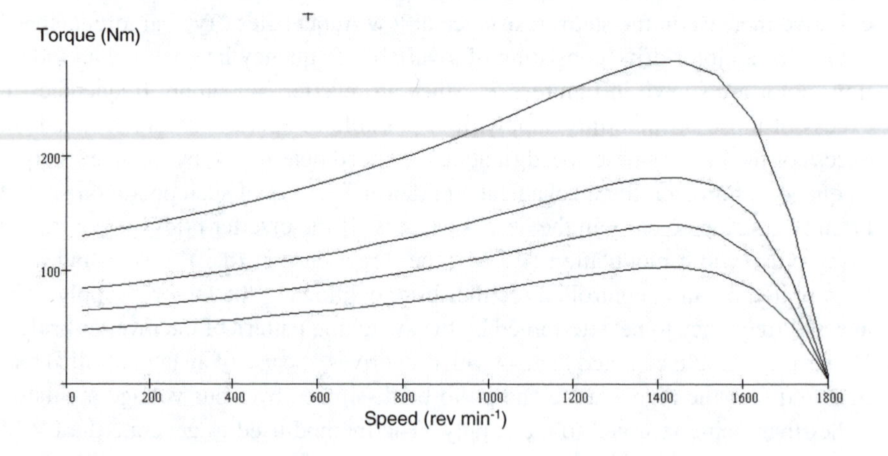

(b) Speed-torque curve, note that the peak torque occurs at the same speed, irrespective of the supply voltage.

Figure 7.5. Operation of a two-pole, three-phase induction motor with a variable voltage, fixed frequency supply. The supply frequency in this case is 60 Hz, giving a synchronous speed of 1800 rev min^{-1}.

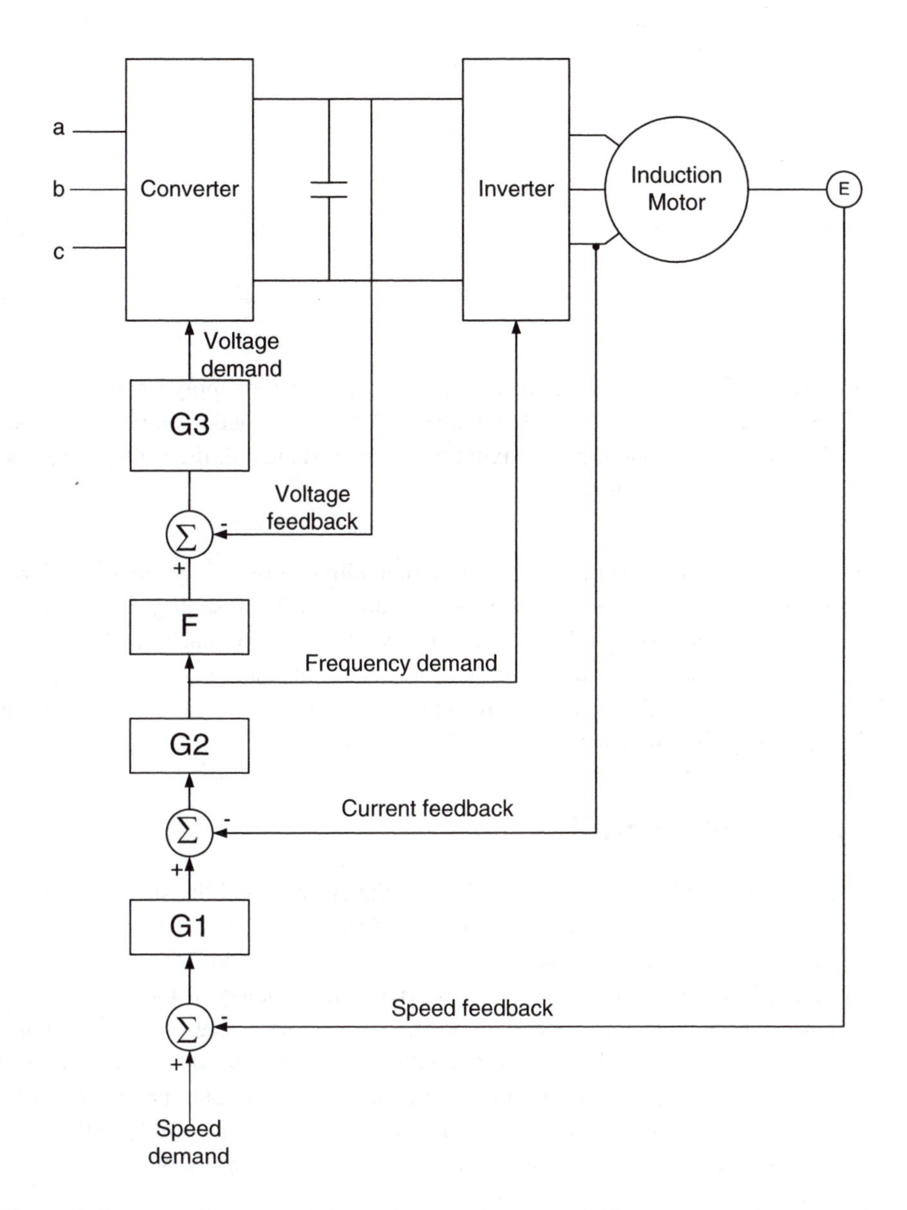

Figure 7.6. Block diagram of the variable-voltage, variable-frequency inverter: F is a function generator that defines the link voltage demand as a function of the inverter frequency; G1, G2 and G3 are gain blocks within the control loops.

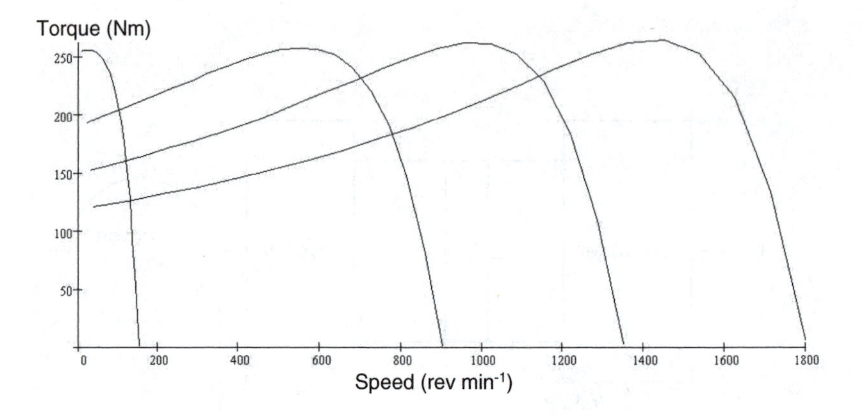

Figure 7.7. Torque-speed characteristics of the motor for supply frequencies of 5, 15, 30, 45 and 60 Hz. The supply voltage has been controlled to maintain constant. It should be noted that to give maximum torque at standstill, the supply frequency needs to be approximately 5 Hz.

minimised. Techniques of selective harmonic elimination using a modified PWM waveform have been receiving considerable attention because they can reduce the harmonic content even further. In the most widely used approach, the basic PWM waveform is modified by the addition of notches. This method does not lend itself to conventional analogue or digital implementation, and so microprocessors are being widely used to generate the PWM waveform.

7.3 Vector control

Under scalar control, the motor voltage (or the current) and the supply frequency are the control variables. Since the torque and the air-gap flux within an induction motor are both functions of the rotor current's magnitude and frequency, this close coupling leads to the relatively sluggish dynamic response of induction motors, compared to high performance, d.c., brushed or brushless servo drives. As will be discussed, a standard induction motor controlled by a vector-control system results in the motor's torque- and flux-producing current components being decoupled. This results in transient response characteristics that are comparable to those of a separately excited motor. Consider the d.c. motor torque equation

$$T = K_t I_a I_f \tag{7.10}$$

where I_a is the armature current, I_f is the field current which is proportional to the air-gap flux, and K_t is the torque constant. In a conventional d.c. brushed-motor control scheme, it is the air-gap flux that is held constant, and the armature current (and hence the torque) is controlled. As the armature current is decoupled from the field current, the motor's torque sensitivity remains at its maximum value during both steady-state and transient operations. This approach to decoupled control is not possible using a scalar-control scheme applied to an induction motor.

In order to give servo-drive capabilities to induction motors, *vector control* has been developed. The rational behind this approach can be appreciated from the phasor diagram of an induction motor's per-phase equivalent circuit (Figure 7.3). The electrical torque can be expressed as

$$T_e = K_T \psi_m I_r \sin\delta \tag{7.11}$$

where ψ_m and I_r are the root-mean-square (r.m.s.) values of the air-gap flux and the rotor current, respectively. If the core losses are neglected, (7.11) can be further simplified to

$$T_e = K_T' I_m I_s \sin\theta = K_T' I_m I_a \tag{7.12}$$

where $I_a (= I_s \sin\theta)$ is the torque component of the stator current (see Figure 7.8). As is readily apparent, this torque equation is now in an identical form to the equation for d.c. motor: I_m is the magnetising or flux component of the stator current, and I_a is the armature or torque component of the stator current, while K_T is a torque constant which is determined by the motor's electromechanical characteristics. In order to vary either I_m or I_a, the magnitude and phase of the supply current must be controlled. The principle of how one current can be independently determined by controlling the current vector can be appreciated by considering Figure 7.9, where the peak value of the current vector, and its phase angle, are independently controlled relative to a predetermined reference frame. The key element in any vector controller is to achieve this in real time as the motor's demanded and actual speed vary under the operational requirements. The requirements of the drive package are summarised in Figure 7.10, where I_m' and I_a' constitute the speed and torque demands to a vector controller; the output of this controller is the current waveform demand to a conventional three phase inverter.

7.3.1 Vector control principles

In a vector controller, the magnitude and the phase of the supply currents must be controlled in real time, in response to changes in both the speed and the torque demands. In order to reduce this problem to its simplest form, extensive use is made of conventional two-axis theory; by the selection of the correct reference frame, the three-phase a.c. rotational problem found in an induction motor can be reduced to a two-axis, stationary d.c. solution. Within the vector controller, the required motor currents are computed with reference to the rotor's frame of reference, while the three phase motor currents are referenced to the stator's frame of reference; to achieve this, a set of transformations must be developed.

If the supply to an induction motor is a balanced, three-phase, a.c. supply, then the conventional two-axis, or d-q, approach to motor modelling can be used to analyse the operation of the induction motor. This approach permits the time-varying

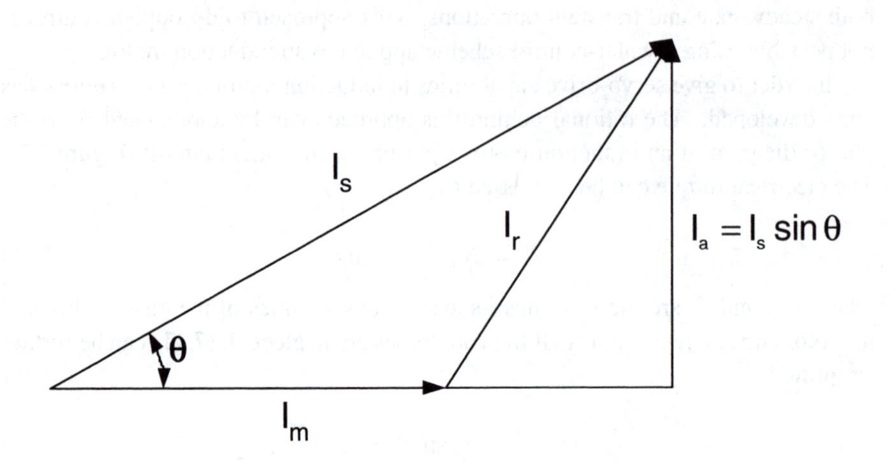

Figure 7.8. The relationship between I_a and I_s as applied to vector control.

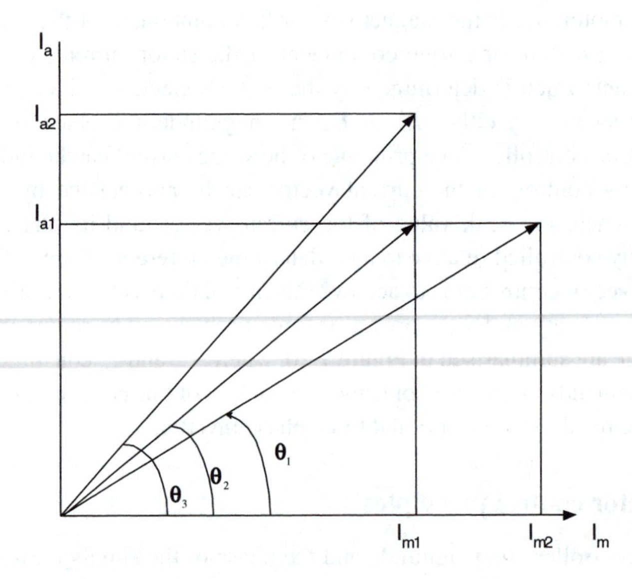

Figure 7.9. The principle of vector control. The values of I_a or I_m can be independently controlled by adjustment of the magnitude of I_s and the angle θ.

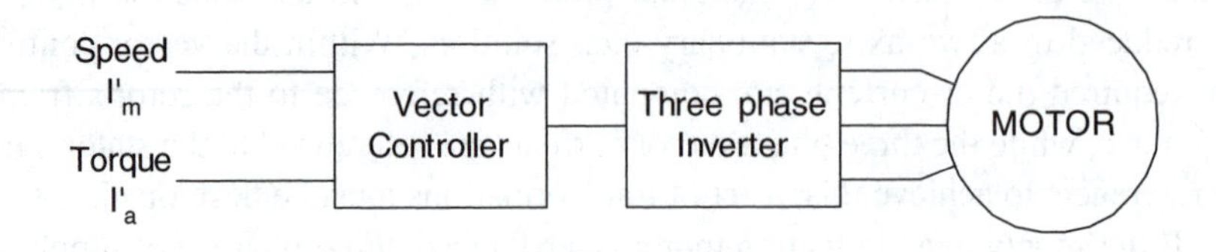

Figure 7.10. The outline of a vector controller.

motor parameters to be eliminated, and the motor variables can be expressed relative to a set of mutually decoupled orthogonal axes, which are commonly termed the direct and the quadrature axes.

The required set of transformations can be developed as follows. Firstly, the transformation between the two stationary reference frames, the three phase $(a\,b\,c)$ frame, and the equivalent two-axis, d_s-q_s frame (see Figure 7.11(a)) is given by the relationship

$$\begin{bmatrix} v_a^s \\ v_b^s \\ v_c^s \end{bmatrix} = \begin{bmatrix} \cos\theta & \sin\theta & 1 \\ \cos(\theta - \frac{2\pi}{3}) & \sin(\theta - \frac{2\pi}{3}) & 1 \\ \cos(\theta + \frac{2\pi}{3}) & \cos(\theta + \frac{2\pi}{3}) & 1 \end{bmatrix} \begin{bmatrix} v_q^s \\ v_d^s \\ v_0^s \end{bmatrix} \tag{7.13}$$

with the inverse given by

$$\begin{bmatrix} v_q^s \\ v_d^s \\ v_0^s \end{bmatrix} = \begin{bmatrix} \cos\theta & \cos(\theta - \frac{2\pi}{3}) & \cos(\theta + \frac{2\pi}{3}) \\ \sin\theta & \sin(\theta - \frac{2\pi}{3}) & \sin(\theta + \frac{2\pi}{3}) \\ 0.5 & 0.5 & 0.5 \end{bmatrix} \begin{bmatrix} v_a^s \\ v_b^s \\ v_c^s \end{bmatrix} \tag{7.14}$$

Two points should be noted about these transformations. Firstly the zero sequence voltage v_0^s is not present because the three-phase supply is considered to be perfectly balanced; secondly, the q^s and d^s axes are considered to be coincident, then θ can be set to zero. The net effect of this is to simplify the mathematical relationships; therefore the speed of any computation is increased.

The second transformation that must be considered is the transformation from the stationary d^s-q^s axes to the corresponding rotational d-q axes. If the d-q reference frame is rotating at the induction motor's synchronous speed, ω_e, relative to the fixed frame (see Figure 7.11(b)), the transformations are given by

$$v_q = v_q^s \cos\omega_e t - v_d^s \sin\omega_e t \tag{7.15a}$$

$$v_d = v_q^s \sin\omega_e t - v_q^s \cos\omega_e t \tag{7.15b}$$

and the inverse relationship is given by,

$$v_q^s = v_q \cos\omega_e t - v_d \sin\omega_e t \tag{7.16a}$$

$$v_d^s = v_q \sin\omega_e t - v_q \cos\omega_e t \tag{7.16b}$$

If $v_a^s = V_m \sin\omega_e t$, and v_a^s, v_b^s, and v_c^s form a balanced three-phase supply, substituted into equation (7.16), will result in $v_q = V_m$ and $V_d = 0$; hence the supply voltage within the stationary frame is transformed to a d.c. voltage within the synchronous rotating reference frame. This approach can be extended to all time-dependent variables within the motor's model; this results in a simplified mathematical model for an induction motor.

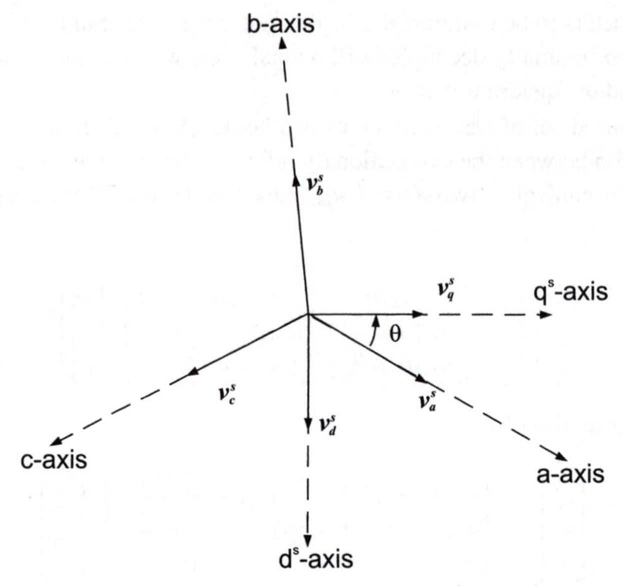

(a) The transformation of the voltages from the $(a_s\ b_s\ c_s)$ axis to the stationary d^q-q^s axes.

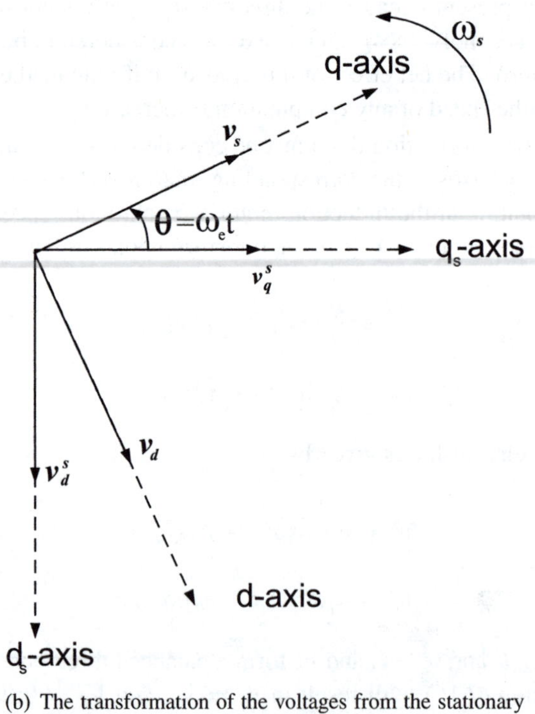

(b) The transformation of the voltages from the stationary d^s and q^s axis to the rotating d-q axes.

Figure 7.11. The principle of d-q transformations as applied to the induction motor.

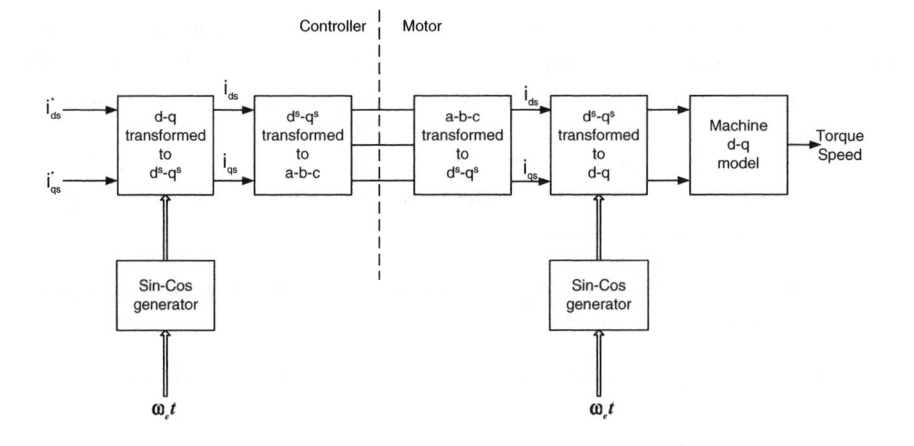

Figure 7.12. The transformations which should be considered in the control of an induction motor, the demand is the *d*-*q* current required to produces the speed and torque.

7.3.2 Implementation of vector control

The theory of vector control discussed above shows that torque control of an induction motor can be performed by the effective decoupling of the flux- and torque-producing components of the stator current. It should be noted that, within an induction motor, the rotor currents and the flux cannot normally be directly measured. The separation of the stator current into flux and torque-producing components can, however, be undertaken by the use of the transformations and the relationships already discussed. This decoupling of the orthogonal field and the armature axes permits high-performance dynamic control, in a similar fashion to the control of d.c. brushed motors. In order to implement the vector control of an induction motor, information, either measured or derived, about the position and magnitude of the currents and the fluxes within the motor is required.

The overview of a vector-controller scheme given in Figure 7.12 allows the various transformations that need to be considered to be located. Those to the right of the motor terminals are within the motor, while those to the left are within the controller and they must be implemented in real time. The inverter is omitted from this block diagram, but it can be assumed to give an ideal motor supply current. It follows, therefore, that for an induction motor to operate under vector control, the values of $\sin \omega_e t$ and $\cos \omega_e t$ need to be determined as part of the overall control strategy.

Vector control can be implemented using either the direct or the indirect control methods. In the direct-measurement scheme, use is made of flux sensors located within the motor to directly measure the flux; this strategy is not easily implemented within industrial applications,where the motor's construction needs both to be as simple as possible and to have the minimum number of interconnections between the controller and motor. In contrast, an indirect vector-control system uses the motor's parameters (the supply current and frequency) and rotational-position measurement to determine the control variables. This indirect strategy cannot be considered to be as accurate as the direct approach.

It can readily be appreciated that an induction motor's vector controller is a highly sophisticated system, and in order to achieve satisfactory control it must be capable of achieving the following:

- Measurement of the rotor position, and then computation of the required transformation in real time.
- Control of the magnitude and phase relationships of the supply current.

7.3.3 Vector control using sensors

Within an indirect vector controller the location of the rotating reference frame must be accurately determined. As noted earlier, the d_s-d_q reference frame is considered to be fixed relative to the stator, while the d-q reference frame rotates at a synchronous speed, ω_e (see Figure 7.13). At any point in time, the angle between the stationary and the rotating frame is θ_e. This angle is given by the sum of the rotor's angular position, θ_r, and the slip angle, θ_s. These angles need to be determined in the implementation of indirect vector control. For decoupled control, the stator's flux component and the torque component are aligned with the rotating d- and q-axes, respectively. As noted earlier, the slip of an induction motor can be determined from the demanded rotor current; this fact is used in indirect vector controllers to determine ω_s, for the demanded rotor current, and then $\sin \omega_s$, and $\cos \omega_s$ can be determined. The values of $\sin \omega_r$ and $\cos \omega_r$ can be measured directly; if the induction motor is fitted with rotary position encoder, both these angles can then be used to determine $\sin \omega_e$ and $\cos \omega_e$, as required by the axes transformations. This approach does, however, require that the parameters of the motor have been accurately determined during manufacture or as part of the commissioning procedure.

Figure 7.14 shows a block diagram of a vector controller which uses this approach; the position and speed-error amplifiers are of a conventional design and they can be implemented in either an analog or digital form; the output is the required torque-producing component of the stator current. The flux component of the stator current is normally maintained at a constant value, unless field weakening is required. The current demands are transformed to the stationary reference frame by the application of equations (7.5) and (7.13), and this is followed by two to three phase transformation to determine the three-phase supply currents. The current demand is used to control a conventional three-phase power bridge, through the use of a local current loop. Vector controllers can be used with a number of different power stages, depending on the application; and there is a case for the use of voltage-sourced transistors or MOSFET inverter bridges because of their fast current control. Current-sourced inverters or the cycloconverters are considered to

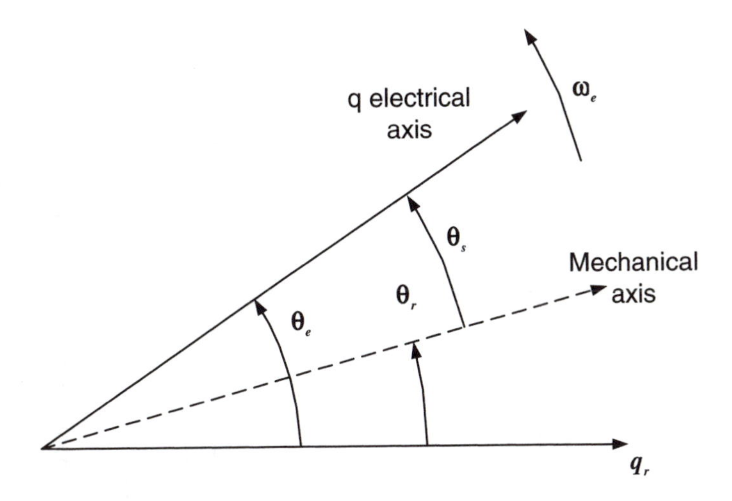

Figure 7.13. The relative angles between the stationary and rotating axes, as used in the computation of the transforms.

be suitable for high-power, low-speed applications. The system described above is a closed-loop system, because of the presence of a rotor encoder. Rotor encoders are required for vector controllers used in those servo applications which require fine control of speed or position down to and through standstill. Recently, a number of manufacturers have developed open-loop vector controllers as replacements for the more conventional variable-frequency inverters. This approach is based on an accurate knowledge of the motor and the system being controlled. The positions of the rotational fields are computed from a knowledge of the supply waveform; the electrical-axis position, of course, rotates at the synchronous speed determined by the supply frequency. In order to give these drives additional flexibility, they are being designed with systems that will measure the motor's characteristics prior to the initial powering up as part of the commissioning process. This is achieved by the use of accurate current measurement and voltage injection techniques, together with the monitoring of the dynamic performance to determine the system variables. Even with these additions, the performance of open-loop vector controllers is unlikely to compete with closed-loop systems in servo applications, but they will prove attractive as replacements for high-performance inverter drives.

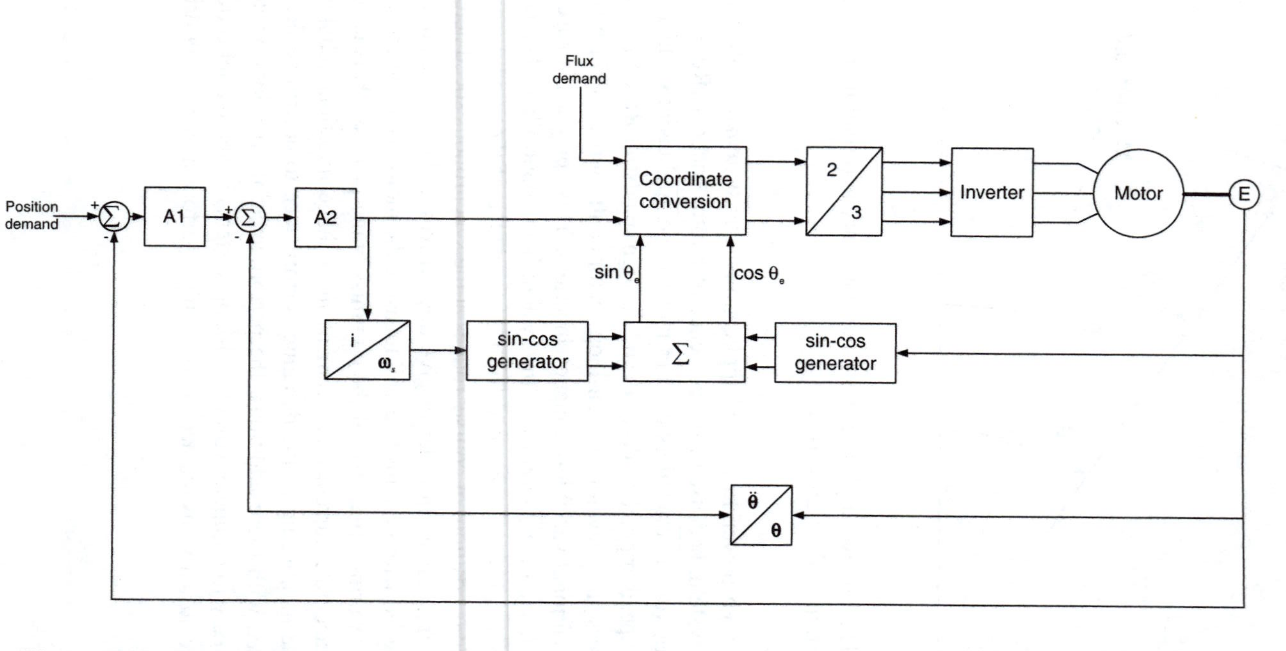

Figure 7.14. A block diagram of a sensor based induction-motor control system operating under vector control.

7.3.4 Sensorless vector control

Currently there is considerable interest in the development of sensorless induction motor controllers. It should be noted that this term is somewhat incorrect – a sensorless controller removes the need for a position or velocity transducer to be located on the motor, and these are effectively replaced by processing the outputs current and voltage sensors to derive the same information. The advantage of this approach are reduced hardware complexity, lower cost and reduction in the size of the induction motor package. The elimination of the sensor cable leads to better noise immunity and increased reliability for example, operating an induction motor at the end of a well string to pump oil to the surface, is an ideal application for a sensorless vector-controlled induction motor.

As has been previously discussed, the vector control of an induction motor requires the estimation of the magnitude and location of the magnetic flux vectors in the machine. For a sensorless controller, using either open or closed loop estimators, the complexity of the algorithms determines the performance of the complete motor-drive system. The elimination of the speed sensor is of particular interest as the mechanical speed is different from the speed of the rotating flux, as shown in equation (7.2). A large number of possible solution have been considered; the paper by Holtz (2002) provides an overview of the available techniques.

In order to illustrate the principles of sensorless vector control we can consider an approach based on MRAC (model reference adaptive control). As discussed in Sections 7.3 and 7.3.2 the current vector has to be determined with reference to a specific coordinate frame that is moving in space. In a MRAC approach the controller contains a model of the machine that is capable of estimating the machine parameters from the motor's line current and voltages.

Figure 7.15 shows the principles of a MRAC based controller. The system contain three elements, a model of the motor, a controller and a conventional current controlled inverter. The model relies on the principle that the flux in the machine can be computed from both the stator and rotor model – in this case the stator model is used as the reference. The rotor model estimates the rotor flux from the measured current and $\hat{\omega}$ or tuning signal. The tuning signal is obtained from a comparison of the flux generated from the stator and and rotor models, and is used in a closed loop to adjust the rotor model. The model provides the estimated speed, and hence the speed error, and rotor flux that are used by the controller to generate the current demand for the inverter. In practice this approach to sensorless control can satisfactorily control the motor' speed almost to standstill.

7.4 Matrix converter

In all the control systems described above, the a.c. supply to the induction motor was generated using a conventional rectifier-inverter or d.c link inverter arrangement. Recently research has been undertaken on an a.c.–a.c. converter that is capable of giving compatible performance to the d.c. link inverter (Wheeler et al.,

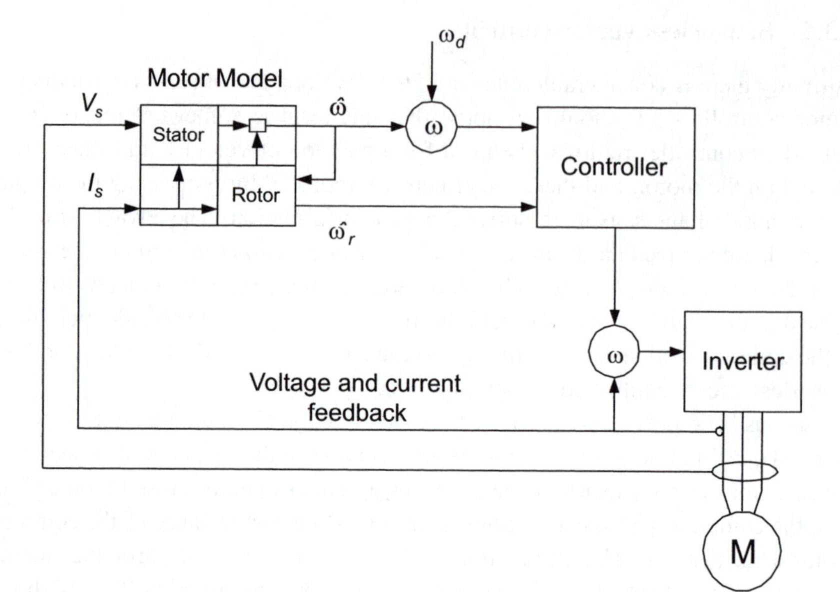

Figure 7.15. A sensorless controller for an induction motor based on the use of a MRAS. The speed demand is ω_d, and the controller is based on the architecture shown in Figure 7.14.

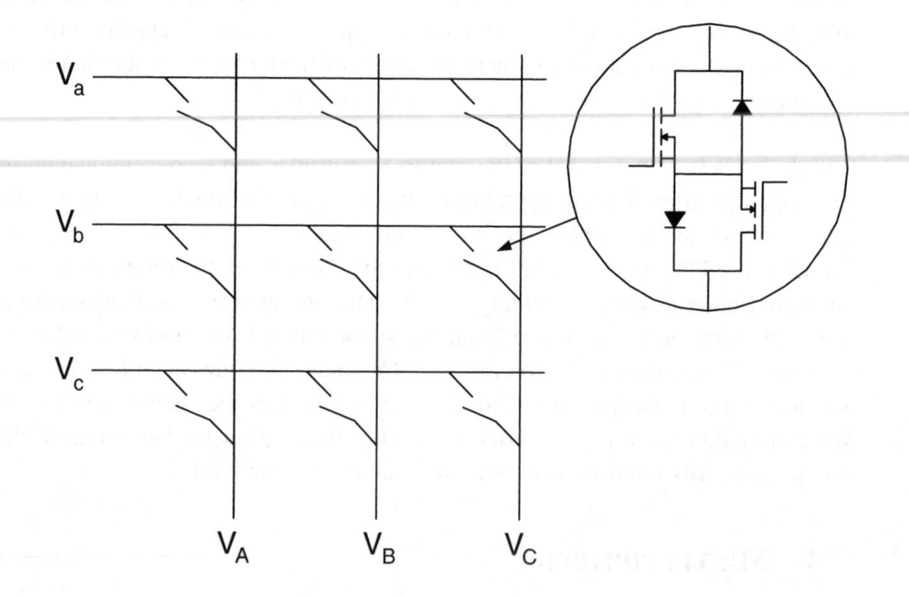

Figure 7.16. Switch layout of a matrix converter. V_a, V_b, V_c are the supply, V_A, V_B, V_C are connected to the load. The load's voltage and frequency is determined by controlling the individual switches.

2002). The matrix converter uses nine bidirectional switches to generate the output waveform. Control of the output waveform is achieved by switching in a predefined pattern, Figure 7.16. The converter's input current is built up from segments of the three output currents, and consists essentially of a supply frequency component, plus a high frequency component that can be removed by filtering. The converter is inherently bidirectional and is capable of operating the induction motor in all four quadrants, with the minimum of energy storage components. In practice the operation is not straightforward due to the lack of free-wheeling paths, and the possibilities of short circuits. The timing of the individual switches is critical, as well as the provision for device protection. In order the generate the required output waveforms, the matrix converter needs to be modulated in response to the demands from the scalar or vector controllers. Work by Sunter and Clare (1996) has demonstrated that the matrix controller can be used to provide servo-grade performance, in particular high-speed reversal, when used as a power controller within a vector controller.

As noted in Wheeler et al. (2002), the matrix converter can be used to provide the high power quality required for vector controllers. The problems with over currents and voltage spikes resulting from the arrangements of the power switches could be minimised by the use of soft commutation techniques.

7.5 Summary

The use of vector-controlled induction motors represents an alternative to the other forms of brushless motors for servo drives. In the selection of a vector-controlled induction motor, the following points need to be considered, particularly when they are being compared with permanent-magnet sine-wave-wound motors:

- Induction motors are inherently more difficult to control.
- Induction-motor drives are typically larger than permanent-magnet sine-wavewound motor-drives, for identical output powers; this is because of the rotor power loss in induction motors. Therefore, a provision may have to be made for forced cooling.
- For the same output torque, the efficiency (which directly dictates the frame size) is lower for induction motors. Permanent-magnet sine-wavewound machines have of higher efficiencies because of the lack of any rotor losses.
- Induction motors can be designed for higher flux densities than those of permanent-magnet sine-wave-wound motors, which are limited by the design of the rotor and its permanent magnets.
- Induction motors cost less than the equivalent permanent-magnet sine-wave-wound motors, due to there simplicity and lack of permanent magnets.

- In induction motors, field weakening is easily achieved over a wide speed range; this is not possible in permanent-magnet sine-wave-wound motors.
- The vector control of induction motors requires a considerable amount of computing power, and while microprocessors are an advantage they are not a necessity. In safety-critical applications, the use of motors incorporating sophisticated microprocessor-based controllers may constitute an undue safety risk.

This chapter has reviewed vector control applied to squirrel-cage induction motors; the resultant characteristics are suitable for servo applications. This development represents significant alternative choice for design engineers. Vector-controlled induction motors are rapidly becoming accepted as the preferred choice for high-power servo applications. Their undoubted advantages and disadvantages need to be critically compared (because of their complexity and hence their cost) when they displace conventional servo drives from any application.

Chapter 8

Stepper motors

The motors discussed so far have been effectively analogue in nature, with the motor's speed being a function of the supply voltage; stepper motors, however, are essentially digital. The rotary motion in stepper motors occurs in a stepwise manner from one equilibrium position to the next, and hence a stepper motor's speed will be a function of the frequency at which the windings are energised. In industrial applications, stepper motors are not widely used as the main robotic or machine-tool drive, but they are widely used as an auxiliary drive (for example within product feed systems, or as a low power end-effector's actuator) or within a computer peripheral (for example within a printer). One area where stepper motors have found widespread use is the drives within small educational robots; this is largely due to their simplicity of control and the low system cost. There are a number of characteristics that make a stepper motor the first choice as a servo drive, including:

- Stepper motors are able to operate with a basic accuracy of ± 1 step in an open-loop system. This inherent accuracy removes the requirement for a positional or speed transducer, and it therefore reduces the cost of the overall system.

- Stepper motors can produce high output torques at low angular velocities, including standstill with the hybrid stepper motor.

- A holding torque can be applied to the load solely with direct-current (d.c.) excitation of the stepper motor's windings.

- The operation of stepper motors and their associated drive circuits is effectively digital, permitting a relatively simple interface to a digital controller or to a computer.

- The mechanical construction of stepper motors is both simple and robust, leading to high mechanical reliability.

8.1 Principles of stepper-motor operation

The essential feature of a stepper motor is its ability to translate the changes in stator winding's excitation into precisely defined changes, steps, of the rotor's position. The positioning is achieved by the magnetic alignment between the teeth of a stepper motor's stator and rotor. There is a wide range of stepper motors on the market, but they are all variations of two basic designs: variable-reluctance stepper motors or hybrid stepper motors. Variable-reluctance stepper motors can be also found as either multistack or single-stack motors. In the variable-reluctance design, the magnetic flux is provided solely by stator excitation, whereas the hybrid design uses the interaction between the magnetic flux produced by a rotor-mounted permanent magnet and that resulting from the stator winding's excitation.

8.1.1 Multistack variable-reluctance motors

The longitudinal cross section of a multistack variable-reluctance motor is shown in Figure 8.1(a). The motor is divided into a number of magnetically isolated stacks, each with its own individual phase winding. The stator of each stack has a number of poles (four in this example), each with a segment of the phase winding; adjacent poles are wound in opposite directions. The position of the rotor relative to the stator is accurately defined whenever a phase winding is excited, where the teeth of the stator and rotor align to minimise the reluctance of the phase's magnetic path. To achieve this, the rotor and the stator have identical numbers of teeth.

As can be seen in Figure 8.1, when the teeth of stack A are aligned, the teeth of stacks B and C are not. Hence by energising phase B after switching off phase A, a clockwise movement will result; this movement will continue when phase C is energised. The final step of the sequence is to re-energise phase A. After these three excitations, stack A will again be aligned, and the motor will have rotated three steps, or one tooth pitch clockwise, in the process to produce continuous clockwise rotation. The sequence of excitation will be A:B:C:A:B:C ... ; and for anticlockwise rotation it will be A:C:A:CB.... The length of each incremental step is

$$\text{step length} = \frac{360}{N R_T} \text{degrees} \tag{8.1}$$

where N is the number of stacks, and R_T is the number of rotor teeth per stack. The motor shown in Figure 8.1 has eight teeth per rotor and three stacks, resulting in a step length of 15°. A higher-resolution motor, with a smaller step angle, can be constructed by having more teeth per stack or by having additional stacks. The use of more stacks will increase the motor length and it will increase the number of individual phases to be controlled, leading to increased system costs.

The flux generated in each pole will determine the torque which is generated. In a multistack motor, the four-pole windings can be connected either in series,

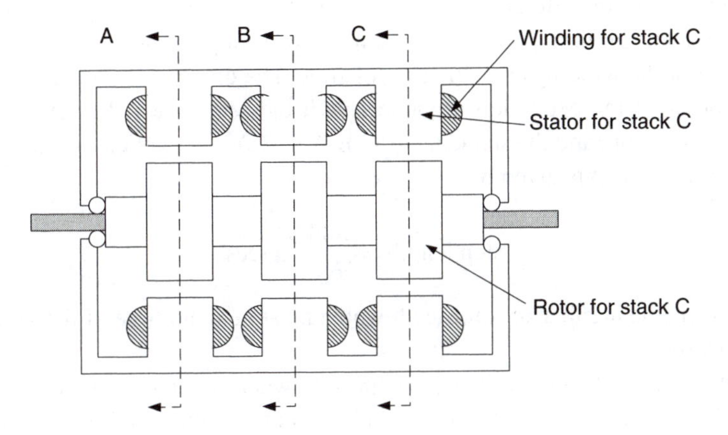

(a) Longitudinal cross section through the motor.

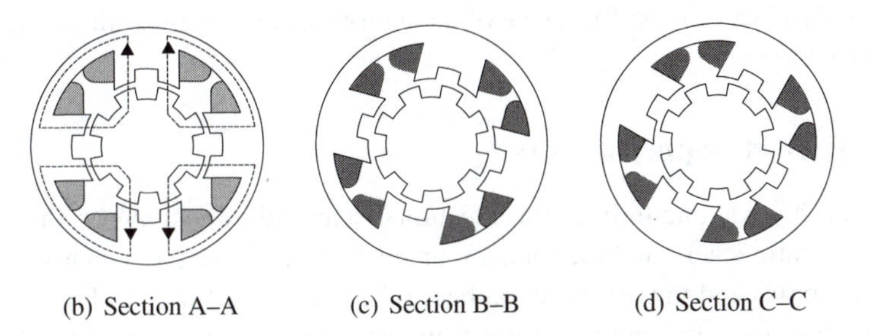

(b) Section A–A (c) Section B–B (d) Section C–C

Figure 8.1. A three-stack variable reluctance stepper motor; the flux path is shown for phase A.

in series-parallel, or in parallel, resulting in different characteristics for the power supply and for the controlling semiconductor switch

8.1.2 Single-stack variable-reluctance motors

The essential difference in construction between multistack and single-stack stepper motors is clearly apparent from Figure 8.2, which shows a longitudinal and a radial cross section of a single-stack motor. The motor consists of only one stack with three independent stator windings; in addition, the number of teeth on the rotor and stator are different. The operation of this form of stepper motor is, in principle, identical to the operation of a multistack stepper motor, with sequential excitation of the windings resulting in rotation. The direction is again determined by the order of the excitation sequence, with the sequence A:B:C:A:B:C... for clockwise rotation, and the sequence A:C:B:A:C:B:A... for anticlockwise rotation. The length of a step is given by

$$\text{step length} = \frac{360}{R_T}\text{degrees} \tag{8.2}$$

where R_T is the number of rotor teeth which must be a multiple of the number of motor phases.

Figure 8.2(a) shows the flux paths present when one motor winding is energised. It is readily apparent that a small amount of flux will leak via the teeth of the unexcited poles, which results in a degree of mutual coupling between the phases and reduces the performance of the motor in comparison with an equivalent multistack motor.

8.1.3 Hybrid stepper motors

Figure 8.3 shows a longitudinal cross section of a hybrid stepper motor; the location of the two stator stacks and the rotor-mounted permanent magnet can also be seen. The stator poles and the rotor are toothed; tyhe motor illustrated in Figure 8.3 has sixteen stator teeth and eighteen rotor teeth, and the teeth at either end of the rotor are displaced by half a tooth pitch relative to each other.

The main flux path is from the rotor magnet's north pole, through the rotor, the air gap and the stator at section X–X, through the back iron, and finally through the stator, the air gap and the rotor at section Y–Y, returning to the magnet's south pole. The motor is wound with two phases, with phase A wound onto poles 1, 3, 5, and 7, and phase B wound onto poles 2, 4, 6, and 8. In addition, the poles of each phase are wound in different directions, resulting in the flux directions which are shown in Table 8.1. For each winding, two different flux directions are possible if the winding is supplied with a bidirectional current.

The interaction between the stator windings and the rotor magnet can be studied by considering the case when phase A is energised by a positive current. Due to the presence of the permanent magnet, the flux in the cross section X–X must flow

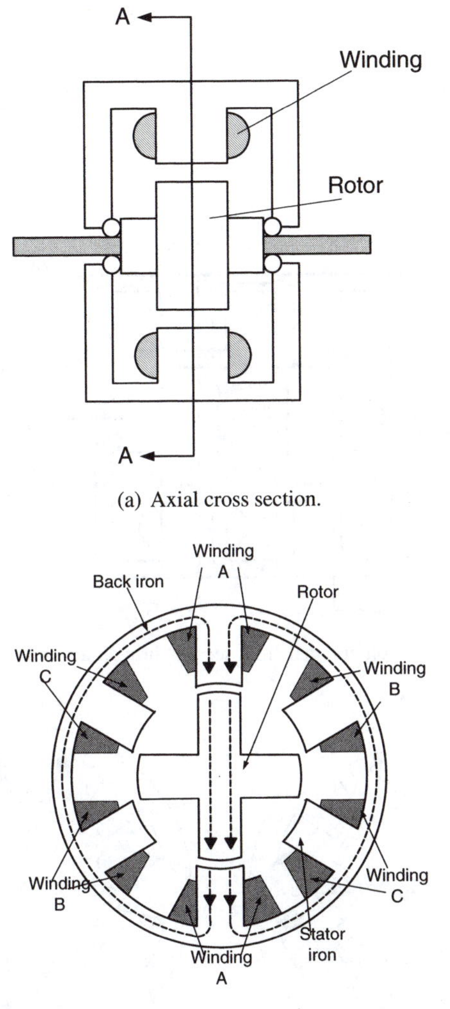

(a) Axial cross section.

(b) Radial cross section.

Figure 8.2. A single stack variable reluctance stepper motor, the flux path for phase A is shown in the radial cross section of the motor.

Table 8.1. The relationship between the radial-field direction and the excitation current for a hybrid stepper motor

Phase	Current direction	Direction of radial field	
		Outwards	Inwards
A	Positive	3,7	1,5
A	Negative	1,5	3,7
B	Positive	4,8	2,6
B	Negative	2,6	4,8

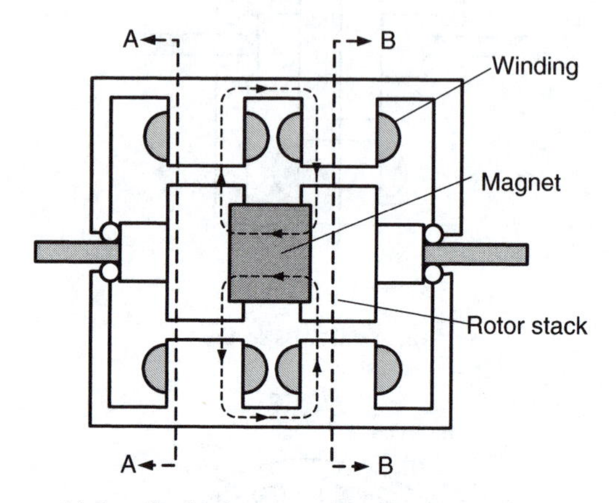

(a) Longitudinal cross section through the motor.

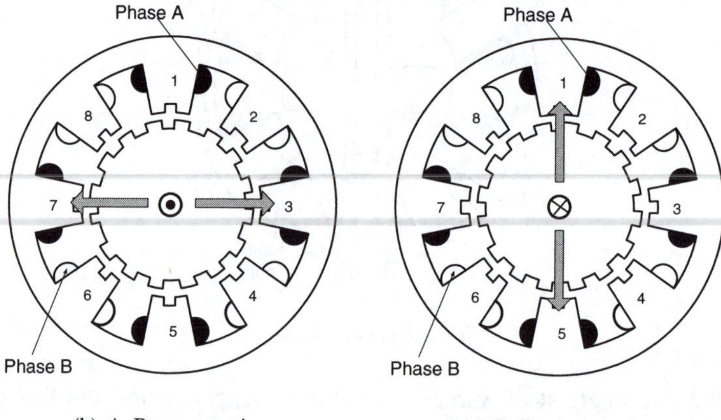

(b) A–B cross section (c) B–B cross section

Figure 8.3. A hybrid stepper motor. The radial cross-section through the stator stack shows the flux path if phase A is energised with a positive current. It should be noted that the view is from the outside of the motor in each case.

radially outwards, resulting in a flux concentration at poles 3 and 7; the opposite situation occurs at the other end of the motor, where the flux flows radially in, and the flux is concentrated in poles 1 and 5. If the magnetic flux is concentrated in certain poles, the rotor will tend to align along these poles to minimise the reluctance of the air-gap. When phase A is energised with a positive current, this will occur under poles 3 and 8 of section X–X, and under poles 1 and 5 of section Y–Y. Continuous rotation of the motor results from the sequential excitation of the two motor phases if the excitation of winding A has just been removed. and if winding B is now excited with a positive current, then alignment of the stator and rotor teeth has to occur under poles 4 and 8 of section X–X and under poles 2 and 6 of section Y–Y; the rotor has to move clockwise to achieve this alignment. Hence a clockwise rotation will require the excitation sequence, A+, B+, A-, B-, A+, B+ ..., and an anticlockwise rotation requires A+, B-, A-, B+, A+, B - . . . . The drive circuit for a hybrid stepper motor requires bidirection-current capability, either by the use of an H-bridge or of two unipolar drives if the motor is wound with bifilar windings.

As with variable-reluctance stepper motors, the step length can be related to the number of rotor teeth, and, as the complete cycle for a hybrid stepper requires four states, the step length is given by

$$\text{Step length} = \frac{90}{R_T} \tag{8.3}$$

where R_T is the number of. teeth on the rotor. In the example shown in Figure 7.5, the step angle is 5°; in practice motors are normally available with a somewhat smaller step length.

8.1.4 Linear stepper motor

The rotary stepper motor, when integrated into a package with a ball screw, is capable of giving incremental linear motor, and is a widely used solution for many low cost applications. However, over recent years the true linear stepper motor has become available. The operation of a linear stepper is in principle no different to a rotary machine. The key components of a linear stepper motor are shown Figure 8.4.

The moving assembly has a number of teeth that are similar to those found on the rotor in a traditional stepper motor, and incorporates two sets of windings and one permanent magnet. From the diagram it can be seen that one set of teeth is aligned with the teeth. As in a rotary stepper motor, energisation of a winding causes the teeth to align. The magnetic flux from the electromagnets also tends to reinforce the flux lines of one of the permanent magnets and cancels the flux lines of the other permanent magnet. The attraction of the forces at the time when peak current is flowing is up to ten times the holding force. When current flow to the coil is stopped, the moving assembly will align itself to the appropriate tooth set, and a holding force ensures that their is no movement. The linear stepper motor controller sets the energisation pattern for the windings so that the motor

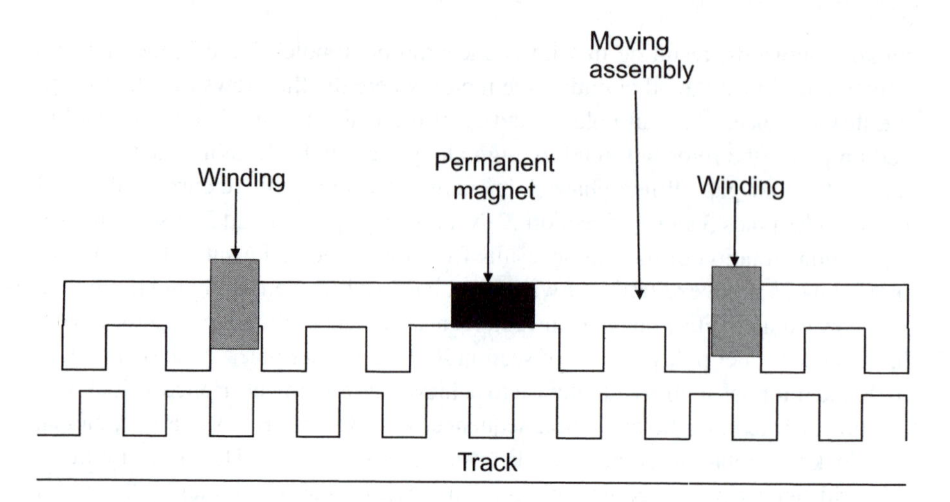

Figure 8.4. Cross section of a linear stepper motor. The motor consists of a stationary track, and a moving assembly incorporating magnets and the windings. As shown in the diagram, only one set of teeth on the moving assembly aligns with the track teeth.

moves smoothly in either direction. By reversing the pattern, the direction the motor travels is reversed.

8.1.5 Comparison of motor types

The previous sections have briefly reviewed a number of stepper motor configurations. Within a motor-selection procedure the various characteristics of each motor type will have to be considered, particularly those relating to the step size, the detent torque, and the rotor inertia:

- Hybrid stepper motors are available with smaller step sizes than variable reluctance motors; hence they are more suitable for limited-movement, high resolution applications. The larger step size of variable-reluctance motors, is more suited to extended high-speed motion, in which the required excitation the drives will be less than for in hybrid motors.

- The permanent magnets of the hybrid motor will produce a continuous detent torque, ensuring that the motor retains its position without the necessity of energising the drive. This is particularly useful for fail-safe applications, for example, following a power failure.

- The rotor's mass in variable reluctance stepper motors is less than its mass in hybrid motors; this ensures that the speed of response to a change in the demand is maximised. As will be discussed later, the inertia determines the mechanical resonance of the drive system the lower is the inertia, the higher is the allowable frequency of operation.

- While a linear motion can be obtained by the combination of a ball screw with any type of stepper motor, giving a low cost linear actuator, the liners stepper motor has a number of performance advantages. However, it should be noted that as with any linear motor, vertical operation can prove problematic.

8.2 Static-position accuracy

The majority of stepper-motor applications require accurate positioning of a mechanical load, for example within a small industrial robot. An externally applied load torque will give rise to positional errors when the motor is stationary, since the motor must develop sufficient torque to balance the load torque, otherwise it will be displaced from its equilibrium position. This error is noncumulative, and it is independent of the number of steps which have been previously executed. As the system's allowable error will determine which motor is selected for a particular application, the relationship between the motor, the drive and the load must be understood.

Figure 8.5 shows the relationship between the generated torque and the rotor position when a single phase is excited. At the point where the rotor and the stator teeth of the excited phase are in total alignment, no torque will be produced. As the rotor is moved away, a restoring torque results. The static-torque-rotor position characteristics repeats with a wavelength of one-rotor-tooth pitch; thus, if the rotor is moved by greater than $\pm 1/4$ tooth pitch, the rotor will not return to the initial position, but it will move to the next stable position. The shape of the curve is a function of the mechanical and the magnetic design of the motor, but it can be approximated to a sinusoidal curve with the peak value determined by the excitation current. If an external load is applied to the motor, the rotor must adopt an equilibrium position where the generated torque is equal to the external load torque. If the load exceeds the peak torque, the position cannot be held. The positional error introduced by an external load can be approximated by

$$\theta_e = \frac{\sin^{-1}(-T_L/T_{pk})}{R_T} \tag{8.4}$$

and this value can be reduced by either increasing the peak torque, T_{pk}, by an increased winding current, or by selecting a different motor with a larger number of rotor teeth.

Another measure of the motor's static-position error is to use the concept of stiffness, which is given by the gradient of the static-torque-position characteristic at the equilibrium position, K. The stiffness is given by the gradient of the torque-position characteristic at the equilibrium point; so, for a given displacement, the load torque that the motor will be able to support is given by

$$T = -K\theta_e \tag{8.5}$$

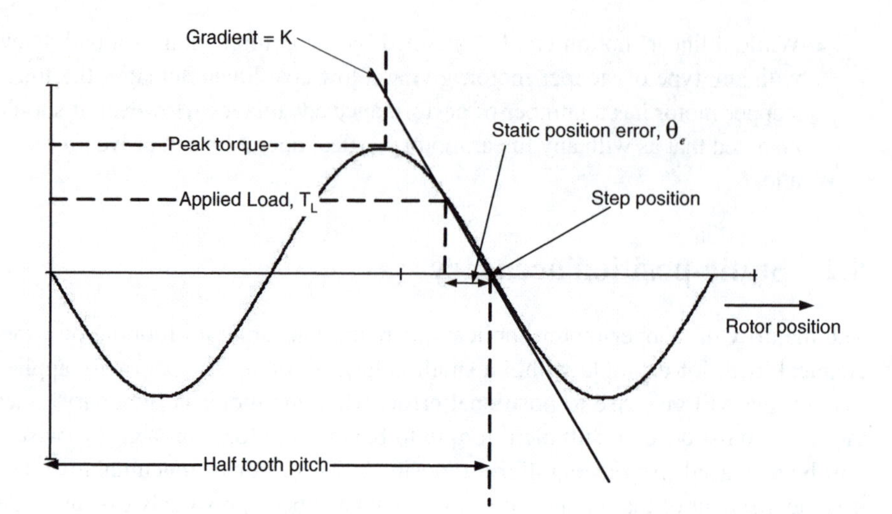

Figure 8.5. Static–torque rotor–position characteristics showing the static position error, θ_e due to the applied load T_L and the motor stiffness, K.

In some motors the torque-position characteristic is shaped to result in a different stiffness for different displacements; in this case, the stiffness which is closest to the expected amplitude must be selected.

Example 8.1

Determine the static position error for a stepper motor with eight rotor teeth, rated at 1.2 Nm, when a load of 0.6 Nm is applied.

The approximate positional error is defined by equation 8.4, hence

$$\theta_e = \frac{\sin^{-1}(-T_L/T_{pk})}{R_T} = \frac{sin^{-1}(-0.6/1.2)}{8} = 3.8°$$

In practice this value is less than that experienced by the actual system, due to the approximations used.

8.3 Torque-speed characteristics

In the application of a stepper motor to a motion-control system, the designer needs knowledge of the motor's torque speed characteristics. This information is supplied by the manufacturers in the form of pull-out characteristics, which show the maximum torque that can be developed at any speed (see Figure 8.6). If the applied load torque exceeds the torque that can be generated by the motor, the motor will pull out of synchronism with the magnetic field, and it will stall. From Figure 8.6, the following points can be noted:

- The motor is capable of operating with a load of T', up to a speed of N' (steps s^{-1}). Above this speed, the motor will not start.
- There are significant dips in the pull-out-torque curve at a number of speeds. These dips are caused by resonance between the motor and the excitation frequency.
- At low motor speeds the phase currents are effectively rectangular. At high speeds, the time constant for the phase current's rise and decay will become a significant proportion of the total available excitation time (see Figure 8.7). Therefore, the effective phase current, and hence the torque which is produced, will be reduced. In addition, as shown in Figure 8.7, high speeds result in an induced stator voltage which also distorts the current waveform. This is particularly marked with hybrid stepper motors because of the presence of the permanent magnet in the rotor.

As shown in Figure 8.7, the phase currents of a stepper motor are almost rectangular at low speeds, allowing the pull-out torque of a motor to be determined from the static-torque-rotor-position characteristics for a particular excitation scheme. The pull-out torque can, within certain limits, be dependent on the driven inertia. With a high load inertia, the pulsating variations of the motor torque will only lead to small variations of the motor speed. Under these conditions, the pull-out torque can be considered to be equal to the average motor torque. If the sum of the motor and the load inertias is low, the motor will stall whenever the load torque exceeds the generated torque.

Since stepper motors are designed to operate in discrete steps, at very slow speeds, the motor will come to rest between each excitation. Due to the dynamics of stepper motors and their loads, the single-step transient behaviour tends to be very oscillatory, and the effects of this have to be considered in the design of an overall system, because they can result in significant accuracy problems in a poorly damped system. As discussed in Section 3.6, the undamped natural frequency of oscillation in a drive system was shown to be

$$f_n = \frac{1}{2\pi}\left(\sqrt{\frac{K}{J_{tot}}}\right) \quad (8.6)$$

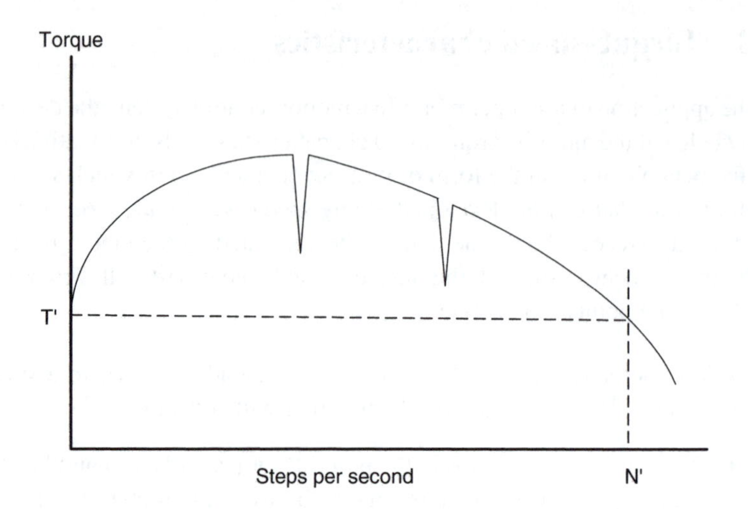

Figure 8.6. Typical pull-out torque-speed characteristics of a stepper motor: the dips in the curve are resonance points. In order to achieve torque throughout the speed range, a value less than the peak torque must be selected.

where K is the stiffness at the rotor position under consideration and J_{tot} is the sum of the motor inertia and the load inertia reflected back to the motor.

This oscillating behaviour can be damped out, if required, for single-step operations by the use of mechanical (that is, viscous) or electrical damping. Excessive vibration of the mechanical system will result in wear, leading to premature mechanical failures.

This resonance behaviour results in a loss of torque at well-defined stepping rates, as shown in the pull-out torque-speed characteristic in Figure 8.6. These stepping rates can be determined from the natural frequency of the system, and they are given by,

$$f_k = \frac{f_n}{k} \quad \text{(for } k = 1, 2, \ldots) \tag{8.7}$$

Hence, if the motor and load have a natural resonance frequency of 120 Hz, the dips in the speed-torque curve will occur at 40, 60, 120, steps s^{-1}.

8.4 Control of stepper motors

The design of a drive system that incorporates a stepper motor should start with consideration of the steady-state performance; the choice of the type and step angle of the stepper motor is dictated largely by the maximum allowable positional error and by the maximum stepping rate which is required. While a stepper motor can be operated under either an open-loop or a closed-loop control system, this chapter will primarily discuss the open-loop approach. Closed-loop stepper-motor

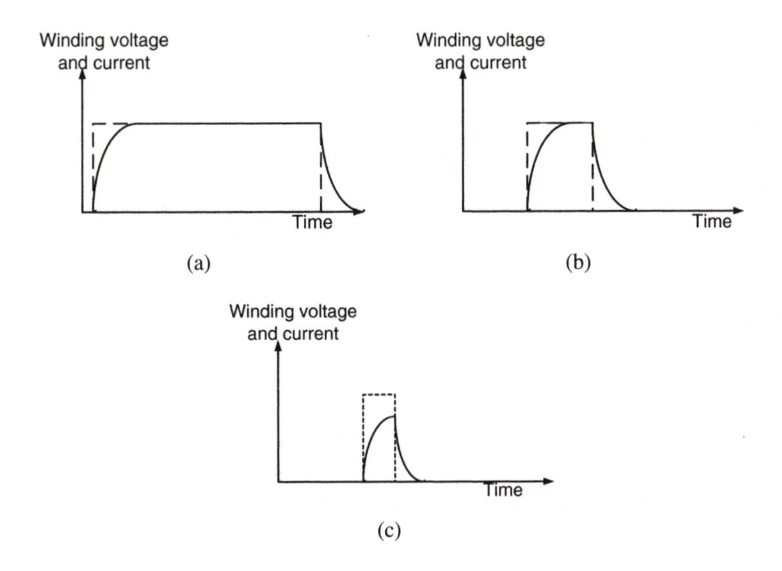

Figure 8.7. Current through a unipolar stepper-motor winding as a function of speed. The applied voltage is shown by the dotted line: (a) low speed, (b) medium speed and (c) high speed.

drives are no different from any other closed-loop drive, which will be discussed in Chapter 10. Due to the inherent operation of a stepper motor, one change of phase excitation will result in the motor moving a specified, and accurately known, distance. The stepper motor's position is controlled by generating a pulse train of known length, which is converted into the correct sequence of winding excitations by a translator, the winding power being switched by the drive circuit. A block diagram of a typical open-loop-stepper drive system is shown in Figure 8.8.

During the design process, information is required on the restrictions that have to be placed on the timing of the pulse train to ensure satisfactory operation. These restrictions can be summarised as:

- The maximum step rate permitted for the required load torque. This can be determined from the motor's pull-out characteristic.

- The motor's transient performance. If the load has a high inertia, the motor's speed must be ramped up to ensure that the motor remains in synchronism with the step demand.

There is no feedback from the motor or load to the controller in an open-loop system, so it is imperative that the motor responds correctly to each incoming pulse because any loss in position cannot be detected and then compensated for. In order to achieve a satisfactory performance from a stepper motor, the operation of the pulse generator should be carefully considered during the design process.

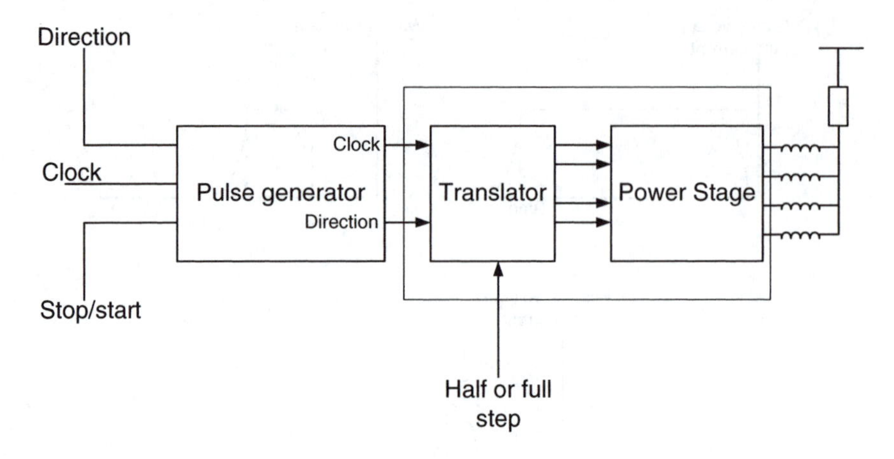

Figure 8.8. A block diagram of an open-loop stepper drive.

8.4.1 Open-loop control

An open-loop position-controller for a stepper-motor generates a string of pulses, at a fixed frequency, until the motor reaches the target position. If the pulse rate is set too high, and the load has a high inertia or static friction, the motor may not be able to accelerate to the required speed without losing steps; or, in an extreme case, it can fail to rotate at all. If the pulse frequency to the motor is ramped up, it will be possible to ensure that, under normal operating conditions, the motor does not lose synchronism. The maximum allowable starting rate for a motor can be determined from a knowledge of the motor and the load. The equation of motion for a system of inertia J_{tot} is given by

$$T_m - T_L = J_{tot} \frac{d^2\theta}{dt^2} \tag{8.8}$$

where T_L is the load torque and T_m is the average output torque of the motor. If this equation solved using the initial conditions $t = 0$, $\theta = \theta_e$ (where θ is the static error due to the load torque), and $d\theta/dt = 0$ then

$$\theta = \frac{(T_m - T_L)t^2}{J_{tot}} + \theta_e \tag{8.9}$$

After one excitation period of length t_p, the rotor will be at a new position θ_f (see Figure 8.9); the maximum allowable initial stepping rate can determined to be;

$$f_{start} = \frac{1}{t_p} = \sqrt{\frac{T_m - T_L}{J_{tot}(\theta_f - \theta_e)}} \tag{8.10}$$

As expected, the lower the load inertia, or the greater the motor torque, the higher the permissible starting frequency. The starting frequency in most applications will be lower than the at-speed frequency. Therefore, an acceleration-deceleration

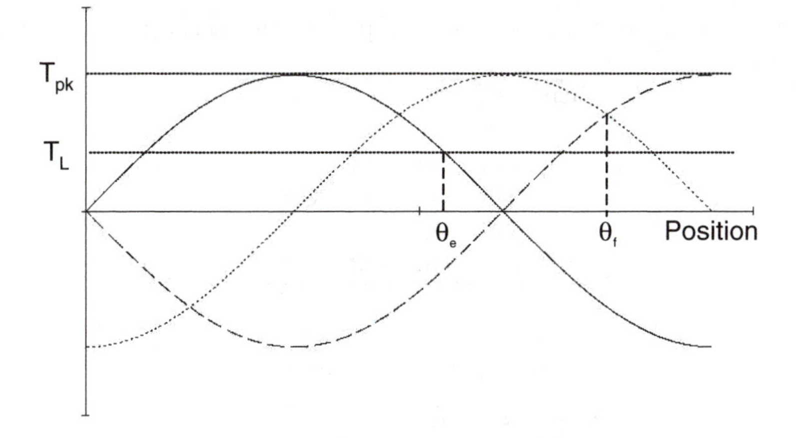

Figure 8.9. The static-torque characteristic for a stepper motor with an applied load. When a phase is energised the motor has to move from θ_e to θ_f without loss of synchronism. The three curves are the individual static-torque characteristics of the individual phases.

capability must be provided; this is normally in the form of a variable-frequency pulse generator. To realise this characteristics, a number of different approaches can be taken, based either on dedicated hardware or on microprocessors.

8.4.2 Translators and drive circuits

The output from the pulse generator forms the input to the stepper-motor's translator and drive circuit. The drive circuit for a stepper motor is normally of a lower rating and complexity than for the motors that have been discussed previously. The function of the translator is to control the excitation of the motor phases in response to the incoming pulses and the required direction of motion; this is achieved through the use of a shift register and a look-up table, which is normally provided within a single integrated circuit. The output sequence for a full step switching pattern is given in Table 8.2.

Since the phase windings of both hybrid and variable-reluctance stepper motors are electrically isolated and controlled by individual drive circuits, the possibility of energising a number of phases simultaneously can be considered. If one winding of a stepper motor is excited, a stable-equilibrium point will occur every rotor-pole pitch at positions A′, B′ and C′ in the case of a three-stack variable-reluctance motor, Figure 8.10.

If, on the other hand, two phase-windings are excited, the resultant torque summation will produce two new equilibrium points, BA′ and CB′, which are midway between the single-winding equilibrium points. Therefore, if the windings of the variable reluctance stepper motor considered earlier in this chapter are excited in

Table 8.2. The full step sequence: the four power-stage outputs are identified in Figure 8.8

Step	A	B	C	D
1	On	Off	Off	Off
2	Off	On	Off	Off
3	Off	Off	On	Off
4	Off	Off	Off	On
5	On	Off	Off	Off

Table 8.3. Half step sequence

Step	A	B	C	D
1	On	Off	Off	Off
2	On	On	Off	Off
3	On	On	Off	Off
4	Off	On	On	Off
5	Off	Off	On	Off
6	Off	Off	On	On
7	Off	Off	Off	On
8	On	Off	Off	On

the sequence A, BA, B, CB, C, AC, A . . ., each excitation change will result in a movement of half its normal step. This approach to stepper-motor control is termed half-stepping. It should also be noted that the peak-torque resultant for multiphase energisation is greater than that which occurs when a single phase is used. This is of particular importance when the number of stacks is greater than three. It is normal practice to energise three or four phases in any one time in the control of a seven-stack stepper motor. Half-stepping operations can be applied to hybrid stepper motors, but due to the bipolar nature of these motors' drives the power capacity of the drive system has to be increased. For a forty per cent increase in the torque, the power supply has to be increased by one hundred per cent.

In half-stepping, the phases are energised at the rated current. If the current in each phase is controlled, it is possible to produce further equilibrium points, leading to further subdivisions of a motor's basic step; this approach is termed *mini-stepping*. While this approach will result in a greater resolution, each phase's current must be individually controlled, leading to additional complexity in the drive system; the switching pattern is given in Table 8.3.

Hybrid stepper motosr have a bipolar-current requirement, whereas variable reluctance stepper motors require a unipolar drive. In a unipolar drive, the output of the translator is directly used to switch the individual phase currents; the power devices are normally MOSFETs. Since the winding current's decay time has an adverse effect stepper-motor performance, it is common practice to add a zener

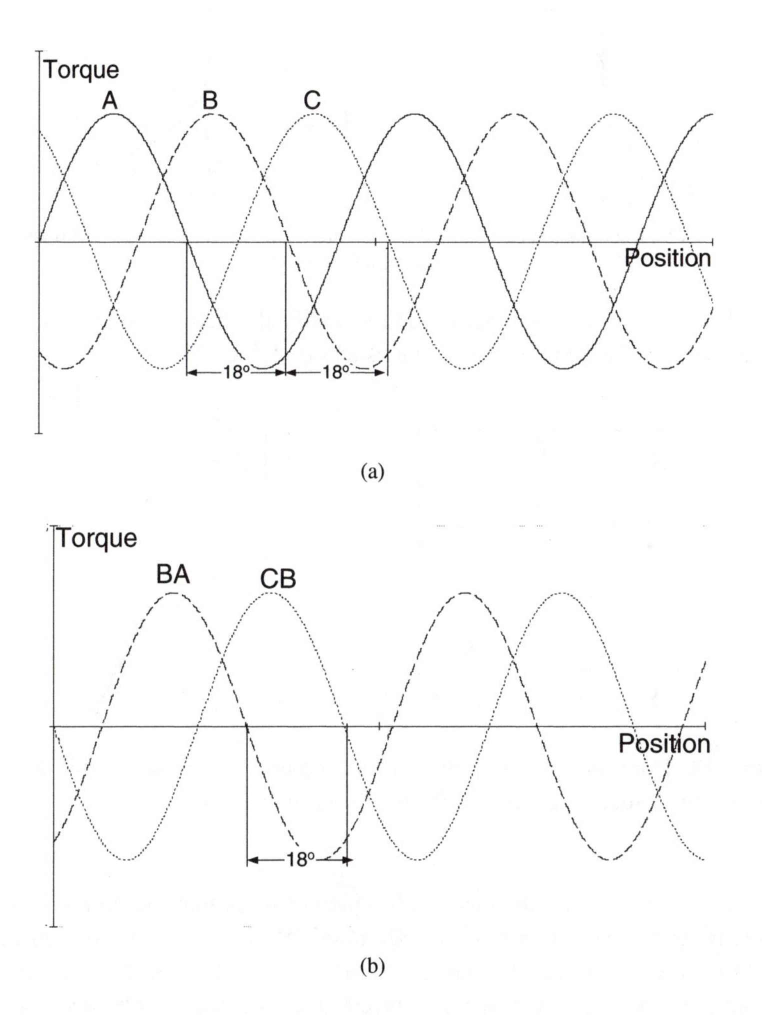

Figure 8.10. Static-rotor rotor-position characteristics when: (a) one phase is excited and (b) two phases are excited. If both curves are combined the step angle is reduced from 18° to 9°.

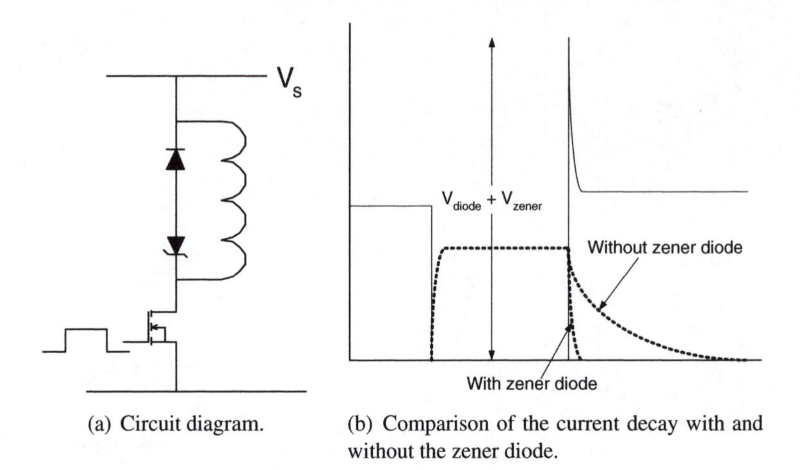

(a) Circuit diagram.

(b) Comparison of the current decay with and without the zener diode.

Figure 8.11. The use of a zener diode to modify the delay characteristics of the winding current, the currents are shown as a dotted line.

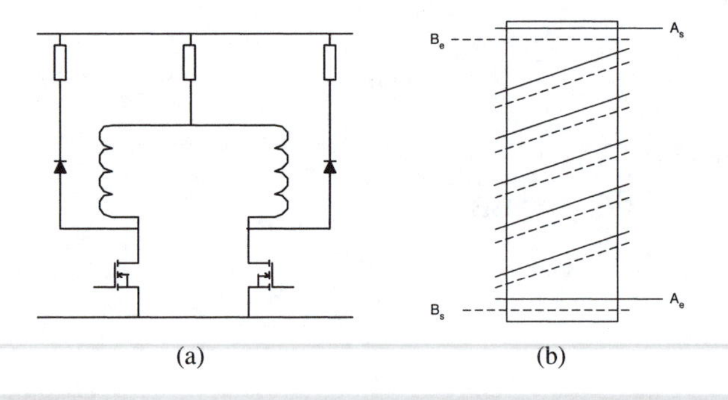

(a) (b)

Figure 8.12. The use of two unipolar drives to control one phase of a bifilar–wound motor: (a) the circuit diagram and (b) the bifilar windings.

diode or a resistor to the flywheel path which ensures that the current decays at an increased rate , (see Figure 8.11). Bidirectional winding currents can be controlled by using an H-bridge, identical to that used in d.c. brushed motors. With this configuration, the free-wheeling current decays more rapidly, because of the opposition of the supply voltage, so it is not necessary to add a resistance or zener diodes to the flywheel path. A different approach is the use of a motor wound with a bifilar winding; this will result in a reduction in the number of switching devices used to control a phase from four to two (see Figure 8.12). Each of the bifilar windings has as many turns as the equivalent winding for a bipolar motor, so the size and cost of the motor increases; but this is counterbalanced by an equivalent reduction in the drive costs. As the two windings share the same pole, there is close magnetic coupling between the coils; this must be taken into account when designing a drive system.

8.5 Summary

This chapter has briefly reviewed the theory and control of a range of stepper motors, and it has been shown that stepper motors are able to provide a low-cost solution to motion-control problems, provided that the limitations of the motors are fully appreciated during the design process.

Chapter 9

Related motors and actuators

The previous chapters have considering the motors and drives that are normally used within the range of applications that have been identified in Chapter 1. However, there are a number of specialist or unconventional motors that can and are being used in an increasing number of applications. These motors may be selected for a wide verity of reasons, both technical and commercial. This chapter considers a number of theses motors and their associated controllers, therefore allowing the design engineer to have an overview of all available technologies. In this chapter the following motors are considered:

- voice coil actuators
- limited-angle torque motors
- piezoelectric motors
- switched reluctance motors
- shape memory alloy, SMA.

While these motors currently have specialist niches in servo drive applications, a range of exciting motors are currently being developed based a wide range of technologies, including electrostatic and micro electromechanical (MEM)technologies, and these will no doubt find their way into more general use over time (Hameyer and Belmans, 1999). Currently this technology is still the research stage, but the applications currently being explored are significant and challenging, and for example include the manipulation of a single DNA molecule (Chiou and Lee, 2005).

9.1 Voice coils

Voice coils or solenoids are ideally suited for short linear (typically less than 50 mm) closed-loop servo applications and both operate on similar principles. In a voice coil, the actual coil moves, while in a solenoid, the iron core moves. Typical

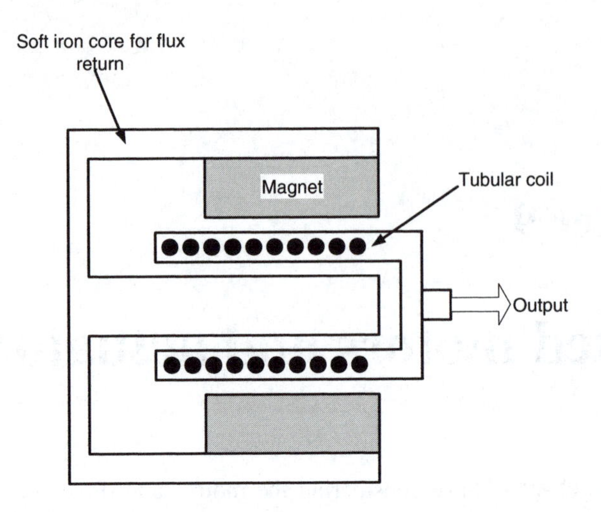

Figure 9.1. The cross section of a voice coil; the dimensions of the air gap has been exaggerated.

positioning applications include direct drives on pick and place equipment, medical equipment, and mirror tilt and focusing actuators. In addition voice coils can also be used in applications where precise force control is required, due to the linear force versus current characteristics.

A voice coil is wound in such a way that no commutation is required, hence a simple linear amplifier can be used to control the actuator's position. The result is a much simpler and more reliable system. Voice coils have a number of significant advantages including small size, very low electrical and mechanical time constants, and low moving mass that allows allows for high accelerations, though this depends on the load being moved.

Voice coil actuators are direct drive, limited motion devices that utilise a permanent magnet field and coil winding (conductor) to produce a force proportional to the current applied to the coil. These non-commutated electromagnet devices are used in linear (or rotary) motion applications requiring a linear force output, high acceleration, or high frequency actuation.

The electromechanical conversion mechanism of a voice coil actuator is governed by the Lorentz force principle; which states that if current-carrying conductor is placed in a magnetic field, a force will result. The magnitude of the force is determined by the magnetic flux destiny, B, the current, i, hence for a winding of N turns, the resultant force is given by

$$F = BLiN \tag{9.1}$$

In its simplest form, a linear voice coil actuator is a tubular coil of wire situated within a radially oriented magnetic field, as shown in Figure 9.1. The field is produced by permanent magnets embedded on the inside of a ferromagnetic cylinder.

The inner core of ferromagnetic material is aligned along the axial centreline of the coil, and joined at one end to the permanent magnet assembly, is used to complete the magnetic circuit. The force generated axially upon the coil when current flows through the coil will produce relative motion between the field assembly and the coil, provided the force is large enough to overcome friction, inertia, and any other forces from loads attached to the coil. For a specific operating displacement of the actuator, the axial lengths of the coil and the magnet assemblies can be chosen such that the force vs displacement curve can be optimised, resulting in the reduction of force at the mid-stroke force being limited to less than 5% of the maximum force.

The sizing and selection of a voice coil actuator is no different from any other linear application, the process defined in Section 3.8.4 can be followed.

9.2 Limited-angle torque motors

Limited-angle torque motors are a range of special-purpose motors that are capable of giving controllable motion up to $\pm 90°$ from their rest position. While brushless motors, as discussed in Chapter 6, have many benefits, they have the penalty of being relatively expensive and complex, if only a limited range of motion is required. The requirement for a limited range of movement can be found in many applications, including the operation of air or hydraulic servo-valves and oscillating mirrors. In addition, their inherent reliability of operation makes a limited-angle torque motors an ideal solution for applications where limited actuation is critical, for example in spacecraft latches, where the only previous solution was to use pyrotechnics.

The basic construction of a limited-angle torque motor is shown in Figure 9.2. While they are broadly similar to brushless d.c. motors, the limited-angle torque motor is a single-phase device, which eliminate the need for the commutation logic and the three-phase power bridge that are found in multiphase machines. The torque motor's winding can be wound in conventional slots or as a toroid over a slotless stator. The rotor in a limited-angle torquer incorporates one or more magnets. The slot-wound limited-angle torque motor has a number of advantages over toroidally wound motors; in particular they have better thermal dissipation and a higher torque constant. However, because of the presence of slots, the output torque ripple and hysteresis losses are greater. The torque ripple can be considered to be zero with toroidally wound motors due to the non-varying reluctance path and the large air gap. In addition the slot-wound limited-angle torque motor exhibits a higher motor constant, K_m, than the corresponding toroidally wound motor, due to the larger number of conductors that are exposed to the magnetic field.

Cogging is essentially zero in toroidally wound limited–angle torque motor, a result of a non-varying reluctance path and relatively large air gap. Toroidally wound armatures, moreover, are typically moulded onto the stator, which protects the windings from damage and holds them in place.

In the selection of a limited-angle torque motor for an application, a number of

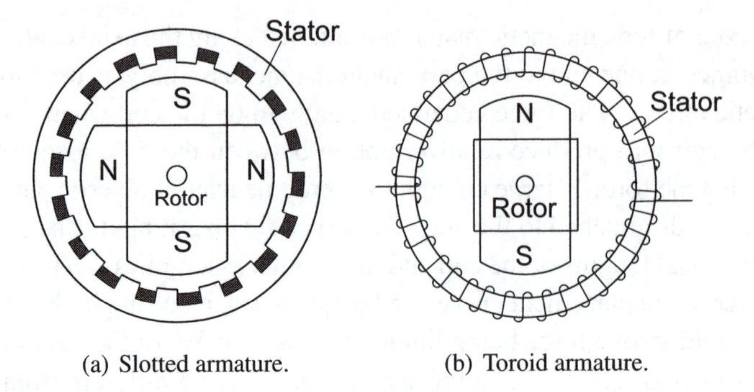

(a) Slotted armature. (b) Toroid armature.

Figure 9.2. Internal construction of limited angle torque motors.

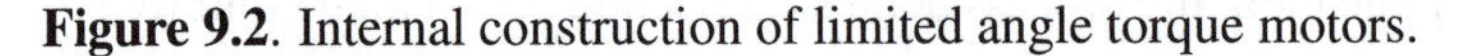

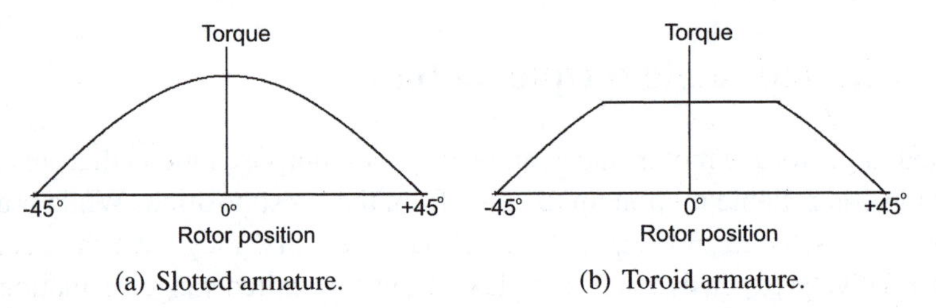

(a) Slotted armature. (b) Toroid armature.

Figure 9.3. Torque-position characteristics for a limited angle torque motor.

parameters shall be considered, including:

- *Peak torque*. As in a conventional motor, this is the torque which is developed at the rated current.
- *Excursion angle*. This is the maximum angle that the rotor can move from the peak-torque position, and it is normally expressed as a plus/ minus value. Figure 9.3 shows typical characteristics for a slot-wound and a toroidally wound motor. In the latter case, the constant-torque region should be noted. Limited-angle torque motors are currently available in ratings from 7×10^{-4} to 0.142 N m, with excursion angles between $\pm 18°$ and $\pm 90°$.

As limited-angle torque motor are single-phase motors, they are easily controlled by single-phase bipolar PWM amplifiers which are identical to those used with brushed d.c. motors. In certain applications, a linear amplifier could be used to increase the bandwidth and to reduce the electrical noise. The limited-angle torque motor produces torque through a rotation angle determined by the number of motor poles. Current of one polarity produces clockwise torque, and vice versa. Manufacturers generally provide a theoretical torque versus shaft-position curve. Typically, the characteristic curve for a slotted armature limited-angle torque motor is represented by a cosine function; that is

$$T = T_p \cos \frac{\theta N}{2} \tag{9.2}$$

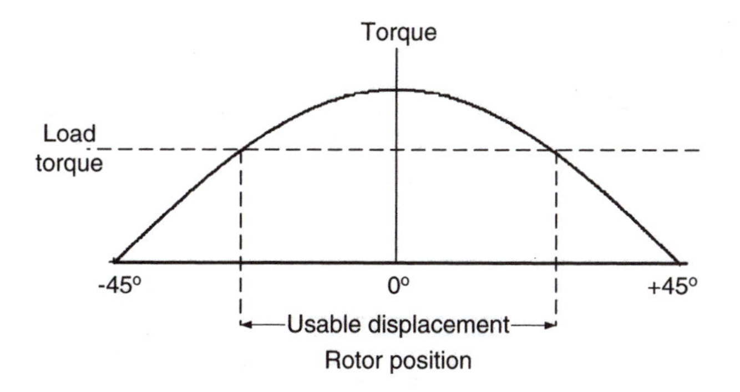

Figure 9.4. The restriction in usable displacement of a limited-angle torque motor as a function of load torque.

where θ is angle of rotation, N is number of poles, and T_p is the peak torque. The general torque characteristic for toroidally wound motors can be represented by a similar curve, but it may also have a flat top.

The selection of a limited-angle torque motor for an application follows an identical route to that of any motor. The process starts with the determination of the application's constraints and of the performance which is required. Once the torque, and the angle over which it is to be applied, has been determined, the suppliers data must be referred to. As the torque-angle characteristic of limited-angle torque motor is sinusoidal, care must be taken to ensure that these devices can produce the required torque throughout the proposed actuation angle, as shown in Figure 9.4.

9.3 Piezoelectric motors

Many specialist applications require motors of extremely high resolution, for example, micropositioning stages, fibre-optic positioning, and medical catheter placement. One motor that can meet these requirements is the piezoelectric motor. When compared to a conventional motors and its associated power train, the piezoelectric motor has a faster response times, far higher precision, inherent brake capability with no backlash, high power-to-weight ratio, and is of smaller size.

The operation of this motor is based on the use of piezoelectric materials where a material is capable of being deformed by the application of a voltage. A range of materials such as quartz (SiO_2) or barium titanate ($BaTiO_3$) exhibit the piezoelectric effect. However in motors normally mass-produced polycrystalline piezoelectric ceramic is used. To produce a suitable ceramic, a number of chemicals are processed, pressed to shape, fired, and polarised. Polarisation is achieved using high electric fields (2500 V/mm) to align material domains along a primary axis. In Figure 9.5(c), a voltage is applied to a piezoelectric crystal to produce a displacement. If the material has a displacement constant of 500 pm V^{-1}, the application

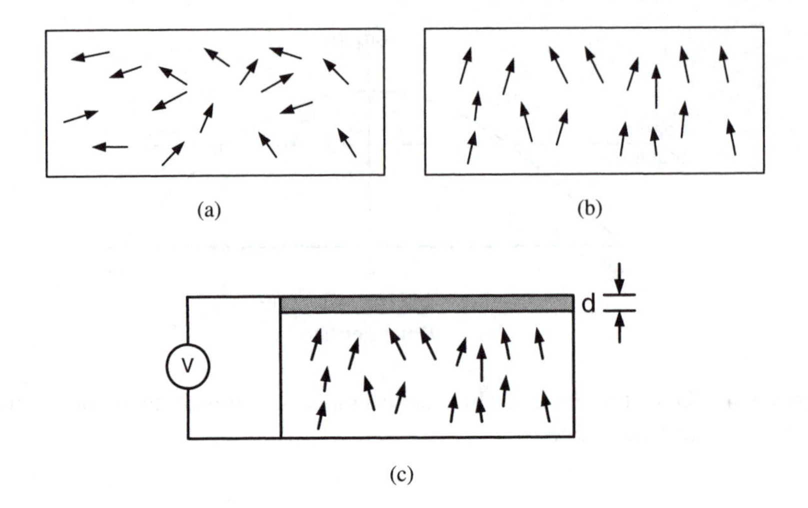

Figure 9.5. The characteristic of a piezoelectric material. (a) shows domains in the the unpolarised material, which align when polarised, as shown in (b). The application of a voltage causes axial displacement, *d*.

of 200 V, will produces an axial displacement of 0.1 μm.

Figure 9.6 shows the basic concepts of a piezoelectric motor. Two piezoelectric crystals are preloaded against a flat wear surface, by way of the motor shoe, to produce a normal contact force. The friction is important in the design of the motor, since the friction force is used to translate the motion of the piezoelectric ceramic into the motor's output. As a positive sinusoidal voltage waveform is applied which increase its thickness, the axial motion imparts a frictional force along the wear strip. When the drive voltage goes negative, the same crystal thickness contacts. This action creates a separation between the motor shoe and the wear strip, allowing the motor to return to its original position without dragging the wear strip backward. As the drive voltage swings positive again, the crystal stroke cycle repeats and the wear strip moves another incremental step to the left.

9.4 Switched reluctance motors

While not originally designed for high-performance servo applications, the switched reluctance motor is making inroads into this area, due to the availability of low-cost digital signal processing. The switched reluctance motor is particularly suitable to a wide range of applications due to the robustness of the mechanical and electrical design.

In a reluctance machine, the torque is produced by the moving component moving to a position such that the inductance of the excited winding is maximised. The moving component is typically the machine's rotor – which can be either internal or external depending on the design – or a linear component in the case of a linear

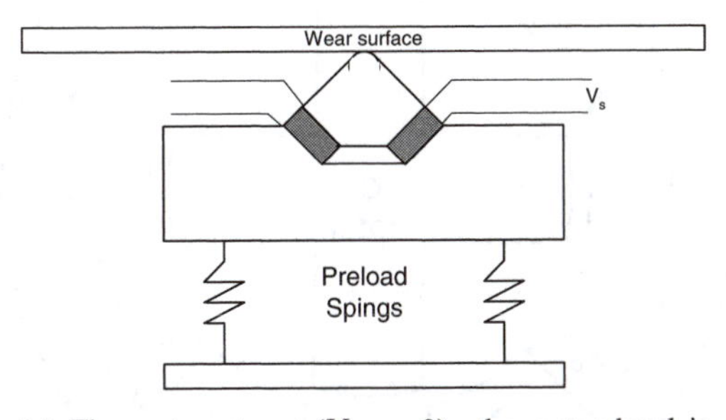

(a) The motor at rest ($V_s = 0$): the motor head is preloaded against the wear surface.

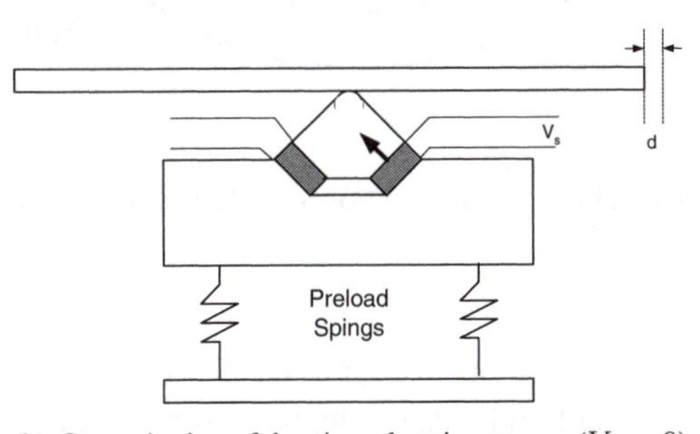

(b) On excitation of the piezoelectric actuator ($V_s > 0$), the head moves against the wear surface, moving the wear surface.

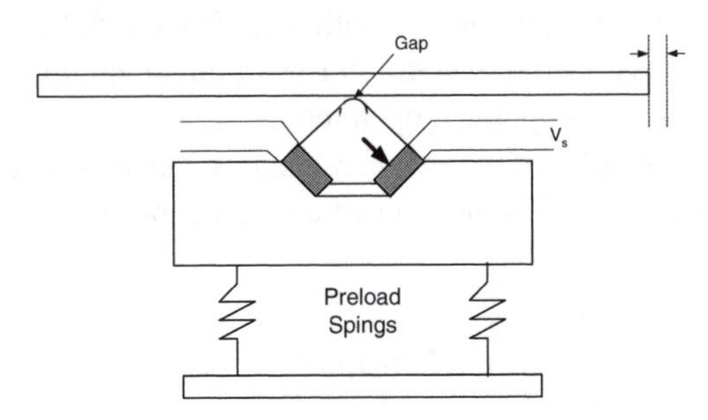

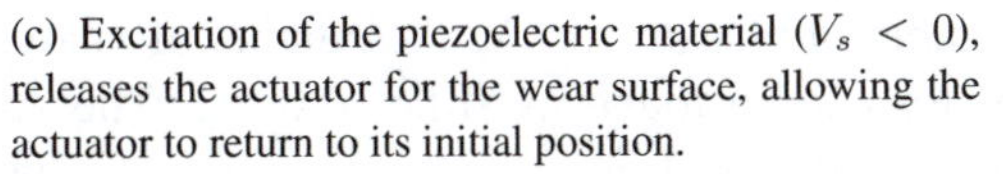

(c) Excitation of the piezoelectric material ($V_s < 0$), releases the actuator for the wear surface, allowing the actuator to return to its initial position.

Figure 9.6. The operation of a piezoelectric motor.

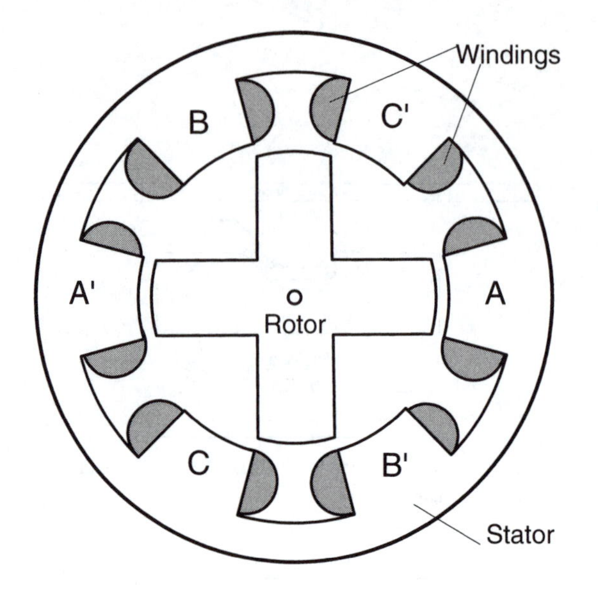

Figure 9.7. The cross section of a switched reluctance motor.

reluctance motor.

The switched reluctance motor is topologically and electromagnetically similar in design to the variable-reluctance stepper motor discussed in Section 8.1.2. The key differences lie in the details of the engineering design, the approach to control, and hence its performance characteristics. The switched reluctance motor is operated under closed loop control, with a shaft mounted encoder being used to synchronise the phase currents with rotor position. In comparison the variable-reluctance stepper motor is operated open loop.

The operating principles of the switch reluctance machine can be considered by examination of Figure 9.7. The number of cycles of torque production per motor revolutions is given by

$$S = m\,N_r \tag{9.3}$$

where m is the number of phases, and N_r the number of phases. A more detailed analysis of the motor can be found in Miller (2001). The voltage equation for a single phase can be calculated in a similar fashion to that used for a brushless motor

$$v = Ri + \frac{d\psi}{dt} = Ri + \omega_m \frac{d\psi}{d\theta} \tag{9.4}$$

where v is the terminal voltage, i is the phase current, ψ is the flux-linkage in volt-seconds, R is the phase voltage, L is the inductance of the phase winding, θ

is the rotor position and ω_m is the rotor's angular velocity. This equation can be expanded to give

$$v = Ri + \omega_m \frac{d(Li)}{d\theta} = Ri + L\frac{di}{dt} + \omega_m i \frac{dL}{d\theta} \tag{9.5}$$

In a similar fashion to a d.c. brushed motor it is useful to consider the terminal voltage v as the sum of three components: the resistive voltage drop, the voltage drop due to the inductance and rate of change of current, and the back e.m.f. term, e

$$e = \omega_m i \frac{dL}{d\theta} \tag{9.6}$$

From equation 9.5 it is possible to calculate the instantaneous electrical power, vi, as,

$$vi = Ri^2 + Li\frac{di}{dt} + \omega_m i^2 \frac{dL}{d\theta} \tag{9.7}$$

which allows the rate of change in magnetic energy to be calculated:

$$\frac{d}{dt}\left(\frac{1}{2}Li^2\right) = \frac{1}{2}i^2\omega_m \frac{dL}{d\theta} = Li\frac{di}{dt} \tag{9.8}$$

The electromagnetic torque generated by the motor can therefore be determined from the instantaneous electrical power minus the resistive voltages drops due and the rate of change of magnetic stored energy:

$$T_e = \frac{vi - Ri^2 - \frac{d}{dt}\left(\frac{1}{2}Li^2\right)}{\omega_m} \tag{9.9}$$

$$= \frac{1}{2}i^2\frac{dL}{d\theta} \tag{9.10}$$

The rate of change of inductance as a function of rotor position is one of the design parameters of the switched reluctance machine. From equation 9.9 it is clear that the torque does not depend on the direction of current flow, however the voltage must be reversed to reduce the flux-linkage to zero. A suitable power circuit for a single winding is shown in Figure 9.8. It is immediately clear that this circuit is far more robust that the conventional PWM bridge shown in Figure 6.5(a), as a line-to-line short circuit is not possible.

The circuit shown in Figure 9.8 is capable of operating the motor as either a motor or a generator, as vi can either be positive or negative, and the power flow is determined by the switching pattern of the power bridge relative to the rotor's position. A block diagram of a suitable controller for a basic switched reluctance motor is shown in Figure 9.9. It is recognised that although this type of drive is simple, and gives adequate performance for speed control, it is incapable of providing instantaneous torque control as required by a servo or similar application.

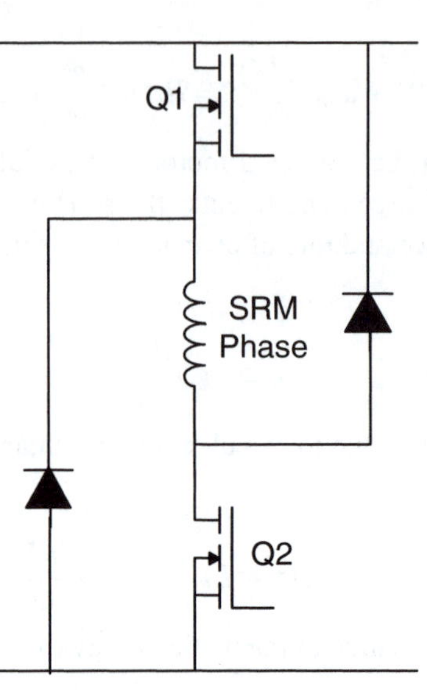

Figure 9.8. A single phaseleg as used in a switched reluctance motor. The current direction is determined by Q1 and Q2, with the respective flywheel diodes.

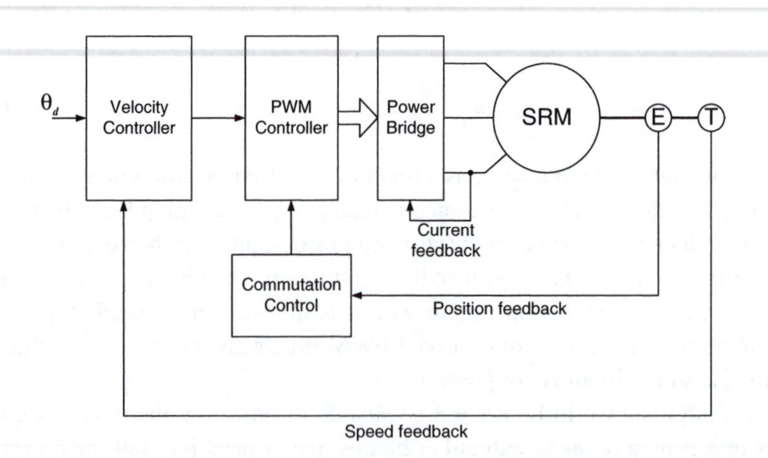

Figure 9.9. A typical controller for a switched reluctance machine operating under velocity control.

To achieve a performance similar to a conventional servo-drive, commutation as a function of rotor position has to be replaced by a control strategy that produces the desired total motor torques by controlling the individual phase currents. The approach taken is very similar in principle to that used to control the a.c. induction motor as discussed in Section 7.3, and in the paper by Kjaer et al. (1997). It is clear that the switched reluctance machine is a very robust machine, that could offer the designer of high performance application an additional choice in the drive selection. While the switched reluctance machine is becoming widely used in high speed applications is has not been seen in the high performance position control application.

9.5 Shape-memory alloy

Shape-memory alloy (SMA) materials have the unique ability to return to a predetermined shape when heated, leading to their uses in a wide range of applications, particularly when micro-actuation is required. This property arises due to a reversible crystalline phase transformation that occurs between the low temperature martensite and high temperature austenite phases. Although the phases have the same chemical composition and atomic order, the two phases have different crystallographic structures. Austenite has a body-centered symmetric structure that exists at high temperature, while martensite has a low symmetric monoclinic structure that stabilises at relatively low temperature (Jones, 2001, p148). When an SMA is cooled from a high temperature, the material undergoes a martensitic transformation from the high temperature austenite. Since the bond energy in the martensite is low, this phase can be easily deformed. In martensite, even after removal of the stress, the strain remains. This residual strain can be recovered by heating the material to the austenite phase, which causes the SMA to return to the original shape. This response is referred to as the shape memory effect. During the martensite–austenite transformation, the SMA exhibits a large force against external resistances.

Position control system using shape-memory alloy wire actuators with electrical resistance feedback has been used in a large number of applications (Ma et al., 2004). A 0.5 mm diameter nickel–titanium alloy (NiTi or Nitinol) wire can lift as much as 5 kg , which is associated with a 5% length recovery. As shown in Figure 9.10, a SMA actuator consists of a length of wire that is preloaded. The applied voltage will heat the wire, hence controlling its length. This illustrates the problems with this type of actuator; the cooling of the wire depends on the ambient temperature, and hence its dynamic performance is poorer than other actuators, but this is more than made up for by its size and simplicity. This high strain property of SMAs offers great potential as actuators in a variety of different applications ranging through micro-robot manipulation, aircraft wingshape control, and microsystem precision control, (Zhang et al., 2004; Ikuta et al., 1998). In all these applications, precise regulation of the actuator is requrired, which can be undertaken by

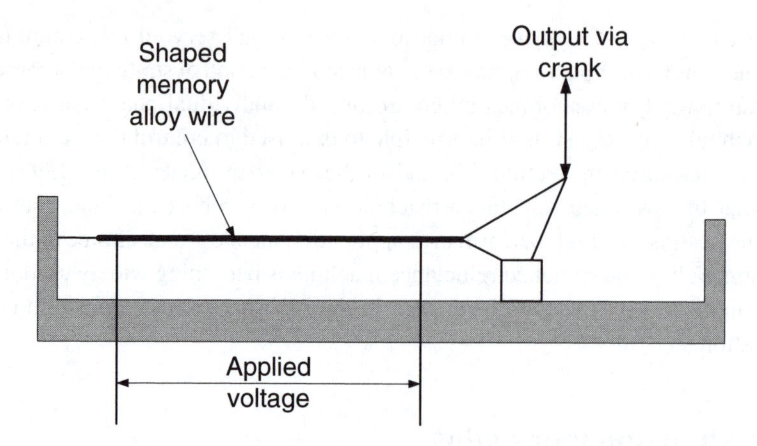

Figure 9.10. The use of a shaped-metal-alloy wire as part of an actuator. The voltage across the wire, and the current through the wire which gives rise to wire heating, is from the output of a conventional position control loop.

controlling the temperature of the wire within a closed loop controller.

9.6 Summary

This chapter has discussed a number of current and future drive systems, which have unique properties. These drives and actuators will give the designer systems with unique characteristics that can be exploited as required.

Chapter 10

Controllers for automation

Previous chapters have discussed the operation and application of a wide range of brushed and brushless motor-drive systems as applied to machine-tool, robotic and similar applications. Chapter 1 considered the overall systems and their broad control requirements, while in Chapter 2, particular emphasis was placed on the criteria to be applied during any selection exercise. Finally, Chapter 3 discussed the mechanical elements of the power train, and the motor-drive selection procedure. In order to complete this overview of modern drive systems, the operation of motor-drive systems integrated with their associated motion controllers was considered in subsequent chapters. This chapter expands this theme and considers the types of controllers which are available to achieve the overall control requirements of an advanced electromechanical system.

The controllers for the applications can be considered to fall into one of the following broad categories:

- *Motion controllers* are able to control the speed and position of one or a number of axes, either individually or when undertaking a coordinated move (i.e. contouring). In addition to closing the servo loop, modern motion controllers may also provide limited data-management facilities, input and outputs channels, communication, and safety circuits required by the machine tool or robot to execute the design function.

- *Multiaxis computer-numerical-control* for machine-tool or robot controllers were developed for particular application areas. Such controllers will contain a number of motion controllers for the axes of the machine tool or robot, together with a system which will generate the required motion profile, and the user and system interfaces. In the machine tool and robotics industries it is the practice for such controllers to be purchased by an original equipment manufacturer, OEM, for installation within their own product.

- *Programmable logic controllers* are capable of controlling the logical operation of a process, and they are capable of interfacing between the user, the external input/outputs, and the motor-drive system.

While the operation and function of PLCs is totally different to the controllers associated with motion control, PLCs can be crucial to overall system control and they have to be considered in any discussion of modern industrial controllers. In addition to reviewing the operation of these controllers, this chapter will provide an introduction to the control theory which is necessary to analyse a closed-loop servo system.

One of the key drivers over recent years has been the development of drives and other industrial systems with networking capabilities. This chapter will consider some of the systems and networking concepts and relate them back to the drives and concepts previously discussed. The Internet and Intranet are redefining how companies operate. It is anticipated that every imaginable kind of device will eventually be networked and, it is expected that pervasive connectivity of sensors will materialise in the industrial world before it happens in the consumer arena. This will transform sensors and related systems from information devices into communication devices.

10.1 Servo control

A generalised representation of a feedback control system is shown in Figure 10.1. The control loop's purpose is to minimise the error between the actual speed, or position, and the demand. The error signal, suitably amplified, is used to generate the velocity or current demand for the drive amplifier. The choice of a drive with either a velocity- or a current-demand input is determined as a function of the controller's stability strategy, and it determines whether the motor should be fitted with a tachogenerator. An external disturbance which acts independently of the system will affect the operation of the system, and can enter the system at any point. The feedback need not be taken from the controlled system; an example of this is the use of a rotary encoder fitted to a lead screw. This assumes that there is a linear relationship between the rotational and the linear position; this is not necessarily the case. Satisfactory operation of the overall system can only be achieved if the motor-drive can produce the required torque, and hence the acceleration that is necessary to follow the required motion profile within the allowable error. If the drive is not capable of matching this basic requirement, there is no way that the overall system can ever meet its specification.

While the control problems which are typically encountered in robotics and machine tools can be simply stated, their full solution is anything but simple because of the additional complications of variable inertial forces, of coupling between axes (in particular with robots), and of gravity. The general route to the development of a control system is first to develop a full dynamic model, and then solve it to obtain the control laws or strategies for the desired performance. In the analysis which is required, both large movements and movements which are associated with the interaction between the workpiece and the mechanical system must be considered. This task has now been made considerably easier with the introduc-

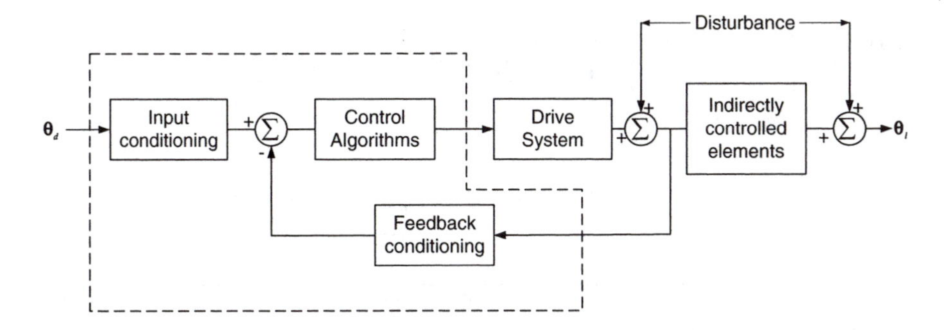

Figure 10.1. A feedback control system, showing the location of the major elements.

tion of a range of computer-based tools, either generic control-simulation packages (of which the best known is perhaps MATLAB and its associated tool-boxes) or packages designed to support a particular manufacturer's motion-control system. The increasing use of digital controllers has permitted the integration of simulation and parameter optimisation in one computer package, which can directly interface with the drive via a communication link and permit real-time configuration of a drive's stability terms.

It is readily apparent from a review of the commercial literature that the majority of modern position-control systems are based on digital processing; however, as an introduction to this overview, it is worthwhile considering an analogue control system, before discussing the implementation of a digital control system. To this end, the analysis of a single axis based on a direct-current (d.c.) brushed motor will be undertaken as a continuous-time system; the equivalent circuit of a d.c. brushed motor has been fully discussed in Chapter 5. The objective of the control loop is to hold the output position, θ_L, as close as possible to the demanded position, θ_L^d. If another motor is used then the control loop, and a transfer function for the drive and the motor, will need to be developed to replace those of a brushed d.c. motor. The block diagram of a simple position-control system is shown in Figure 10.2. In a practical system any gearing and nonlinearities will have to be considered, since these factors would modify the overall loop-transfer functions.

As a first step to the analysis, the transfer function, using Laplace transforms, between the motor's terminal voltage and the output position can be determined to be

$$\frac{\Theta_L(s)}{V(s)} = \frac{K_t}{s(sR_aI_{tot} + R_aB + K_eK_t)} \tag{10.1}$$

where K_e and K_t are the motor's voltage and torque constants, R_a is the armature resistance, I_{tot} is the total inertia of the system, B is the system damping constant, V is the motor terminal voltage, and Θ_L is the angular position of the output shaft. The motor's armature inductance has been neglected, because the motor's electrical

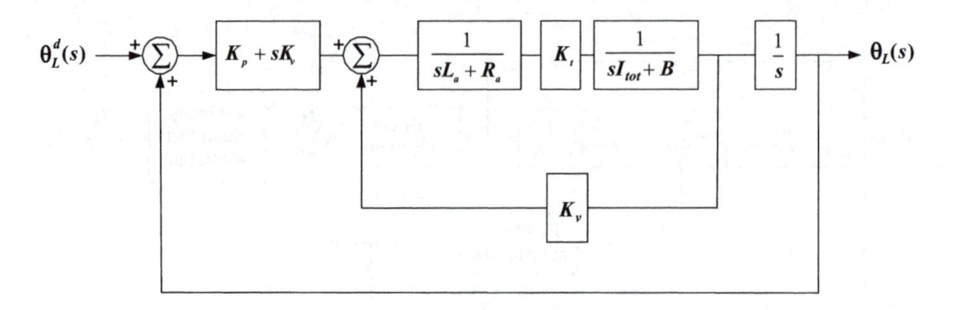

Figure 10.2. The block diagram for a closed-loop controller.

time constant is negligible compared to the system's mechanical time constant.

In a proportional closed-loop servo system, the motor's terminal voltage is directly proportional to the angular difference between the required and the actual position; therefore the motor's supply voltage can be expressed in the form

$$V(s) = K_p[\Theta_L^d(s) - \Theta_L(s)] = K_pE(s) \tag{10.2}$$

where K_p is the proportional gain, including the power-amplifier transfer function. The position feedback signal is derived from an encoder mounted on the load shaft, which will require suitable conditioning (as discussed in Chapter 4). In the full-system model the gain and/or conversion factor will be added to the transfer functions. If the analysis of this control system is undertaken in the conventional manner, it can be shown that the open-loop transfer function is

$$\frac{\Theta_L(s)}{E(s)} = G(s) = \frac{K_t(K_p + sK_v)}{s(sR_aI_{tot} + R_aB + K_eK_t)} \tag{10.3}$$

where K_v is the servo amplifier's derivative feedback gain, which is added to improve the system's response when following a trajectory generated as a polynomial function. The open-loop transfer function, $G(s)$, results in a closed-loop transfer function for the system of

$$\frac{\Theta_L(s)}{\Theta_L^d(s)} = \frac{G(s)}{1 - G(s)} = \frac{K_t(K_p + sK_v)}{s^2R_aI_{tot} + s(R_aB + K_eK_t + K_tK_v) + K_tK_p} \tag{10.4}$$

The transfer function of the motor-drive and its controller is a second-order system, with a zero located at $-K_p/K_v$ in the left-hand of the s-plane. Depending on the location of this zero, the system can have a large positional overshoot and an excessive settling time. In a machine-tool or robotic application, this possibility of an overshoot should be considered with care because it could lead to serious collision damage if it becomes excessive.

In the analysis of a system, the effect on the load's position of an externally applied load or disturbance must be fully considered. In this example, an external torque of $T_D(s)$ is applied to the system (this could be from the gravitational and

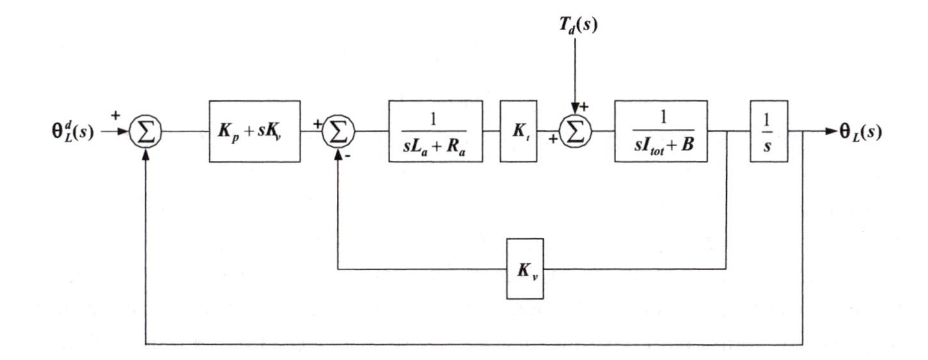

Figure 10.3. Block diagram of a closed-loop control of a single joint with disturbances.

centrifugal forces in a robot or from the cutting forces in a machine tool); see Figure 10.3. The closed-loop transfer function can be rewritten with reference to the disturbance as

$$\left.\frac{\Theta_L(s)}{T_d(s)}\right|_{\Theta_L^d(s)=0} = \frac{K_t(K_p + sK_v)}{s^2 R_a I_{tot} + s(R_a B + K_e K_t + K_t K_v) + K_t K_p} \tag{10.5}$$

To consider the overall performance of the system, it is possible to combine, by superposition, the transfer function relating the demanded position to the output position the transfer function with relating the load torque to the output position, to give the following transfer function:

$$\Theta_L(s) = \frac{K_t(K_p + sK_v \theta_L^d(s) R_a T_D(s))}{s^2 R_a I_{tot} + s(R_a B + K_e K_t + K_t K_v) + K_t K_p} \tag{10.6}$$

Once the closed-loop transfer equations have been developed the performance of the control system can be investigated. In this second-order system, the quality of the performance is based on a number of criteria, including the rise time, the system's steady-state error, and the settling time. The characteristic equation of a second-order system can be expressed in the form

$$s^2 + 2\zeta\omega_n s + \omega_n^2 = 0 \tag{10.7}$$

where ω_n is the undamped natural frequency and ζ is the damping ratio. If this equation is related to the closed-loop poles of equation (10.6) it can be shown that

$$\omega_n = \sqrt{\frac{K_t K_p}{I_{tot} R_a}} \tag{10.8}$$

and

$$\zeta = \frac{R_a B + K_t K_e + K_t K_V}{2\omega_n I_{tot} R_a} \tag{10.9}$$

In a determination of the servo loop parameters, the nature of the application must also be taken into account. In particular, within a manipulator application, it is not possible to have an undamped response to a step input or a possible collision could result, leading to either a critical or an overdamped system; therefore, ζ has to be greater than or equal to one. However, if the system is considerably overdamped, the response to a change in the demand may be so poor as to make the system useless. An additional constraint that should be considered is the relationship of the undamped natural frequency to the vibrational characteristics of the mechanical structure. It is normally recommended that a servo loop's undamped natural frequency is no more than half that of the mechanical structure's resonant frequency. The derivation of the natural frequency for a shaft was discussed in Section 3.6.2. While the detailed derivation for a manipulator system is far more complex, it can be shown that for a single joint the natural frequency is given by

$$\omega_n = \sqrt{\frac{K_s}{I_{tot}}} \tag{10.10}$$

where K_s is the effective stiffness of a joint; $K_s\theta_m(t)$ opposes the inertial torque of the motor, so that

$$I_{tot}\alpha_m(t) + K_s\theta_m(t) = 0 \tag{10.11}$$

where α_m is the acceleration of the motor.

A further critical factor in the consideration of a servo system is the steady state error, e_{ss}, which should be as close to zero as possible in a robotic or a machine-tool application. The error within the system, $e(t)$, can be defined as the difference between the actual and the demanded position:

$$e(t) = \theta_L^d(t) - \theta_L(t) \tag{10.12}$$

For a step input of magnitude X (that is, $\theta_L^d(t) = X$), and if the disturbance input is unknown, then the steady state error of the system can be determined by the use of the final-value theorem. This theorem states that the steady-state error, e_{ss}, is given by

$$e_{ss} = \lim_{t\to\infty} e(t) = \lim_{s\to 0} sE(s) \tag{10.13}$$

Using the overall transfer function, the steady-state error for a step input (X/s) can be determined as

$$e_{ss} = \lim_{s\to 0}\left[\frac{[s^2 R_a I_{tot} + s(R_a B + K_t K_v)]X/s + nR_a T_D(s)}{s^2 R_a I_{tot} + s(R_a B + K_e K_t + K_t K_v) + K_t K_p}\right] \tag{10.14}$$

which simplifies to

$$e_{ss} = \lim_{s \to 0} \left[\frac{R_a T_D(s)}{s^2 R_a I_{tot} + s(R_a B + K_e K_t + K_t K_v) + K_t K_p} \right] \tag{10.15}$$

This shows that the steady-state positional error for a step input in a second order system is a function of the external disturbance. In a fuller analysis, the disturbance torque can be determined if it is the result of gravity loading and centrifugal forces; however other disturbances, such as friction, are difficult to analyse. If the determination of the steady-state error is repeated for a ramp input, it can be shown to be dependent on the ramp constant and on the load disturbance. If this analysis is conducted on a multijointed robot, where a large proportion of the disturbance is from adjacent joints, it will rapidly become apparent that the full analysis is complex and that it requires the use of control-simulation packages.

10.1.1 Digital controllers

In drive systems, there has been an almost complete shift towards the use of digital systems rather than analogue systems; this results in systems with a number of significant benefits. When a digital processor is used within a servo controller, the data will processed at specific intervals, leading to sequential, and discrete, data acquisition and processing activities. In the analysis of a sampled-data system, the data can be transformed from the continuous s-domain to the discrete z-domain by the application of the relationship $z = e^{sT}$, where T is the sampling period. The transfer functions in the z-domain have similar properties to those in the Laplace s-domain. Before considering the implementation of digital-control systems, the advantages of these microprocessor-based systems should be highlighted:

- The use of low-cost microprocessors reduces the parts count within the controller; therefore the system reliability can be increased without a comparable increase in cost.

- Digital control provides a highly flexible system which allows the implementation of a wide range of functions, including non-linear functions.

- Due to the digital nature of the controller there will be no component variations as a function of temperature and time (in contrast to analogue systems), hence the gain and the bandwidth will not be subject to drift.

- The ability of a digital control loop to accept the control values digitally allows easy modification of the stability terms in real time.

- Once the control parameters have been determined they can be used in an identical controller, for example, during maintenance, with the assurance that the system's response will be not be affected.

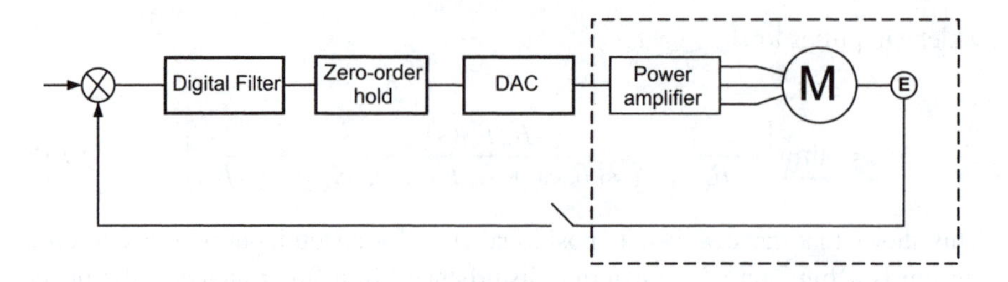

Figure 10.4. Block diagram of a digital-control system, including signal conditioning.

A block diagram of a complete digital position or speed loop is shown in Figure 10.4: it should be noted that the complete system will contain analogue elements, particularly the motor-drive. The servo loop will be implemented in a local microprocessor or microcontroller, with a position or speed demand generated by a profile generator. The feedback that is required is provided by an encoder or a resolver fitted to the load or to the motor, depending on the application. The operation of the system can be summarised as follows. At a predetermined, but constant interval, the output of the digital-to-analogue converter is updated; so the motor changes position. At the same time, the motor's position is determined from the encoder; this is compared with the demand, to determine the error signal. The resultant digital value is applied to the digital filter to provide a new value for the digital-to-analogue converter. A zero-order hold (ZOH) ensures that the output is held constant between the samples. The effects of the sampling period on signals is discussed in Section 4.1.3. The optimisation of the digital filter is not different in any way to the process undertaken with an analogue servo system. As with all digital systems, the time between the samples is finite, which imposes a limitation on the accuracy on the system. While this will not cause significant problems for robotic or machine-tool applications, it is seen as a limitation on the synchronism of very high-speed drive systems. In addition, it should be noted that changes in the sampling rate affect the transient response, and not only the damping characteristics; but these changes may also turn a stable system into an unstable system.

The determination of a digital compensation element within a closed-loop control system can be undertaken using digital-control theory. The analysis of digital-control systems requires the use of a z-transform, which is handled in a similar manner to a Laplace transform: the relationships for a number Laplace and z-transforms is given in Table 10.1. In order for the system to be stable, the roots of the system's digital transfer function have to lie within a unit circle in the z-plane. As with the s-plane analysis, the steady-state error can be determined by the application of

$$e_{ss} = \lim_{t \to \infty} = \lim_{z \to 1}(1 - z^{-1})E(z) \tag{10.16}$$

Table 10.1. Laplace and *z*-transforms; T is the switching period

$x(t)$	$X(s)$	$X(z)$
$\delta(t) = \begin{cases} 0 & t = 0 \\ 1 & t = Kt, K \neq 0 \end{cases}$	1	1
1	$\frac{1}{s}$	$\frac{z}{z-1}$
t	$\frac{1}{s^2}$	$\frac{zT}{(z-1)^2}$
$1 - e^{-at}$	$\frac{1}{s(s+a)}$	$\frac{z(1-e^{-at})}{(z-1)(1-e^{-at})}$
e_{at}	$\frac{1}{s+a}$	$\frac{z}{z-e^{-at}}$
$\sin\omega t$	$\frac{\omega}{s^2+\omega^2}$	$\frac{z \sin\omega T}{z^2 - 2z \cos\omega T + 1}$
$\cos\omega t$	$\frac{s}{s^2+\omega^2}$	$\frac{z(z-\cos\omega T)}{z^2 - 2z \cos\omega T + 1}$

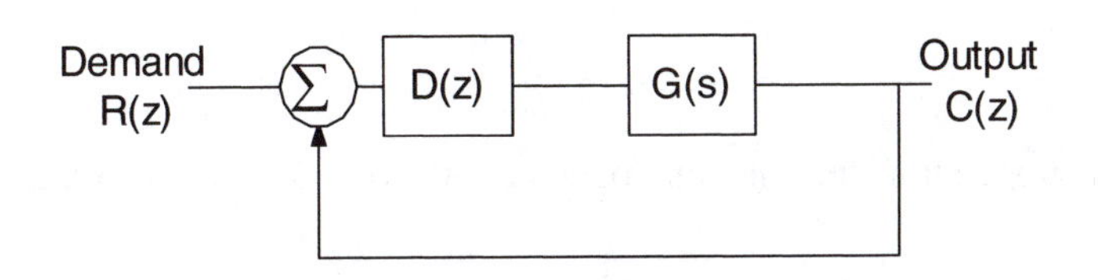

Figure 10.5. An elemental digital–control loop.

Example 10.1

Consider the control loop shown in Figure 10.5, where $D(z)$ is the transfer function of the digital controller, and $G(s)$ is transfer function of the analogue element of the loop, typically a motor-drive system. $D(z)$ is assumed to act directly on the difference between the demand and the actual output, hence determine $G(z)$ to maintain stability.

The overall transfer function will be largely defined by the time constants of the motor and of the load; its z-domain representation is as follows:

$$G(z) = ZOH \times Z[G(s)] \tag{10.17}$$

$$= Z\left[\frac{1 - e^{-sT}G(s)}{s}\right] \tag{10.18}$$

$$= (1 - z^{-1})Z[G(s)] \tag{10.19}$$

Using these relationships, and given the required characteristics of the overall loop characteristics (typically those of the gain and the frequency), a suitable digital transfer controller can be determined. The closed-loop transfer function of this digital system can be written as

$$\frac{C(z)}{R(z)} = \frac{G(z)D(z)}{1 + G(z)D(z)} \tag{10.20}$$

where $D(z)$ can be assumed to be of the form

$$D(z) = \frac{K(z - A)}{(z + B)} \tag{10.21}$$

where K is the system gain and A and B are the coefficients of the digital filter. If the transfer function for the process is taken to be

$$G(s) = \frac{1}{s(s - 1)} \tag{10.22}$$

then the analogue transfer function, together with the ZOH, can be written as

$$G(s) = \frac{C(s)}{R^*(s)} = \frac{1 - e^{-sT}}{s^2(s + 1)} \tag{10.23}$$

Expanding this expression, and using Table 10.1,

$$G(z) = \frac{[ze^{-T} - z - zT] + [1 - e^{-T} - Te^{-T}]}{(z - 1)[z - e^{-T}]} \tag{10.24}$$

If the switching period $T = 1$, the $G(z)$ becomes

$$G(z) = \frac{0.372 + 0.26z}{(z-1)(z-0.37)} \tag{10.25}$$

Equations (10.21) and (10.25) can be combined to give the denominator of equation (10.20). The pole at z = 0.37 in equation (10.25) will cancel if A is equal to 0.37. This leaves B and K to be determined, permitting optimisation of the response.

10.1.2 Advanced control systems

It is current industrial practice to treat each axis of a multiaxis system as an individual servo mechanism. This approach models the varying dynamics of a system inadequately, because it neglects the motion and therefore the changes of the configuration within the system, particularly those changes that occur in manipulators. These changes can be significant, and they may render the conventional-control-strategy approach ineffective. The result of this approach is a reduction in the servo's response speed and in damping, which limits the speed and the precision of the system. Any significant gain in performance requires consideration of more efficient dynamic models of sophisticated control techniques, and of the use of advanced computer architectures. With advances in real-time computing, the implementation of a range of advanced techniques is now possible; techniques based on either adaptive control or on artificial-intelligence approaches (for example, fuzzy logic or neural networks) are of particular interest.

Among the various adaptive-control methods being developed, model-reference adaptive control is the most widely implemented. This concept is based on the selection of an appropriate reference model and on an adaptation algorithm that is capable of modifying the feedback gains of the control system. The adaptation algorithm is driven by the errors between the reference-model outputs and those of the actual system. As a result of this approach, the control scheme only requires moderate computation, and it can therefore be implemented on a low-cost microprocessor. Such a model-reference adaptive control algorithm does not require complex mathematical models of the system dynamics, nor does it require an a priori knowledge of the environment of the load. The resultant system is capable of giving good performance over a wide range of motions and loads.

10.1.3 Digital signal processors

It is clear that a majority of motor drive system requires a degree of digital signal processing – the introduction of which has been made easier by the development of

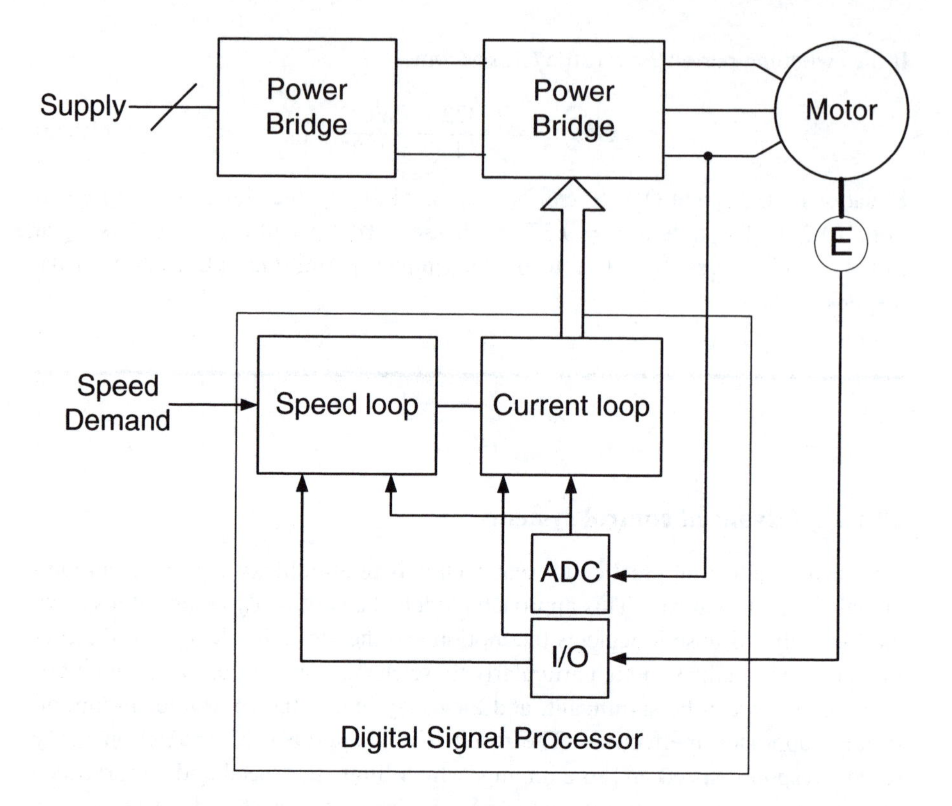

Figure 10.6. The outline of a digital signal processor based motor controller.

the digital signal processor or DSP. In practice a DSP is a powerful microprocessor that is capable of processing data in real time. This real-time capability makes a DSP suitable for applications such as the sensorless control of brushless motors. As has been discussed in earlier sections speed and torque control requires the real time solution of the electromechanical and electromagnetic relationships within the system. Figure 10.6 shows the outline of a brushless motor controller based on a DSP.

The control system includes the following:

- A DSP integrated circuit that executes speed and current algorithms to control the motor using data from the fitted position sensor, which can be either a resolver, an encoder, or a Hall effect sensor depending on the motor type. In addition current feedback is provided from one of the motor's phases.

- Analogue and digital converters to process the position and current data into a digital number of a suitable format for use by DSP. Multi-channel ADCs are used for simultaneous sampling to maintain correct phase information, and can be integrated within the DSP.

10.2 Motion controllers

In many instances several motor-drives are simultaneously controlled by a single supervisory controller, as discussed for CNC machine-tool and robotic applications in Chapter 1. The supervisory controller undertakes a wide range of high-level tasks, such as the generation of position, velocity, and acceleration profiles, together with a range of housekeeping functions, including data management, communications, and operation of the user interface. The choice of controller strategy depends on the number of axes and on the degree of coordination between the axes; possible options include the following.

Axis controllers

A multi-axis controller is capable of controlling a number of motion axis simultaneously. The implementation can be undertaken with the use of a number of single-chip microcontrollers on a single printed-circuit board. A microcontroller is a microprocessor with additional memory (both random-access memory (RAM) and programmable read-only memory (PROM)), together with analogue-to-digital (A/D) converters, digital-to-analogue (D/A) converters, and communications ports fabricated into one package. A number of companies supply customised devices that incorporate motion-control algorithms; all that the users have to supply are the equation parameters and the limiting values. With the increasing power and ease of programming of industry-standard personal computers, a range of motion-control cards using these devices have been developed. They have configured standard expansion sockets which permit a system to be put together with the minimum of effort. Cards are available which can control up to eight axes. By placing more than one axis on a card, multiaxis interpolation and contouring or coordinated motion between axes can be easily undertaken. The boards are normally available in a number of bus configurations giving the system designer considerable flexibility. A typical three-axis card can operate in either independent or vector-positioning modes, and they have the ability to contour up to speeds of 500 000 encoder counts per second.

In a number of cases it is possible for one drive to act as a master unit, while the other drives are directly synchronised to maintain a zero position error between themselves and the master drive (see Figure 10.7). This approach is used to replace mechanical gearboxes and transmission shafts in, for example, textile machinery.

Machine tool controllers

There are a considerable number of multi-axis controllers on the market, many of which are dedicated machine-tool controllers. In the simplest terms they can be considered to consist of a number of axis controllers and the overall system-control computer.

A typical unit is capable of controlling up to five axes and the machine spindle.

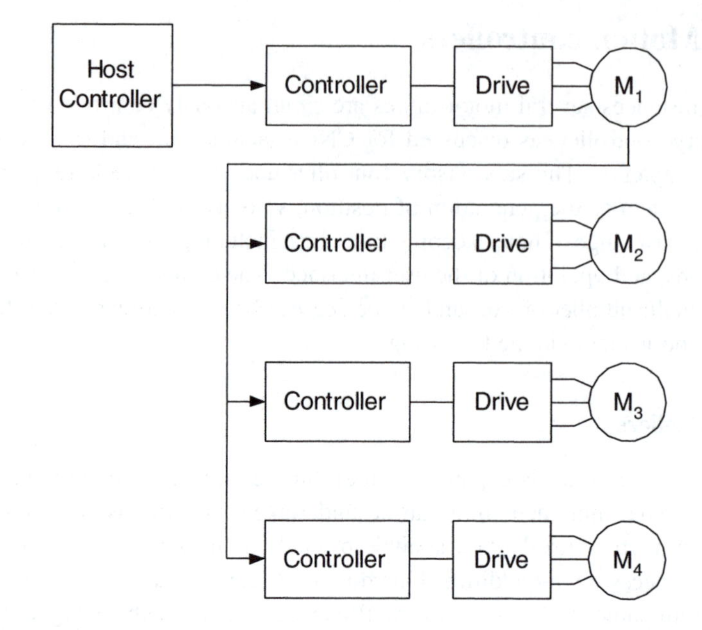

Figure 10.7. The use of a number of single-axis controllers to provide an electronic line shaft. The motors M_2 to M_4 operate with a fixed positional relationship to M_1. The relationship is determined by the individual axis parameters.

The control system can store up to 12 000 command lines, and it can be programmed using a keyboard or directly using information from a designer's computer aided-design (CAD) package. The requirements of such controllers have been discussed previously; however, controllers which are specifically designed for machine-tool applications can have a number of additional features. One feature that should be noted is the provision of a set of industry standard program command codes (see ISO 6983: *Numerical control of machines – Program format and definition of address words*) for machine-tool functions. These codes range from simple on-off commands (for example, spindle-drive on, or coolant on) to a number of canned cycles. A canned cycle is effectively a preprogrammed subroutine, for example, cutter compensation, peck drilling, or boring. The power of this approach is shown in Figure 10.8 where a number of holes have to be drilled in a flat plate. By the use of the correct code, G81, the machine tool will control the drill axis from the starting position to the base of the hole, and it will then return to the starting position. The canned cycle will attempt to repeat itself using the next block of code that contains an X, Y, or Z word, until it is cancelled by the G80 code. The file shown in Table 10.2 will control a CNC machine to drill a set of eight holes as shown.

With the increasing use of CAD packages, it is not uncommon for the design to be directly converted to the machine-tool program, using a suitable post processor which has the ability to optimise the machining process and to minimise the

Table 10.2. Example of the use of the G-codes to drill eight holes. Note that after the first four holes the depth changes from 1 to 0.5.

n100	g90 g0 x0 y0 z0	Co–ordinate home and set absolute position mode
n110	g1 x0 g4 p0.1	
n120	g81 x1 y0 z0 r1	Initiate the canned drill cycle
n130	x2	
n140	x3	
n150	x4	
n160	y1 z0.5	Change the depth being drilled
n170	x3	
n180	x2	
n190	x1	
n200	g80	Turn off the canned cycle
n210	g0 x0	Rapid move to home position
n220	y0	
n220	z0	
n220	m2	End of programme

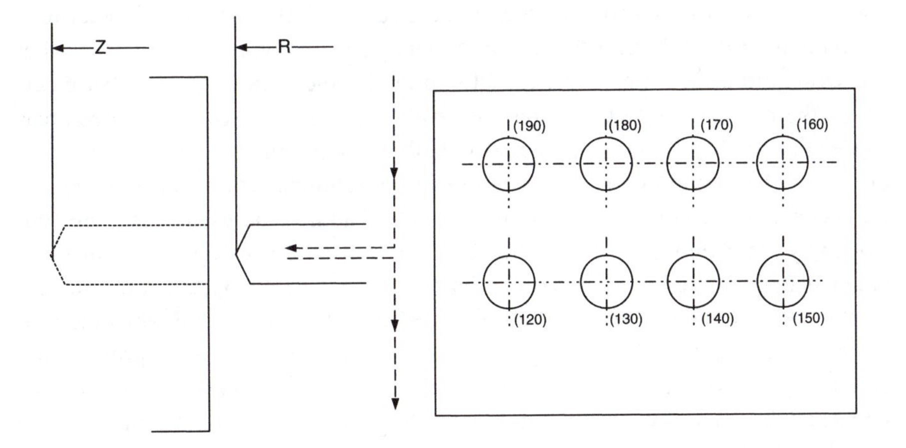

Figure 10.8. A drilling cycle using the canned cycle, G81. The cycle advances the drill to position R, and then feeds at drill speed to Z, and then retracts back to R. The controller then moves the drill to the next whole position, prior to the drilling being repeated. The numbers in brackets refer to the programme line responsible for drilling the hole.

production time.

10.3 Programmable logic controllers

The development of programmable logic controllers (PLCs) was driven primarily by the requirements of automotive manufacturers who constantly changed their production line control systems to accommodate their new car models. In the past, this required extensive rewiring of banks of relays – a very expensive procedure. In the 1970s, with the emergence of solid-state electronic logic devices, several auto companies challenged control manufacturers to develop a means of changing control logic without the need to totally rewire the system. A PLC is designed to be relatively 'user-friendly'. In a PLC based system, push-buttons, limit switches, and other conventional components can be used as input devices to the PLC. Likewise, contactors, auxiliary relays, solenoids, indicating lamps can be directly wired as output to a PLC.

Many industrial processes consist of a considerable number of interrelated activities which have to be performed in a predetermined and fixed sequence. Consider the manufacturing cell shown in Figure 10.9, which comprises a robot and its controller, two machine tools, conveyors, and a parts store. While the sequencing could be undertaken by the robot or either of the machine-tool controllers, there will be considerable advantages (particularly in the speed of computation) in using an overall sequencing controller which is based on a separate programmable logic controller. The PLC will receive inputs from the robots, from the machine-tool controllers and from sensors fitted within the cell; and its program will determine the outputs to the robot, to the machine tools, to the conveyors, and to the other process equipment. In essence, the PLC provides the logic sequence that determines the process. A PLC is a purpose-built computer consisting of three areas (see Figure 10.10): processing, memory (both the program and the working memory), and the input/output interface. As in conventional computer architecture, these elements are connected to common data and address buses, and they are controlled by the central processing unit (CPU). For program storage, use is made of either battery-backed CMOS RAM (complementary metal-oxide semiconductor, random-access memory) or by PROM. A PROM can only be used when the program development has been completed and no further changes in the program are anticipated. A separate area of RAM is provided as a working memory; this can be backed by a battery to aid fault finding after a system failure. Compared with the requirements of present-day personal computers, the memory requirements of a PLC are quite modest; a memory of 64K will hold up to 1000 instructions, which is adequate for most applications. Although similar to personal computers in terms of their hardware, a number of specific features of PLCs make them suited to industrial control applications, including the following.

- The input and output channels can be wired directly from the PLC to external systems without any additional interfacing.

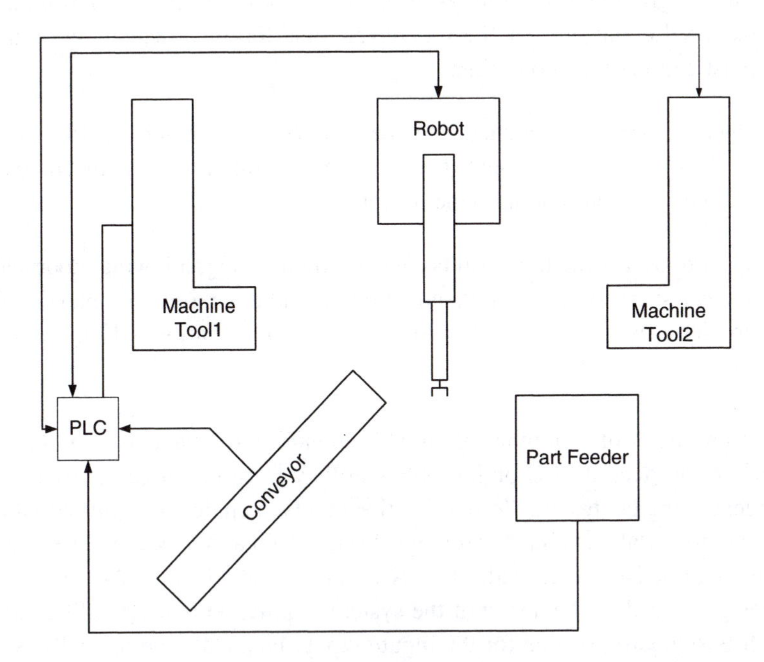

Figure 10.9. An overview of a manufacturing cell, showing how the various elements can be linked via a PLC.

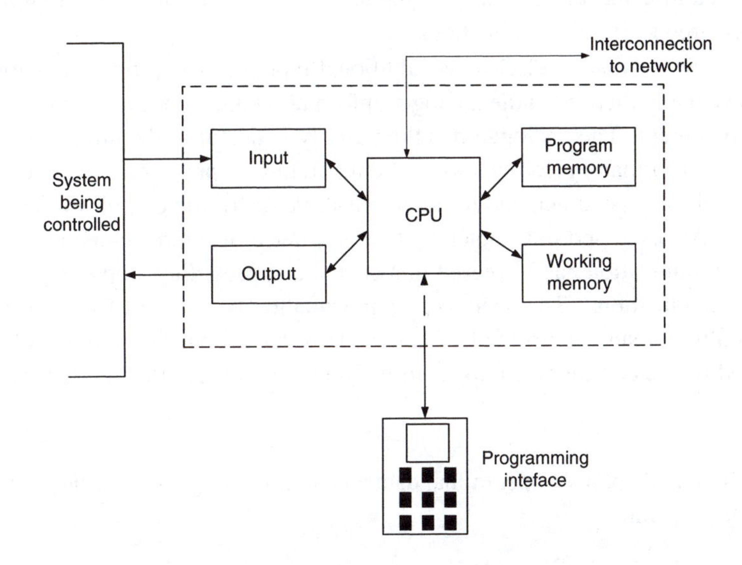

Figure 10.10. The internal structure of a PLC.

- The programming is undertaken in a ladder logic form, either from a local hand-held controller or from a conventional terminal depending on the sophistication of the controller.

- Once the program is loaded it can be permanently stored in the PLC, so once the PLC is commissioned it becomes an integral part of the machine and it will become transparent to the operator.

- The program structure permits easy reprogramming, allowing modifications and revisions to be incorporated with minimal down time compared with the problems associated with a wiring change when a hardwired logic system is used.

The provision of a number of suitable input/output channels is a significant element of the overall PLC philosophy, and they form the direct interface to the equipment being controlled. To provide this ability, a choice of input/output packages is commercially available, including inputs rated at 5 V d.c. or 24 V d.c., and outputs rated at 24 V d.c. and 100 mA or 110 V a.c., 1 A, or 240 V a.c., 2 A. The designer needs to be aware of the system requirements when a PLC is specified. It is standard practice for the input/output channels to be optically isolated to provide a high degree of isolation, but relays or triacs are normally used for the higher-power a.c. outputs. The isolation provided will protect a PLC from external switching transients up to $\pm$1500 V. In a small self-contained PLC, the inputs and outputs are physically located in the same casing and they tend to have the same rating. In a modular PLC, a range of input/output modules of different ratings can be configured to suit the application.

The more advanced PLCs have additional input/output features; in particular, they have the ability to handle analogue information and communications with remote computers. The analogue-handling ability is useful in the direct monitoring of process information (for example, of temperature or pressure), with the outputs being used as a speed demand to the motor-drive or for the control of the heating element. As discussed in Chapter 1, the importance of intermachine and factory-wide communication has increased in recent years, providing ample opportunities for PLC applications. The need to pass information between PLCs and other devices within a manufacturing plant has ensured that all but the simplest PLCs are provided with a communications facility. The main uses for PLC communications are:

- Remote display of operational data and alarms using either printers or visual display units.

- Data logging for the archiving or quality control.

- Passing program changes, either process parameters or the resident program.

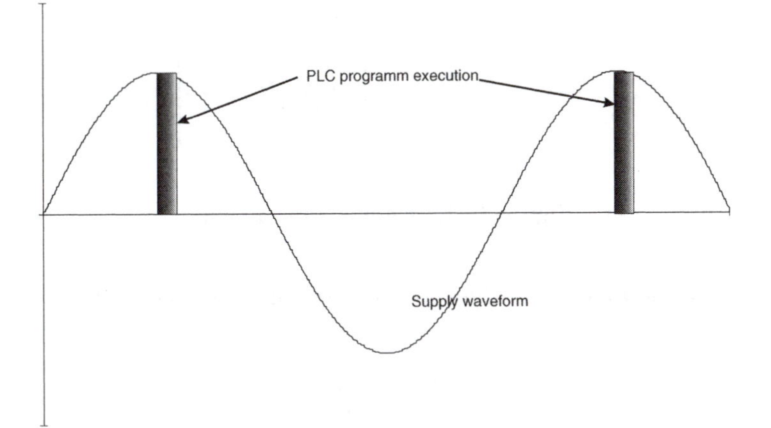

Figure 10.11. The relationship between the supply waveform and the execution of the PLC program.

- Linking the PLC into a computer hierarchy, which may contain many other PLCs and computers, to form a true computer-integrated-manufacturing (CIM) facility. The programming of a communications channel is effectively no different from the programming of the logic sequences discussed in the next sections.

The programming unit of a PLC can be provided as a separate removable unit, whose functions are dependent on the size and complexity of the PLC. For small PLCs a simple keyboard will suffice, while larger systems can use a separate personal computer to provide simulation and other supporting software.

The internal operation of a PLC must be considered before its programming is discussed. Figure 10.11 shows the principle of the operation: each execution of the program cycle is initiated at the peak of the PLC's supply waveform. Therefore, in practice, a PLC effectively remains idle for a considerable portion of time. In theory this may lead to errors in the logic if the output of a logic step is used as an input to a logical relationship earlier in the ladder sequence. With careful programming this effect can be minimised; and, in any case, the logic will be re-executed within 20 ms on a 50 Hz system. This will not cause any problems in most applications. However, if the logic becomes highly interdependent, the possibility of catastrophic failure will increase; hence such applications will need to be considered with care, and in extreme cases the PLC should be replaced with other control systems, and the programming should be undertaken with considerable rigour.

10.3.1 Combinational-logic programming

There are two basic approaches to programming a PLC, a combinational-logic approach or a sequential approach. The former can be demonstrated by considering

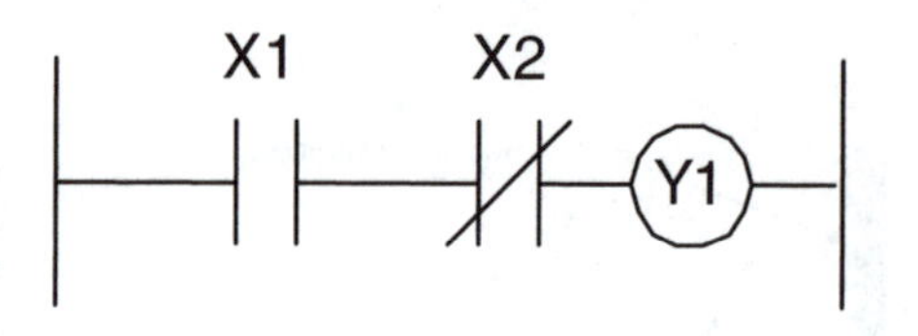

Figure 10.12. Combinational logic.

Table 10.3. Instructions for the logic shown in Figure 10.12

LD	X1
ANI	X2
OUT	Y1

the ladder rung shown in Figure 10.12, where the output Y1 will become active only if the input X1 is on, and the input X2 is off; the symbols used are identical to those used in conventional relay logic. The instruction sequence necessary to program a rung of this ladder into the PLC is shown in Table 10.3 (where the instruction LD identifies the start of a new rung of the ladder, ANI is the inverted AND instruction, and OUT identifies the output channel being controlled). In these sequences, X is an input channel and Y is an output channel which is being sampled or controlled.

It is possible to nest a number of networks, as shown in Figure 10.13, by the use of AND or OR commands. The network shown can be written in Boolean algebra as

$$Y1 = [(X0 + X1 + X2) \cdot X3 \cdot X4] + X5 \tag{10.26}$$

and it can be programmed using the sequence shown in Table 10.4.

In addition to pure logic manipulation, a PLC will normally incorporate programmable timers and counters to increase its flexibility. An example of an application that incorporates a timer is shown in Figure 10.14; the logic provides an extension of 20 seconds in Y1, following the falling edge of the input pulse at X0.

Table 10.4. Instructions for the logic shown in Figure 10.13

LD	X0
OR	X1
OR	X2
AND	X3
AND	X4
OR	X5
OUT	Y1

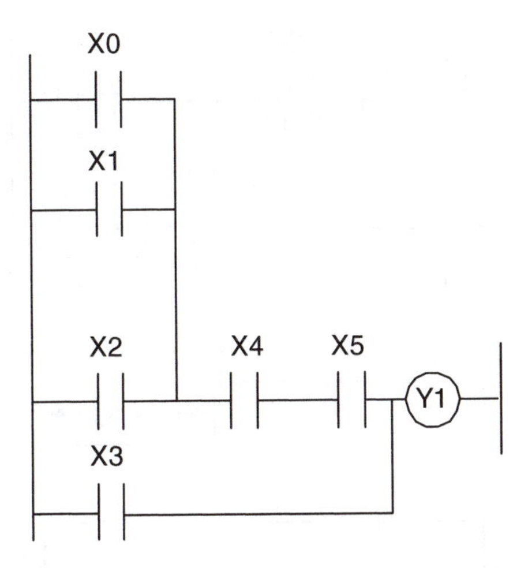

Figure 10.13. Combinational logic extended through nesting.

Table 10.5. Instructions for the pulse extender shown in Figure 10.14

LD	XO	Input
OR	Y1	
ANI	T450	
OUT	Y1	Output
ORI	X1	
AND	T450	
K20		Timer period

The operation is as follows: when X1 is activated, the output Y1 changes state; in addition, a latch is placed across the input. The timer is inhibited by the normally closed contact X0. When the input is deactivated the timing cycle is initiated, and, following timeout, the output of T450 becomes active, breaking the connection to Y1 and the timer and releasing the output and resetting the timer. The instructions for this example are shown in Table 10.5.

Timers can also be used to provide delays, control program sequencing, and to generate programmable mark/space-ratio pulse trains. In addition, a PLC will also contain blocks to provide down counters, auxiliary relays, and other features that will assist the program development.

10.3.2 Sequential-logic programming

The outputs in a sequential-control strategy do not only depend on the present inputs, they also depend on the sequence of the previous events; so some form of memory is required to implement this approach. Sequential problems can be solved using conventional logic networks. However, modern PLCs have a number

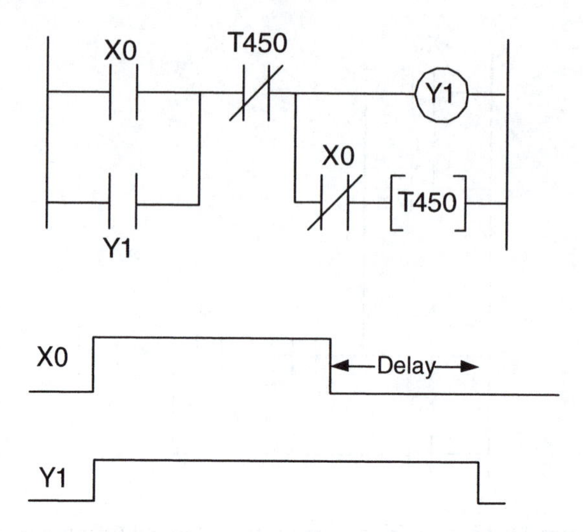

Figure 10.14. The use of a PLC timer, to extend a pulse.

of features, particularly shift registers and drum timers, that considerably simplify the program structure. As with digital electronics, sequential-logic requires a different approach than is used in the design of combinational logic. The key steps are:

- The functionality of the process needs to be carefully and fully described.
- The description then needs to be developed into a fully descriptive functional diagram.
- The function diagram is converted to the required Boolean logic.
- The Boolean logic is converted to a ladder diagram and then into required code for the PLC.

The design process can be considered using the manufacturing process shown in Figure 10.15. A conveyor belt is used to move components towards a manufacturing process. A limit switch, LSI, which can be either mechanical or optical, detects the arrival of the component and it causes the belt to stop. Ram 1 will then move the component into the process until the switch LS2 is activated. On activation, this ram is retracted clear of the process and the conveyor; the third switch, LS3, is then operated. At the same time, the manufacturing process will be initiated. On completion of the process, the components ejection, by Ram 2, from the process is signalled by a fourth limit switch, LS4. On clearing the process the belt is restarted, and the next component is moved up to the first limit switch, LS1. The flowchart for this process is shown in Figure 10.16, and the process can be expressed by four states with the following Boolean relationships:

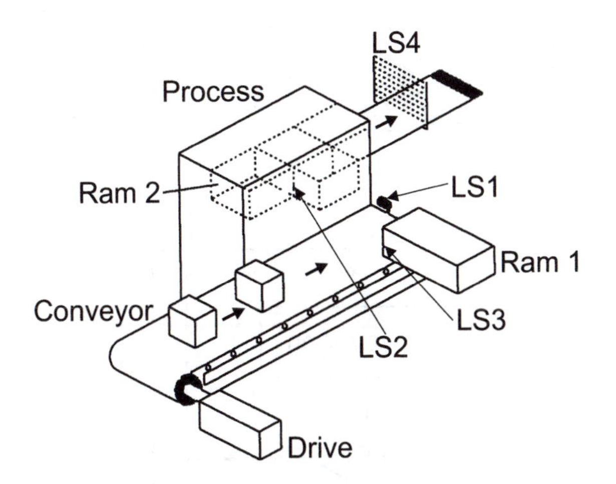

Figure 10.15. A process line that is controlled by sequential-logic.

$$\begin{aligned} S1 &= S4 \cdot LS4 + S1 \cdot \bar{S2} && \text{Belt forward} \\ S2 &= S1 \cdot LS1 + S2 \cdot \bar{S3} && \text{Ram 1 extend} \\ S3 &= S2 \cdot LS2 + S3 \cdot \bar{S4} && \text{Ram 1 return} \\ S4 &= S3 \cdot LS3 + S4 \cdot \bar{S1} && \text{Process initiated} \end{aligned} \qquad (10.27)$$

In order to initiate a state (for example State 1, the movement of the belt), the previous state (in this case State 4) has to be valid and a limit switch (LS4) has to be activated. State 1 is held until State 2 by LS1. It should be noted that one state is entered before the previous state is exited; in practice this is only a transitory phase and it should not cause any conflict in the operating process. Using this information, the ladder diagram can be constructed (see Figure 10.17), which in turn can be coded for the PLC. One problem that needs to be addressed is to ensure that the sequence is always started in State 1. This is achieved by the addition of a start control relay, M105. On powering up, the system is therefore forced into State 1. On entering State 3, the control relay M105 is latched via M103 and it is isolated. The functions of the control relays and the inputs are summarised in Table 10.6.

For a practical implementation, additional steps must be included, for example, interfacing of the control relays to the conveyor drive and any required pneumatic or hydraulic control valves, the implementation of safety circuits, and process control and manual-override provision. While this effectively discrete approach is satisfactory for small systems, it can be simplified by the use of PLC functions such as shift registers, control relays, step ladder functions drum times and sequences. The operation of these programming blocks will be detailed in the manufacturer's literature. While these examples can be considered to be relatively trivial, they do show the power of modern PLCs. As discussed earlier, the range of facilities provided

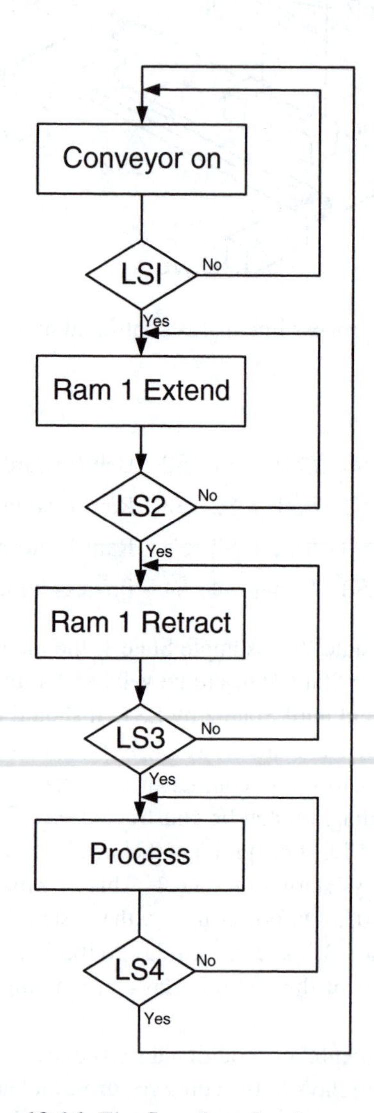

Figure 10.16. The flowchart for the process line.

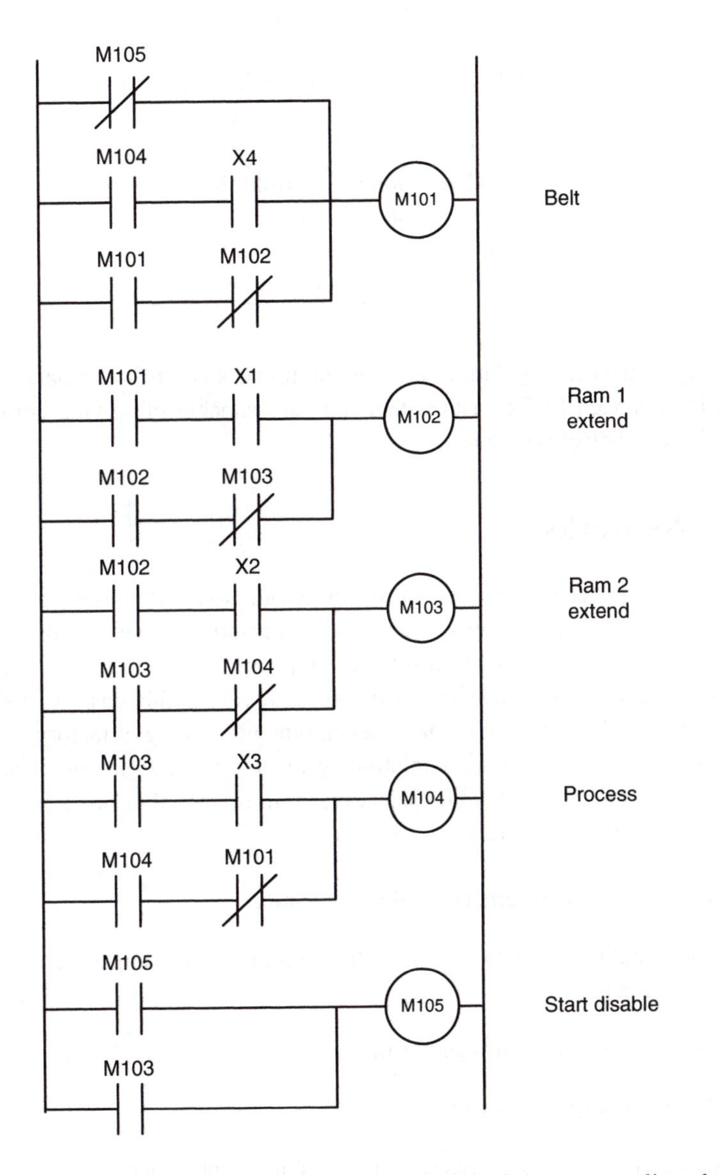

Figure 10.17. The ladder diagram for the operation of the process line shown in Figure 10.15.

Table 10.6. Key PLC assignments for the process shown in Figure 10.15

Element	Function
X1	Input: limit switch
X2	Input: limit switch, LS2
X3	Input: limit switch, LS3
X4	Input: limit switch, LS3
M101	State 1 control relay
M102	State 2 control relay
M103	State 3 control relay
M104	State 4 control relay
M105	Start control relay

for analogue data manipulation, continuous process control, and data communication in advanced PLCs ensures that they are capable of solving complex and sophisticated control problems.

10.4 Networks

The information technology revolution in automation technology is opening up a wide range of savings potentials in the optimisation of system processes and makes an important contribution towards improved use of resources. Industrial communication systems have assumed a key function in this respect: it should be recognised that that networking goes beyond the interfacing of factory systems, to interface with design and documentation systems to form virtual organisation and other related systems. The actual selection of an industrial network is based on a number of criteria including:

- The detailed requirements of the application.
- The capabilities of the network technologies, including speed and timing requirements.
- Integration with existing equipment.
- Availability of components.
- Costs of installation, devices, training, and maintenance.

Within the industrial networking domain the lack of standardisation is the major problem in inter-computer communication. The large number of signal characteristics and topologies is just part of the problem. Manufacturers of many machine-tools or robotic systems, have developed their own instruction sets, memory access systems, and I/O access systems. A machine language program that

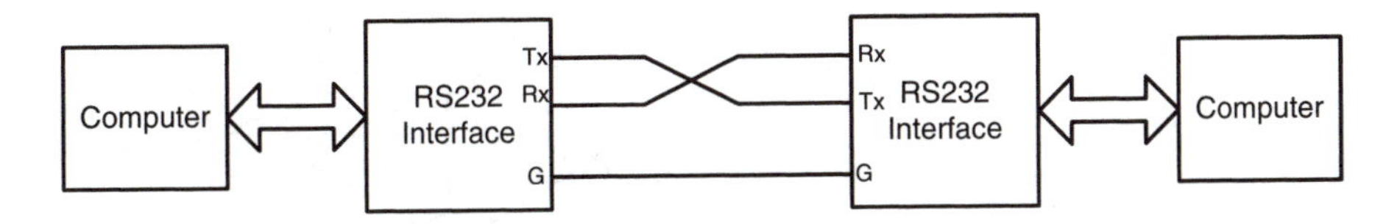

Figure 10.18. The key features of a RS 233 communication link. G is ground, Tx is data transmission, and RX is data receive.

works on one system will not necessarily be usable on a system provided by a different manufacturer. However, as will be discussed later in this section, sets of standards for industrial systems are evolving, in part driven by manufacturers themselves. Before discussing standards, an overview of the basic network concepts are required.

The exchange of data over a network can be either asynchronous or synchronous, depending on the protocol being used. In asynchronous communications, each data word is sent as a separate message. Asynchronous communication is often adequate if only two computers are connected together. Online programming and monitoring is often undertaken asynchronously with a single computer directly connected to a PLC, robot, or other controller. Synchronous communication data messages consist of many data words that are preceded by a header containing information about the data 'packet', and followed by a footer containing error-checking information. Synchronous communication is appropriate where large amounts of data are to be transmitted quickly. Local area networks (LAN) use synchronous data communications. The RS 232 standard is perhaps the most widely accepted serial communication standard. The RS 232 standard dictates that a binary 1 should be sent as a negative 3 to 12 volts, and a binary 0 at plus 3 to 12 volts. The standard also specifies a 25 pin connector, even though only 3 pins are essential (ground, transmit, and receive), Figure 10.18. RS 485 is a standard with growing acceptance. It uses the improved grounding of the RS 422 standard, but allows the connection of multiple computers via the same set of four conductors. It can be used, therefore, in applications such as in in-plant local area networks. Most PLC local area networks use the RS 485 standard.

For long-distance communication (over 2 km), the RS232 standards is not sufficient. One commonly-used solution is to convert an RS 232 output into AC signals via a modem that can then be carried by standard telephone lines. In addition serial binary data can also be coded into light. Fibre optic networks use light frequency and amplitude modulation techniques similar to those used in modems. The Fibre Digital Data Integration (FDDI) standard offers potential data transmission rates well into the hundreds of Mbps. One of the significant advantages of a fibre optic link is its resilience to EMI, which is a common cause of many problems within an industrial network.

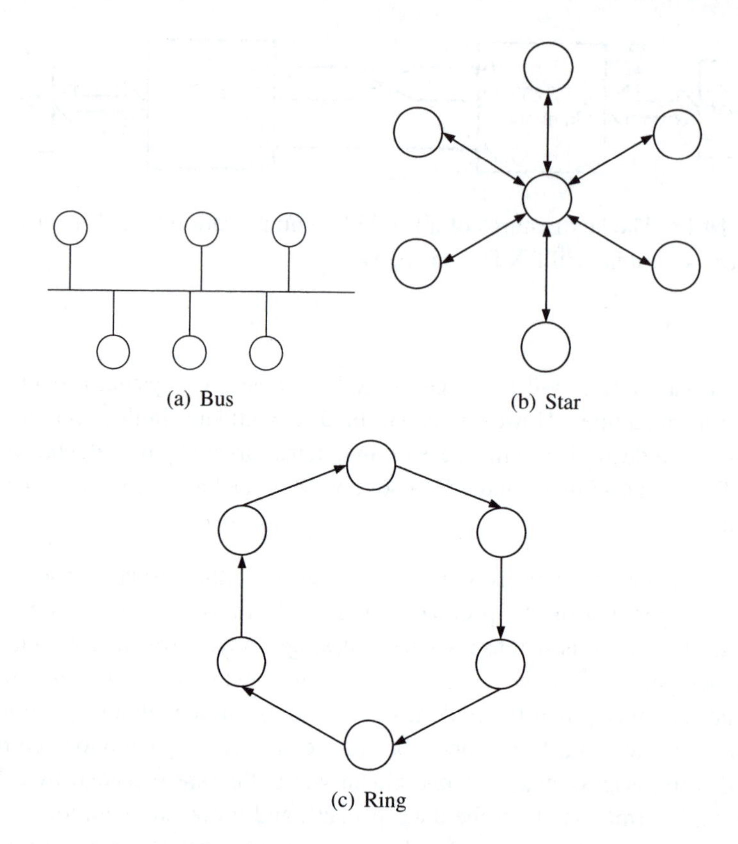

Figure 10.19. Communication network architectures.

10.4.1 Network architecture

There are several architecture options for interconnection of cell or robot controllers and central computers within the industrial environment.

- In bus networks, Figure 10.19(a), all nodes share the same communication bus, hence techniques are necessary to prevent collisions when more than one system tries to send data simultaneously. One method of allowing shared access is the Carrier Sense Multiple Access with Collision Detection (CSMA/CD) bus standard. A node that wishes to send first listens and if the line is free, transmits. While transmitting, the sender listens to ensure that no other computer has tried to send at the same time. If such a collision does occur, the transmission is immediately terminated and the sender waits for a short time before retrying the transmission. A different approach is to use a token passing bus networks, where each computer on the network gets its turn (token) in a pre-assigned sequence, and can transmit only at that time.
- In a star network topology, Figure 10.19(b), each link connects only two nodes, with some nodes acting as message centres or hubs, accepting and

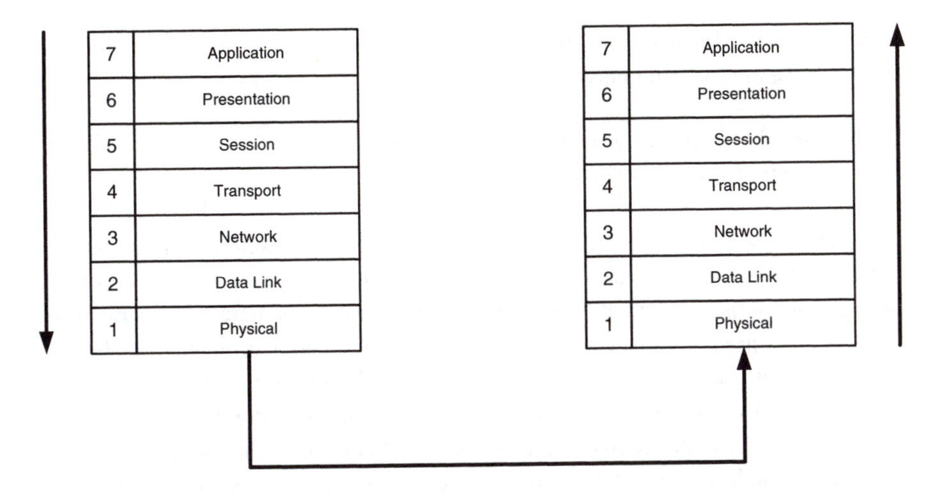

Figure 10.20. The OSI network model, showing the flow of data from one application, through the network, to a second application.

relaying messages to other nodes. Due to the computing load, network hubs may not have any computing time left for other functions.

- In ring networks, Figure 10.19(c), messages are passed from one computer to the next, in one direction, around the ring. The computer that is communicating passes control to the next computer in the ring when it has finished its communication. Computers waiting to originate communications must wait until they receive control.

To ensure correct communication, both ends of the link need to adopt a common protocol. The protocol specifies such factors as speed, data coding method, error checking, and handshaking requirements, and implies that the form of the data must be useful at both ends of the communication.

To define standards use is made of the Open Systems International (OSI) model. The OSIs seven layers break up communication services into four application service layers and three network service layers, Figure 10.20, summarised below:

- Layer 1 – Physical: Encodes and physically transfer messages.
- Layer 2 – Data link: Breaks up and reassembles messages, together with error detection and correction at the bit level.
- Layer 3 – Network: Routing of message packets.
- Layer 4 – Transport layer: Splits the data into messages to be transported.
- Layer 5 – Session layer: Establishes sessions between machines.

- Layer 6 – Presentation layer: Converts data formats between application representations and network representations.
- Layer 7 – Application layer: Provides a uniform layer that abstracts the behaviour of the network.

In the model an 'application' is taken as meaning any source of a request for network access at a computer. An industrial control program would be considered 'applications'. Commands from the 'application' to the network's application services layer are recognised and responded to by the applications layer. The presentation layer translates data forms as required. The session layer at the sending computer works with the session layer at the receiving computer to transmit data only as fast as it can be received. The transport layer ensures that data is not lost or damaged in the lower layers. The four application service layers must exist at each node.

The three network service layers of programs may be provided by telecommunication companies, or may be built into a local area network. The network layer handles routing of messages via shared networks, and multiplexes messages so that a network can be shared by many 'applications'. The data link layer adds addresses to outgoing messages, examines all network data for messages addressed to this node, and does basic error checking. The physical layer inserts messages onto the shared network conductors and receives all message traffic, converting between binary data and whatever form data is in when being moved via the network.

There are situations where not all seven layers are required. Eliminating some reduces the hardware and software cost. If, for example, both computers can use the data in the same form, then the data translation function of the presentation layer is unnecessary. If the node's only function is to connect one type of network to another (e.g., token ring to CSMA/CD), then only the lower 2 or 3 layers are necessary.

10.4.2 Industrial networking

Communication in automation is becoming increasingly direct, horizontally, as well as vertically through all hierarchy levels, Figure 10.21. At sensor/actuator level the signals of the binary sensors and actuators are transmitted via a sensor/actuator bus. At this level, a low-cost technique, through which data and a 24-volt power supply for the end devices are transmitted using a common medium, is an important requirement. The data are transmitted purely cyclically. At field level the distributed peripherals, including I/O modules, measuring transducers, drive units, valves and operator terminals require communication with the automation systems via real-time communication systems. At automated cell level, the programmable controllers such as PLC and IPC communicate with each other. The information flow requires large data packets and a large number of powerful communication functions. Smooth integration into company-wide communication

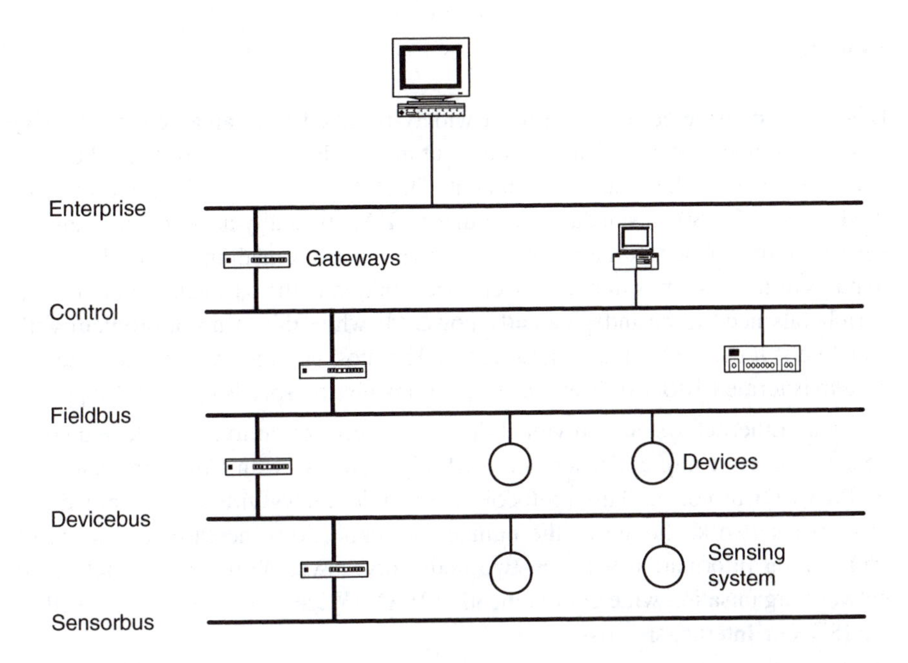

Figure 10.21. The hierarchy of an industrial network, ranging from the enterprise network for corporate information to the interconnection of individual sensors through a Sensorbus.

systems, such as Intranet and Internet via TCP/IP and Ethernet are now becoming important requirements.

A number of commercial products are available at each of the levels of the hierarchy, as listed in Table 10.7. Each product has a clearly defined standard, some of which are open standards, and are supported by a number of manufacturers. It is not uncommon for a systems manufacturer to provide a range of interface cards for their products, hence allowing its integration into a manufacturing system irrespective of the bus standard being used. In order to show the key features at each level of the hierarchy an overview of a number of systems are given below.

Table 10.7. Some of the currently available networking standards

Application	Product
Enterprise bus	Ethernet
Control bus	HSE, ControlNet
Fieldbus	Foundation Fieldbus, Profibus PA
Device bus	DeviceNet, Profibus DP, Interbus-S
Sensor bus	CAN, ASI, Seriplex, LANWorks

Ethernet

Ethernet is considered to be the most widely-installed local area network (LAN) technology, and as such is found throughout most industrial organisations, but normally restricted to the office environment. The Ethernet protocol is formally specified in the IEEE 802.3 standard. An Ethernet LAN typically uses coaxial cable or special grades of twisted pair wires as the transmission medium. One of the key features of the system is that no power is transmitted in the bus, hence all hubs and peripherals need to be independently powered; while this is not a problem with computers, it can be for industrial sensors. The most commonly installed Ethernet system is termed 10BASE-T and provides transmission speeds up to 10 Mbps.

In an Ethernet system individual devices are connected to the cable and compete for access using a Carrier Sense Multiple Access with Collision Detection (CSMA/CD) protocol. This protocol in principle allows virtually unlimited devices per network, though as the number of connections increases the speed of transmitting information between two nodes decreases. With an Ethernet based network organisation wide communication WAN (Wide Area Network) as well as via ISDN or Internet are possible.

In addition to the basic system two additional standards are available, Fast Ethernet or 100BASE-T, which provides transmission speeds up to 100 megabits per second and is typically used for LAN backbone systems, or Gigabit, which Ethernet provides an even higher level of backbone support at 1000 megabits per second.

DeviceNet

DeviceNet, (DeviceNet, 2005), is a digital, multi-drop network that is capable of operating between industrial controllers and I/O devices, and is considered to be the de facto standard in the US semiconductor industry. In the DeviceNet architecture each device and/or controller is considered to be a node on the network. One of the important features of this technology is that power is provided on the network. This allows devices with limited power requirements to be powered directly from the network, reducing connection points, physical size and cost. DeviceNet conforms to the OSI model, and as such is an open standard.

DeviceNet uses Controller Area Network or CAN for its communication layer. The CAN protocol is defined by ISO 11898-1 and comprises the data link layer of the seven layer ISO/OSI reference model. CAN provides two communication services, either the sending of a message or the requesting of a message (remote transmission request, RTR). The CAN is capable of providing, a multi-master hierarchy, which permits the development of intelligent and redundant systems. An important feature is that if one network node fails the network is still able to operate.

PROFIBUS

PROFIBUS (PROFIBUS, 2005) or the Process Field Bus is based on the IEC 61158 standard and is mainly used at field level with capabilities to operate down to the sensors, or up to the production levels. A number of variants have been designed for factory operation (PROFIBUS DP), motion control (PROFIdrive), process operation (PROFIBUS PA) or safety applications (PROFIsafe). In PROFIBUS DP and PROFIdrive the communication is based on RS485, and has the capability of speed up to 31.25 Kbps up to distances of 1900 m, and uses a logical token ring with a master/slave paradigm.

In using PROFIBUS a drive system can be controlled remotely, for example setting speeds; the actual closed loop speed control is carried out by the drive itself. The application shown in Figure 10.7 is a typical application where the synchronising features of the PROFIdrive can be used. Instead of the drives being hardwired, each drive communicates to a central controller of the bus. The closure of the loop over the bus requires the maintenance of synchronisation, this is achieved by *clock cycle synchronism*, where the exchange of data is controlled by a master clock. This will allow a drive control to be decentralised, hence the example considered in Figure 10.7 could consist of a number of PROFIBUS enabled drives communicating over a common network, instead of a hard-wired controller as shown.

10.4.3 SCADA

SCADA is the acronym for supervisory control and data acquisition, a computerised concept for gathering and analysing real time data across large industrial systems. SCADA systems are used to monitor and control a plant or equipment in industries such as telecommunications, water and waste control, energy, oil and gas refining and transportation. A SCADA system gathers information, such as where a leak on a pipeline has occurred, transfers the information back to a central site, alerting the home station that the leak has occurred, carrying out necessary analysis and control, such as determining if the leak is critical, and displaying the information in a logical and organised fashion. SCADA systems can be relatively simple, such as one that monitors environmental conditions of a small office building, or incredibly complex, such as a system that monitors all the activity in a nuclear power plant or the activity of a municipal water system.

10.5 Summary

This chapter concludes the examination of the application of modern drive systems to machine-tool and robotic systems. This brief examination of control techniques and controllers should have given designers an appreciation of the power which is available in modern systems. Irrespective of the controller and the drive, the overall performance of a system is as only as good as the weakest link in the chain.

Designers therefore must have a clear understanding of the performance specification prior to entering the design and development process. In certain aspects, the controller is the key element, because it provides a direct interface with the user; topics such as user interfaces have not been considered here, but they do need to be reviewed as part of any design process. The selection of a controller, its programming, and its interface to the system should be undertaken with care.

Units and conversion factors

The tables below give the definitions and conversion factors for quantities which are widely found in the design of motion control systems. Unless otherwise specified, ounces (oz) and pounds (lb) are considered to be *units of force*.

Length

SI units		Conversion factors	
km	kilometre	1 m	$= 10^{-3}$ km
m	metre		$=10^{3}$ mm
mm	millimetre		$=10^{6}\mu$m
μm	micrometre		$=10^{9}$ nm
nm	namometre		
		1 in	= 25.4 mm
		1 ft	= 0.3048 m
		1 mil	$= 10^{-3}$ in
			= 25.4 μm

Area

SI units		Conversion factors	
m^2	square metre	1 m^2	$= 10^{6}$ mm^2
mm^2	square millimetre		= 10.764 ft^2

Volume

	SI units	Conversion factors	
m^3	cubic metre	$1\ m^3$	$= 1000\ l$
l	litre		$= 10^9\ mm^3$
mm^3	cubic millimetre		$= 6.1024 \times 10^4\ in^3$

Linear velocity

	SI units	Conversion factors	
$km\ s^{-1}$	kilometres per second	$1\ m\ s^{-1}$	$= 10^3\ mm\ s^{-1}$
$m\ s^{-1}$	metres per second		$= 3.6\ km\ h^{-1}$
$mm\ s^{-1}$	millimetres per second		$= 39.370\ in\ s^{-1}$

Linear acceleration

	SI units	Conversion factors	
$m\ s^{-2}$	metres per second squared	$1\ m\ s^{-2}$	$= 39.370\ in\ s^{-2}$

Plane angle

	SI units	Conversion factors	
rad	radian	1 rad	$= 57.296°$
rev	revolution	1°	$= 1.7453 \times 10^{-2}$ rad
°	angular degree		$= 2.7778 \times 10^{-3}$ rev
′	angular minute		
″	angular second	1 rev	= 6.2832 rad

Angular velocity

	SI units	Conversion factors	
$rad\ s^{-1}$	radians per second	$1\ rad\ s^{-1}$	$= 0.15915\ rev\ s^{-1}$
$rev\ s^{-1}$	revolutions per second		$= 9.5493\ rev\ min^{-1}$
$rev\ min^{-1}$	revolutions per minute	$1\ rev\ min^{-1}$	$= 1.6667 \times 10^{-2}\ rev\ s^{-1}$
$°\ min^{-1}$	degree per minute		$= 6°\ s^{-1}$
			$= 0.10472\ rad\ s^{-1}$
		$1\ rev\ s^{-1}$	$= 360°\ s^{-1}$
			$= 6.2832\ rad\ s^{-1}$

Angular acceleration

	SI units	Conversion factors	
rad s^{-2}	radians per sec per sec	1 rad s^{-2}	= 0. 15915 rev s^{-2}
rev s^{-2}	revolutions per sec per sec		= 9.5493 rev min^{-2}
$^{\circ}s^{-1}$	degrees per sec per sec	1 rev s^{-1}	= 360° s^{-2}
			= 6.2832 rad s^{-2}

Mass

	SI units	Conversion factors	
g	gram	1 kg	= 10^3 g
kg	kilogram		= 2.2046 1b (*mass*)

Force

	SI units	Conversion factors	
N	newton	1 N	= 0.22481 lbf
kgf	kilogram force		= 7.2330 poundals

Torque

	SI units	Conversion factors	
N m	newton metre	1 N m	= 0.7376 lb ft
			= 8.85075 lb in
			= 141.612 oz in

Moment of inertia

	SI units	Conversion factors	
kgm^2	kilogram metre squared	1 kg m^{2}	= 10^7 gcm ^{2}
g m^2	gram centimetre squared	1 oz in s^2	= 7.0616×10^3 kg m^2

Energy

	SI units	Conversion factors	
J	joule	1 J	= 1Ws
kW h	kilowatt hour		= 2.7778×10^{-7} kWh
			= 9.4781×10^{-4} Btu
		1 kW h	= 3.6×10^6 J
			= 3.14121×10^3 Btu

Power

	SI units		Conversion factors
W	watt	1 W	= 1 Js^{-1}
kW	kilowatt		= 0.73756 lb ft s^{-1}
		1 hp	= 735.5 W

Rotational power

Derived as the product of torque and the angular speed:
power (P) = torque (T) $\times$ angular speed (n)

$P = T\,n$	W; Nm; rad s^{-1}
$P = 0.10472\,T\,n$	W; Nm; rev min^{-1}
$P = 7.3948 \times 10^{-4}\,T\,n$	W; oz in; rev min^{-1}

Derived motor constant – torque

	SI units	Conversion factors
NmA^{-1}	newton metres per amp	1 NmA^{-1} = 141.612 oz in A^{-1}

Derived motor constants – voltage

	SI units	Conversion factors
V $r.p.m^{-1}$	volts per revolutions per minute	1 V $r.p.m.^{-1}$ = 9.55×10^{-3} V $rad^{-1}s^{-1}$

Derived motor constants – damping

	SI units	Conversion factors
N m rad^{-1}s	Newton metres per radian per second	1 N m rad^{-1} = 104.72 Nm $krpm^{-1}$

Bibliography

Ambrose, R., Aldridge, H., Askew, R., Burridge, R., Bluethmann, W., Diftler, M., Lovchik, C., Magruder, D., and Rehnmark, F. (2000). Robonaut: NASA's space humanoid. *IEEE Intelligent Systems and their Applications*, 15(4):57–63.

Arkin, R. C. (1998). *Behaviour–Based Robotics*. MIT Press, Cambridge, MA.

Barrett Technology (2005). BarrettHand. 139 Main Street, Kendall Square Cambridge, MA.

Bose, B. K. (1987). *Power Electronics and AC Drives*. Prentice–Hall, Englewood Cliffs, NJ.

Chiou, C. H. and Lee, G. B. (2005). A micromachined DNA manipulation platform for the stretching and rotation of a single DNA molecule. *Journal of Micromechanics and Microengineering*, 15(1):109–117.

Crowder, R. M. (1991). An anthropomorphic robotic end effector. *Robotics and Autonomous Systems*, 7:253–268.

Crowder, R. M. and Smith, G. A. (1979). Induction motors for crane applications. *IEE Journal of Electric Power Applications*, 2(6):194–198.

DeviceNet (2005). DeviceNet. http://www.odva.org/.

Hameyer, K. and Belmans, R. (1999). Design of very small electromagnetic and electrostatic micro motors. *IEEE Transactions on Energy Conversion*, 14(4):1241 –1246.

Hendershot, J. R. and Miller, T. J. E. (1994). *Design of Brushless Permanent–Magnet Motors*. Magna Physics Publishing and Clarendon Press, Oxford UK.

Holland, O. (2003). Exploration and high adventure: the legacy of grey walter. *Philosophical Transactions: Mathematical, Physical and Engineering Sciences*, 361(1811):2085–2121.

Holtz, J. (2002). Sensorless control of induction motor drives. In *Proceedings of the IEEE*, volume 90, pages 1359–1394.

Howse, M. (2003). All electric aircraft. *IEE Power Engineering*, 17(4):35–37.

Ikuta, K., Tsukamoto, M., and Hirose, S. (1998). Shape memory alloy servo actuator system with electric resistance feedback and application for active endoscope. In *IEEE International Conference on Robotics and Automation*, volume 1, pages 427–430. IEEE.

Jacobsen, S., Wood, J., Knutti, D., and Biggers, K. (1986). The Utah/MIT dextrous hand: Work in progress. In Pham, D. and Heginbotham, W., editors, *Robot Grippers*, pages 341–389. IFS.

Jones, I. P. (2001). *Materials Science for Electrical and Electronic Engineers*. Oxford University Press.

Kassakian, J. G., Wolf, H.-C., Miller, J. M., and Hurton, C. J. (1996). Automotive electrical systems circa 2005. *IEEE Spectrum*, pages 22–27.

Kjaer, P. C., Gribble, J. J., and Miller, T. J. E. (1997). High-grade control of switched reluctance machines. *IEEE Transactions on Industrial Applications*, 33(6):1585–1593.

Klute, G., Czerniecki, J., and B, H. (2002). Artificial muscles: Actuators for biorobotic systems. *International Journal of Robotic Research*, 21(4):295–309.

Kyberd, P., Evans, M., and te Winkel, S. (1998). An intelligent anthropomorphic hand, with automatic grasp. *Robotica*, 16:531–536.

Ma, N., Song, G., and Lee, H. (2004). Position control of shape memory alloy actuators with internal electrical resistance feedback using neural networks. *Smart Materials and Structures*, 13:777–783.

Miller, T. J. E. (1989). *Brushless permanent-magnet and reluctance motor drives*. Oxford Science Publications, Oxford, UK.

Miller, T. J. E. (2001). *Electronic Control of Switched Reluctance Machines*. Newnes Power Electronic Series. Newnes, Oxford.

Nise, N. (1995). *Control System Engineering*. Addison-Wesley.

Okamura, A., Smaby, N., and M, C. (2000). An overview of dexterous manipulation. In *Proceedings 2000 IEEE International Conferance on Robotics and Automation*, pages 255–262., San Francisco. IEEE.

Papadopoulos, E. G. and Chasparis, G. C. (2002). Analysis and model-based control of servomechanisms with friction. In *Proceedings of the 2002 IEEE/RSJ International Conferance on Intelligent Robots and Systems*, pages 2109–14, Lausanne, Switzerland.

Paul, R. P. (1984). *Robot manipulators: mathematics, programming and control.* Artificial Intelligence. MIT Press, Cambridge, MA.

Pons, J., Ceres, R., and F, P. (1999). Multifingered dexterous robotic hand design and control. *Robotica*, 17:661–74.

Pratt, G. A. and Williamson, M. M. (1995). Series elastic actuators. In *IEEE International Conference on Intel ligent Robots and Systems*, volume 1, pages 399–406.

PROFIBUS (2005). PROFIBUS. http://www.profibus.com/.

Robinson, D. W., Pratt, J. E., Paluska, D. J., and Pratt, G. A. (1999). Series elastic actuator developmement for a biomimetic walking robot. In *IEEE/ASME international conference .On advanced Intelligent Mechatronics*, pages 563–568.

Rolt, L. T. C. (1986). *Tools for the job: a history of machine tools to 1950.* HMSO, 2nd edition.

Salisbury, J. K. (1985). Design and control of an articulated hand. In Mason, M. T. and Salisbury, J. K., editors, *Robot hands and mechanics of manipulation*, pages 151–167. MIT Press, Cambridge, MA.

Sen, P. C. (1989). *Principles of electric machines and power electronics.* John Wiley, New York.

Shell, R. L. and Hall, E. L., editors (2000). *Handbook of Industrial Automation.* Marcel Dekker, Inc.

Sunter, S. and Clare, J. (1996). A true four quadrent matrix converter induction motor drive with servo performance. In *27th Annual IEEE Power Electronics Specialists Conference*, volume 1, pages 146–151.

Waldron, K. J. and Kinzel, G. L. (1999). *Kinematics, Dynamics, and Design of Machinery.* John Wiley and Sons, Inc.

Waters, F. (1996). *Fundementals of Manufacturing for Engineers.* UCL Press, London.

Wheeler, P., Rodriguez, J., Clare, J., Eppringham, L., and Weinstein, A. (2002). Matric conveters: A technology review. *IEEE Transactions on Industrial Electronics*, 49(2):176–288.

Zhang, H., Bellouard, Y., Burdet, E., Clavel, R., Poo, A., and Hutamacher, D. (2004). Shape memory alloy microgripper for robotic microassembly of tissue engineering scaffolds. In *Proceedings. ICRA '04. 2004 IEEE International Conference on Robotics and Automation*, volume 5.